The Rocket Men
Vostok & Voskhod, The First Soviet Manned Spaceflights

Springer
London
Berlin
Heidelberg
New York
Barcelona
Hong Kong
Milan
Paris
Santa Clara
Singapore
Tokyo

Rex Hall and David J. Shayler

The Rocket Men

Vostok & Voskhod,
The First Soviet Manned Spaceflights

Springer

Published in association with
Praxis Publishing
Chichester, UK

Rex Hall
Education Consultant
Council Member of the BIS
London
UK

David J. Shayler
Astronautical Historian
Astro Info Service
Halesowen
West Midlands, UK

SPRINGER–PRAXIS BOOKS IN ASTRONOMY AND SPACE SCIENCES
SUBJECT *ADVISORY EDITOR*: John Mason B.Sc., Ph.D.

ISBN 1-85233-391-X Springer-Verlag Berlin Heidelberg New York

British Library Cataloguing-in-Publication Data
Hall, Rex
  The rocket men: Vostok & Voskhod, the first Soviet manned
  spaceflights. - (Springer–Praxis books in astronomy and
  space sciences)
  1. Manned space flight – History   2. Astronautics – Russia
  (Federation) – History
  I. Title   II. Shayler, David J.
  629.4'5'0947

  ISBN 1-85233-391-X

Reprinted 2002

Printed by MPG Books Ltd, Bodmin, Cornwall, UK

Copy editing and graphics processing: R.A. Marriott
Cover design: Jim Wilkie
Typesetting: BookEns Ltd, Royston, Herts., UK

Printed on acid-free paper supplied by Precision Publishing Papers Ltd, UK

Konstantin E. Tsiolkovsky (1857–1935), the 'Father of Cosmonautics'

'In our country, [it] has become possible only at this time when our whole industrious nation, every working man and woman in our Soviet land have all together set out to turn mankind's dream of conquering the heights beyond the clouds into reality. Today I am very certain that my other dream, interplanetary travel, which I have proved possible theoretically, will also come true. For forty years I have worked on jet-propelled engines and have thought that it would be several centuries yet before we could take a pleasure trip to Mars. But times are changing. I believe that many of you will witness the first flight beyond the atmosphere ...'

An extract from a speech by Konstantin. E. Tsiolkovsky, recorded in 1933 – a mere 24 years before Sputnik 1, and just 28 years before Yuri Gagarin took that first flight

Table of contents

Foreword

When Gagarin's *'Off we go!' (Poyekhali!)* was heard through the noise of the rocket engines on 12 April 1961, it became perfectly clear to everyone that the impossible had taken place at that moment. The heroic deeds of all those who participated in the creation of the rockets and performing the first spaceflights were among the greatest. The 108 minutes of that first spaceflight by Yuri Gagarin have been written into the chronicle of Blue Planet Earth, and will forever remain in the memory of humanity.

I trained and worked with all the cosmonauts involved in the Vostok and Voskhod series of missions. I served on back-up crews for Vostok 3, Vostok 4, Vostok 5, and Voskhod 1. I also was assigned to a number of prime crews before seeing the programme end. It was exciting to be part of that dream to explore space on behalf of the people of the Earth. I went on to fly two space missions in 1969 and 1976.

I worked for thirty years in the cosmonaut group at the Yu Gagarin training centre, starting as a student cosmonaut (1960), and rose to be the cosmonauts' group commander. Several generations of cosmonauts in our country and abroad were in training for space missions within this period. Gradually they were improving their professional skills and personal characters. But most important is that everyone keeps warm memories of those times that were devoted to our objectives and dreams.

Now daily TV programmes begin with Gagarin's *'Off we go!' (Poyekhali!)*, and it proves that Yuri is alive in everyone's memory and that he will live forever!

Boris Valentinovich Volynov
Twice Hero of the Soviet Union
Pilot Cosmonaut of the USSR

Zvezdny Gorodok, December 2000

Сегодня, вглядываясь в события 12 апреля 1961 года, когда сквозь гул двигателей ракеты-носителя донеслось задорное гагаринское «Поехали...», особенно четко понимаешь, что тогда произошло невозможное, и тем более величественным представляется подвиг всех, кто участвовал в создании космической техники и осуществлении первых пилотируемых космических полетов. 108 минут первого космического полета Юрия Алексеевича Гагарина вписаны в летопись «голубой планеты» Земля и навсегда останутся в памяти человечества.

Я проходил подготовку со всеми космонавтами по программам «Восток» и «Восход», был дублером экипажей космических кораблей «Восток 3», «Восток 4», «Восток 5», «Восход 1». Также я был включен в несколько основных экипажей, прежде чем программа была закрыта. Я был рад принимать участие в исследовании космоса от лица всего человечества. Позже два космических полета в 1969 и 1976 годах.

Мне доставляет удовольствие, что сейчас можно рассказать о программе «Восток» и «Восход». Книга рассказывает о высочайшем мастерстве, фантазии, мужестве и большой работе огромного количества людей, обеспечивающих космические полеты, которыми было доказано, что люди могут жить, творчески работать и решать определенные задачи в космосе.

В нашем отряде космонавтов ЦПК им. Ю.А. Гагарина я прошел путь длиною в 30 лет: от слушателя-космонавта (1960 г.) до командира отряда космонавтов. За это время несколько поколений космонавтов нашей страны и других стран готовились к космическим полетам, существенно повышая профессиональный уровень и совершенствуя личностные качества, но самое главное каждый человек навсегда сохранил добрую память о том времени, преданность нашему делу, верность традициям. Сейчас с гагаринского «Поехали!» начинаются телевизионные программы – значит, Юрий жив в памяти каждого и останется жить навсегда!

Борис Валентинович Волынов
Дважды герой Советского Союза
Летчик-космонавт СССР
Звездный городок, декабрь, 2000 г.

Authors' preface

It is a rare privilege to be able to witness the dawn of a new era or a milestone in human history. Early in the twenty-first century we will celebrate the centenary of powered flight – a mode of transport that is now an everyday event. In a similar way, human spaceflight has also come to be accepted as routine. It is easy to forget that it was only four decades ago that this mode of transport began, as we heard the news that a young Soviet pilot called Yuri Gagarin had been launched into Earth orbit. His name and his achievement rang around the world at a time of political and military tension. The 108-minute flight of Vostok 1 stands as one of the key events in the development of mankind.

Although the name and the flight of Gagarin have become well known, much of the build-up to and many of the events following his mission were shrouded in secrecy during the era of the Cold War. We recall here the build-up to Gagarin's mission, the role of the Soviet design bureaux, and the key personalities that drove the programme to success.

We present an account of the development of the spacecraft, the launch vehicle, and the huge launch complex, along with the mostly untold story of the men and women who were Gagarin's colleagues. There was a core of around 50 pilots and engineers who were selected for spaceflight training to provide crews for the first Soviet spaceflights through the mid-1960s; but only eight manned missions were flown between 1961 and 1965, and just 11 cosmonauts reached space. The story and the fate of the unflown members within the cosmonaut team is also told, along with their contribution to the success of the Soviet manned programmes of the early 1960s.

This book examines in detail the achievements of the first decade of spaceflight, an era which included:

- The FIRST artificial satellite (Sputnik)
- The FIRST orbiting of a live animal (Sputnik 2: dog Laika)
- The FIRST recovery of live animals from orbit (Korabl-Sputnik 2)
- The FIRST flight in space by a human (Vostok: Gagarin)
- The FIRST cosmonaut to spend a day in space (Vostok 2: Titov)

- The FIRST dual manned launch. (Vostok 3 and Vostok 4)
- The FIRST extended duration flights over 24 hours (Vostok 3, 4, 5 and 6)
- The longest solo spaceflight in history (Vostok 5: Bykovsky)
- The FIRST flight in space by a woman (Vostok 6: Tereshkova)
- The FIRST multi-crewed flight (Voskhod)
- The FIRST non-pilots to fly in space (Voskhod: Feoktistov and Yegorov)
- The FIRST space walk (EVA) (Voskhod 2: Leonov)

The evolution of Soviet human spaceflight developed from the beginning of the twentieth century, from a lifetime's study by one man through the pioneering decades of the 1930s–1950s, when a group of talented individuals formed the nucleus of what became known as cosmonautics. Understanding the development and importance of the Soviet design bureau infrastructure (which, in the 1940s and 1950s, built a new network of hardware and facilities) explains how the Soviets were able to launch humans on their first steps into the void of space. Interwoven with this development is a complex web of politics and secrecy that surrounded these years and the many individuals – the key members of the State Commission – who were the leading designers in their various fields. Finally, we explain how the genius of one man conceived and brought about the reality of space exploration in the Soviet Union, from pages of theory to international headlines.

The Soviets developed the concept of serial production of spacecraft, using the basic design of the Vostok craft, and we review all these variants, used not just for manned flight, but also for military and civilian unmanned missions. Although Vostok-type manned missions ended in 1965, the unmanned versions continued to fly for the next thirty years, covering the whole history of spaceflight until the beginnings of the International Space Station. This concept of spacecraft production was not adopted by the Americans until the communication satellite buses of the 1980s, and for certain military systems.

This book also recalls the evolution of the cosmodrome, and the development of the R-7 launch vehicle that evolved from the first Soviet ICBM and is still used to launch Soyuz and Progress missions to the International Space Station, some fifty years after it was first designed. Details of the unflown Vostok missions are also presented, along with design studies that laid the foundations for the Soyuz spacecraft and the space stations of the late 1960s and beyond.

In compiling this work, we have drawn on the knowledge of a number of people in the West who have been 'sleuthing' the Soviet programme for many years. The text also includes information drawn from Russia, which has been made available since the break-up of the Soviet Union in 1991. Many of the pictures come from the personal photograph albums of people who worked on these programmes and still live in Star City. For the first time, we have also had access to declassified material – released under the Freedom of Information Act – from the American Intelligence community relating to this period. These documents place into context the thinking that drove America to send its citizens into space and to commit itself to go to the Moon within a decade.

The first spaceflights occurred in the decade of the late 1950s and early 1960s, at a time when a generation began to embrace a sense of adventure, daring and risk. Space explorers became modern-day heroes of the New World of technology and exploration, with many photographs from the period becoming lasting images of a 'can do' age. This new beginning was echoed by the words of Gagarin as he left Earth on that first trip: *'Poyekhali!'* – *'Off we go!'*.

Rex Hall   David J. Shayler
London, England   West Midlands, England

Spring 2001   www.astroinfoservice.co.uk

Acknowledgements

In the era of the Soviet Union, Radio Moscow News and its science and engineering programme broadcasts, *Soviet Weekly*, Novosti booklets and Tass press releases were the only indication of a new mission and its purpose, and then always after it had launched. Few details were forthcoming, and this led to individuals undertaking personal research and investigation to learn more. This was the exact opposite of following the American programme, where a wealth of data was accessible, especially through the civilian space agency NASA.

The primary Western sources of information at this time were the weekly British magazine *Flight International*, and the American magazines *Aviation Week*, *Time* and *Newsweek* (which, in the compilation of this book, were used for their mission coverage for the period 1961–1966). Over the years the British Interplanetary Society took the lead in Soviet space reporting, with articles appearing in *Spaceflight* and *Journal of the British Interplanetary Society* – especially the Soviet Specials and the annual Technical Forums.

Contact with other Soviet space watchers around the world led to the network of what Mike Cassutt called 'space sleuths', and it is to this group that we extend our thanks for their continued exchange of information over the last three decades. This group includes Mike Cassutt, for the Who's Who series; Phil Clark, for his deep knowledge of Soviet hardware and technology; Bart Hendrickx, for his Kamanin Diary translations and other material; Brian Harvey, for his own extensive and detailed research; Gordon Hooper, for his two-volume *Soviet Cosmonaut Team* (GRH Publications, 1990); Nick Johnson, for his series of annual reports; Neville Kidger, for his detailed mission research; Jim Oberg, for his series of articles on the Soviet programme which appeared in *Flight International*, *Spaceflight* and *Space World*; and Bert Vis, for his interviews and photographs.

A special mention should be made of the work of the Kettering Group (based at Kettering Grammar School) and the inspiration of Geoff Perry, who showed us what 'space sleuthing' could reveal; William Sheldon, for his pioneering work in the classification of Soviet launch vehicles; and Marcia Smith, for the series of Library of Congress reports on the Soviet space programme.

This book could not have been produced without the additional support of Lynn

Kelterborn, for her valuable input into the draft manuscript; Phil Clark, for his technical amendments and comments; Mike Shayler, for his computer editing skills in preparing the final submitted draft; Roger Launius (NASA historian), Patrick Moore and Brian Harvey, for their helpful comments and support of the initial proposal; Bob Marriott, for his skills as copy editor, and for spending many hours scanning and preparing the illustrations; and Clive Horwood, Chairman of Praxis, for his professional support and enthusiasm for the whole project.

The book also benefited from the splendid Foreword by Colonel Boris Volynov, and the assistance of Elena Esina, curator of the museum in the Cosmonauts' House at Star City.

Illustrations

The photographs used in this book came primarily from the archives of the authors, but special thanks should be given to Eduard Buinovsky and to those at Star City who loaned us photographs but requested anonymity; and to Jim Harford, and Korolyov's daughter, Natalya. We must thank Ralph Gibbons, Charles Vick and David Woods for allowing us to use their line drawings, which so well illustrate Soviet space technology.

Reference sources

The authors used their own personal archives as well as the following publications for primary reference material. Two of the foremost sources which deserve special mention were Lieutenant Colonel William Barry's thesis *Missile Design Bureaux and Soviet Manned Space Policy 1953–1970* (University of Oxford, 1995), and Asif Siddiqi's *Challenge to Apollo: The Soviet Union and the Space Race 1945–1974* (SP-2000-4408, NASA 2000). References in English are listed here. A list of Russian language sources is available upon request.

Books
1961   *Cosmonaut Yuri Gagarin, First Man in Space*, Wilfred Burchett and Anthony Purdy (Panther)
1963   *Spaceflight Today*, Ken Gatland (Iliffe)
1964   *Spacecraft and Boosters*, Ken Gatland (Iliffe)
1965   *Encyclopedia of Manned Spacecraft*, Ken Gatland (Blandford)
1966   *Soviet Space Exploration: The First Decade*, William Shelton (Barker)
1967   *The Soviet Encyclopedia of Spaceflight*, ed. G.V. Petrovich (Mir, Moscow)
1968   *From Sputnik to Space Station*, Yu Zaitsev (Moscow)
1969   *Soviet Rocketry: First Decade of Achievement*, Michael Stoiko (David & Charles)
1970   *Russians in Space*, Evgevy Riabchikov (Doubleday, UK edition)
        The Kremlin and the Cosmos, Nikolas Daniloff (Knoff)
        The Soviet Manned Space Programme: Cosmonauts in Orbit, Gene and Clare Gurney (Watts)
1973   *Soviets in Space*, Peter Smolders (Lutterworth)
1975   *'It is I, Seagull': Valentina Tereshkova, First Woman in Space*, Mitchell

Sharpe (Crowell)

1978   *Our Gagarin*, Yaroslav Golvanov (Progress, Moscow)

1980   *Handbook of Soviet Manned Spaceflight*, Nicholas Johnson (ASS Publications)

1981   *Red Star in Orbit*, James Oberg (Harrop)
        Where All Roads to Space Begin, I. Borisenko and A. Romanov (Progress, Moscow)

1984   *First Man in Space: The Life and Achievement of Yuri Gagarin* (Progress, Moscow)

1985   *The Soviet Manned Space Programme*, Philip Clark (Salamander)
        Uncovering Soviet Disasters, James Oberg (Random House)

1986   *Almanac of Soviet Manned Spaceflight*, Dennis Newkirk (Gulf)

1987   *Valentina: First Woman in Space*, A. Lothian (Pentland)
        Russian Space History, Sotheby's sale catalogue

1996   *The New Russian Space Programme*, Brian Harvey (Wiley–Praxis)
        Russian Space History, Sotheby's sale catalogue

1997   *Korolyov*, James Harford (Wiley)
        The First Manned Spaceflight: Russia's Quest for Space, Vladimir Suvarov and Alexander Sabelnikov (Nova)

1998   *Starman*, Jamie Doran and Piers Bizony (Bloomsbury)

US space programme of the era:

1960–66 *Aeronautics and Astronautics* (SP-400 reference series)

1963   *Mercury: A Chronology*, James Grimwood (NASA SP-4001)

1969   *Project Gemini: A Chronology*, Grimwood, Hacker & Vorzimmer, (NASA SP-4002)
        Vanguard: A History, Constance McLaughlin Green and Milton Lomask (NASA SP-4202)

1983   *Lockheed U2*, Jay Miller (Aerofax)

1993   *Moonshot: The Inside Story of America's Race to the Moon*, Alan Shepard and Deke Slayton with Jay Barbee and Howard Benedict (Virgin Books)

1999   *The Race*, James Schefter (Century Books)

Reports

The Soviet Space Programme National Intelligence Estimates for 1959 (NE 11-6-59, 21 July 1959) declassified 8 September 1993, and 1962 (NE 11-1-62, 5 December 1962) declassified 1997, and requested by Peter Pesavento.

Foundation of Space Biology and Medicine, Volume III, NASA, 1975.

Soviet Space Programs, 1971–1975, US Library of Congress, Charles Sheldon (Co-ordinator), 1976.

Starry Trip of Yuri Gagarin: Documents about the First Flight of a Man in Space. *Izvestia TsK KPSS* magazine, No.5, 1991 (English translation).

From the Development of History of the Vostok Spacecraft, B.V. Rauschenbach, IAA Paper 91-686, 1993.

A Brief History of Baikonur, J. Villain, IAA Paper 94-IAA.2.1.614, 1994.

History and Creation of the Russian Space Suits, G.I. Severin, IAF Congress Paper
AA-99-IAA.2.1.07, 1999.
BBC International Space Reports (various dates).

Articles
Soviet High Altitude Pressure Suit Development, 1934–1955, Major Charles L.
Wilson, USAF MC, Aerospace Medicine, September 1965.
The Tyuratam Enigma, Dino A. Brugioni, *Air Force* magazine, March 1984
From Foundation Pit to Outer Space, R. Yanbukhtin, *Sputnik* magazine, 1991.
History and Development of Biomedical Investigations on the Soviet Manned Space
Programme, John Uri, GE Government Service, 1992.

Web sites
We made use of several key web sites, including the following:

Sven Grahn	http://www.users.wineasy.se/svengrahn/
Jonathan McDowell	http://hea.www.harvard.edu/QEDT/jcm/space/jsr/jsr.html
Sergey Voevodin	http://www.mcs.net/~rusaerog/sergeyv/VSA.html
Mark Wade	http://www.astronautix.com/

Note
Unless otherwise stated, all times are quoted in Moscow Time (GMT + 3 hrs),
regardless of time of year. All measurements are in Imperial units wherever possible,
in keeping with the theme of the era covered.

List of illustrations and tables

First spacecraft and first cosmonauts

First man and first day

First group flights and first woman

First crew and first EVA

The legacy

Conclusion

Tables

Front cover
On 18 March 1965, Alexei Leonov became the first person to walk in space – just one of many milestones accomplished by eleven Soviet cosmonauts between April 1961 and March 1965. The eleven, shown together in the lower photograph, were the first of their nation to fly into space: (*left to right*) Vladimir Komarov, Konstantin Feoktistov, Yuri Gagarin (first man in space), Alexei Leonov, Gherman Titov, Valery Bykovsky, Valentina Tereshkova (first woman in space), Pavel Popovich, Pavel Belyayev, Boris Yegorov and Andrian Nikolayev.

Back cover
The launch of the first manned spacecraft took place from the Soviet Union on 12 April 1961. On that day, in just 108 minutes, Yuri Gagarin, onboard Vostok, completed one orbit of the Earth. The event became one of the most significant milestones in human history – the day man left Earth for the first time.

Prologue

THE FIRST – YURI ALEXEYEVICH GAGARIN – 12 APRIL 1961

09.58 am Moscow Time (MT): 'Attention! This is Radio Moscow speaking. This message is being transmitted by all radio stations in the Soviet Union:

'The world's first satellite spaceship, Vostok, with a man on board, was put into orbit round the Earth. The pilot is Major of the Air Force, Yuri Alexeyevich Gagarin, a citizen of the Union of Soviet Socialist Republics. After successful launching in the multistage space rocket, the satellite ship, having attained orbital velocity and separated from the last stage of the carrier-rocket, has begun free orbital flight round the Earth.'

This communiqué, issued by the Soviet news agency TASS, also stated that two-way radio communications had been established and that Gagarin was being observed by means of telemetry and onboard TV. Following launch at 09.07 am MT, preliminary data indicated that the orbit was inclined at 65° 4' to the equator, with a maximum altitude (apogee) of 187.5 miles, a minimum altitude (perigee) of 108.5 miles, and a duration (period) of 89.1 minutes. 'Comrade Gagarin, the space pilot, withstood the period of acceleration satisfactorily and at present feels quite well. The systems ensuring necessary life conditions in the cabin of the spaceship are functioning normally. The flight of the Vostok with comrade Gagarin on board continues.'

As the Vostok flew over South America, Tass added that Gagarin had reported that the flight was proceeding normally and that he was feeling fine. At 10.15 am, while flying over Africa, the cosmonaut commented that he was 'feeling no ill effects from weightlessness.' In a statement issued at 10.25 am, Tass revealed that 'the flight around the globe had been carried out in accordance with the preset programme. The deceleration system was switched on, and the spaceship with Major Gagarin, the space pilot on board, began to descend from orbit and land in a predetermined area of the Soviet Union.'

The next report from Tass stated that 'The Soviet spaceship Vostok made a safe landing in a predetermined area of the Soviet Union on 12 April 1961, at 10.55 Moscow Time. After landing, Major Gagarin commented: "Please report to Party and Government, and to Nikita Sergeyevich Khruschev in person, that landing went off normally. I am all right and have no injuries or bruises." The accomplishment of a manned spaceflight holds out vast prospects for man's conquest of space', Tass concluded.

Congratulatory messages from the Central Committee of the Communist Party of the Soviet Union (CPSU), the Presidium of the Supreme Soviet of the USSR, and the Soviet Government were issued. Addressed to 'The communist party and the peoples of the Soviet Union; to the peoples and governments of all countries; to the whole world of progressive mankind', it stated that 'The Soviet Union proudly reports a new era in human progress. A great event has taken place. For the first time in history, man has accomplished a space flight [and] after circling the globe, safely returned to the sacred soil of our country, the Land of Soviets. The first man to have penetrated into space is a Soviet man, a citizen of the Union of Soviet Socialist Republics. It is an unparalleled victory of man over the forces of nature, an immense achievement of science and technology, and a triumph for the human mind. It has led off man's flights into space. This feat, which will live through the ages, is an embodiment of the genius of the Soviet people and the great might of socialism.'

The communiqué continued to note with 'deep satisfaction and legitimate pride' that this new era had been ushered in by the country of victorious socialism – the Soviet Union. The statement recalled that Tsarist Russia was 'a backward country,' and to dream of such accomplishments in such feats of progress, or of competing with more technically or economically developed countries, would not have been possible before the October 1917 revolution.

In blazing the first trail into space, it was clear, according to the Soviets, that their nation had surpassed all other countries of the world. After the launch of the first intercontinental ballistic missile, the first satellite, the first probes to the Moon, and after launching and recovering the first living creatures in orbit around the Earth, this was a significant milestone. 'The triumphant flight of a Soviet man around the Earth in a spaceship was a victory crowning our exploration of space.'

When Gagarin launched, it was 01.07 in the morning in Washington DC. Within 30 minutes, the USAF tracking network had recorded a launch from the Tyuratam launch complex located in Kazakhstan, Soviet Central Asia. The object tracked in orbit was large enough to contain a man, but the White House decided to wait for the official Soviet announcement before making any statement. The official announcement from the Soviets came at 02.00 am Washington time.

The American intelligence network had known for several days that the Soviet launch was imminent. The anticipation was frustrated by America's own attempt to put the first man in space. American astronaut Alan B. Shepard had hoped to ride Mercury Redstone 3 on a 15-minute sub-orbital hop into the Atlantic on 24 March, but this had been delayed to 25 April at the earliest, due to politics and extreme caution.

The Central Intelligence Agency (CIA) had expected a Russian launch on 9 April and advised President Kennedy of the anticipated launch. On that day – a Sunday –

View of Earth from a Vostok spacecraft (note the open shade).

the young President was to throw the first pitch of the opening game of the Washington Senators baseball team, and then, watching the game from his spectator box, would await the breaking news. But the news never broke, and when the latest intelligence reports indicated that the launch would occur during the night of 11–12 April, Kennedy asked not to be woken, and that the news, if it occurred, be given in the morning.

In fact, the *Morning Star* – the Communist Party newspaper published in London – had said that a launch had occurred on 7 April and that the pilot, Lieut-Col V. Ilyushin, was injured during the flight. Ilyushin, the son of the famous aircraft designer, was in hospital, but his injuries were due to a car crash. As usual, the reporters had put two and two together to obtain three.

As news of the flight reached the world's media, the White House not only kept it quiet, but had also not informed NASA. At 4 am, the spokesman of the Mercury astronauts, Lieut-Col Shorty Powers, was woken by a telephone call from a reporter asking for the astronauts' reaction to the news. Annoyed about being disturbed so early, and still half asleep, Powers told the reporter 'We're still asleep down here', and slammed down the telephone. After reading the morning headlines, Powers began to regret his statement, as the papers reported: 'As the Soviets sent a man into space, a NASA spokesman had revealed that the US was still asleep!'

Astronaut Al Shepard had two 'personalities'. The 'Mr Nice Guy' was 'Smilin' Al', while the alter ego was known as the 'Icy Commander'. It was an early morning call to his motel room at Cape Canaveral that woke the Icy Commander, with the news that Russia had put a man in space. Having almost dropped the telephone, he could not rid himself of the thought that had the Americans not been so cautious, he could, three weeks earlier, have been the first man in space. There were risks, but he was a test pilot, and risk was part of the game. Given the chance, he would have made that flight and beaten the Russians. 'We had 'em by the short hairs, and we gave it away,' he reflected years later. Fellow astronaut John Glenn was more direct and honest: 'They just beat the pants off us, that's all.'

For America, it was another body blow, like the orbiting of Sputnik, the first satellite, in 1957. Repeated demonstrations of 'superior Communist technology' was not what the new Kennedy administration needed. Five days after Gagarin flew, the pre-dawn invasion of Communist Cuba by Cuban exiles (sponsored by the CIA and supported by an American aircraft carrier) was a disaster. The invading forces were held and overwhelmed on the beaches by the army of Fidel Castro. The Americans were forced to withdraw, and the invasion was over almost as soon as it had begun. In the world's press, President Kennedy was smeared for his decisions concerning the Bay of Pigs fiasco, which continued to haunt him and his administration.

Then on 25 April, Mercury Atlas 3 was launched, only to be destroyed by the range safety officer 40 seconds later, when the vehicle failed to roll and pitch to the correct attitude. It had been planned as an unmanned sub-orbital test, but was changed to an orbital attempt two days after Gagarin flew. The explosion and loss of the vehicle delayed the use of Atlas for Mercury manned orbital missions.

A bold step was needed to restore American confidence in the young President, and for Kennedy, space seemed to be the answer. He asked where such a feat could best be achieved and be seen as an impressive challenge: a laboratory in space, a flight around the Moon, or a manned lunar landing?

On 5 May 1961, Al Shepard became the first American in space (but not in orbit) and America rejoiced at the success. Twenty days later – just five weeks after Gagarin had led the way, and with only 15 minutes in the American spaceflight log book – Kennedy laid down his challenge to the Soviet Union and the American nation: 'I believe that this nation should commit itself to achieving the goal, before this decade is out, of landing a man on the Moon and returning him safely to Earth.' The Arms Race that had evolved into the Space Race was now to be a Moon Race.

The Mercury astronauts could not believe what they were hearing. At the presentation of awards at the White House following the Shepard flight, they had overheard NASA Administrator James Webb, and Director of Project Mercury, Robert Gilruth, discussing, with Kennedy, post-Mercury manned spaceflight plans, including flights to the Moon. Kennedy assured the astronauts that they were only talking about the idea; but now, days later, here they were with a very public commitment to actually carrying it through.

Each of the seven astronauts offered their own comments on the speech, but it was Deke Slayton who realised that although each of them would still be young enough to make the trip, 'If it was not for the Russians, we wouldn't be going anywhere.'

First dreams, theories and pioneers

TSIOLKOVSKY TO WORLD WAR II

The Father of Cosmonautics

> 'Mankind will not remain on the Earth forever, but in the pursuit of light
> and space, we will, timidly at first, overcome the limits of the atmosphere
> and then conquer all the area around the Sun.'

Thus wrote Konstantin E. Tsiolkovsky in a letter to the engineer Boris Vorobiev, dated 12 August 1911. Tsiolkovsky has long been known as the father of Soviet space exploration (cosmonautics). He was the *first* to advance the idea of reactive engines (rockets) for propulsion in the vacuum of space; the *first* to recognise that artificial satellites and orbital stations were possible; the *first* to propose that rockets should be propelled by liquid oxygen and liquid hydrogen; the *first* to calculate the precise accelerations that were necessary to place a vehicle in orbit or break free of the Earth's gravitational forces; and one of the *first* to address the physiological questions of living in space.

Tsiolkovsky was born on 17 September 1857, at Izhevskoye, in the Ryazanskaya Province, Ukraine. His father was a former forester and amateur inventor, and the family was poor. When he was two years old he contracted scarlet fever, which left him almost totally deaf; but despite learning difficulties whilst attending school, he was a passionate reader, becoming interested in mathematics and physics, and later in mechanics and aeronautics. After failing his entry examinations to the Moscow Technical College in 1873, he used the next three years to teach himself a range of mathematical and scientific subjects, and made extensive use of the technical college library. In 1876 he returned home to support his family, working as a private tutor in Vyatak, where he spent his spare time developing his own workshop. In 1879 he earned a teaching post at Borovsk, near Kaluga, where his writings and theories came to the attention of the Physical and Chemical Society at St Petersburg, which elected him to full membership in 1884.

As a teenager, Tsiolkovsky became fascinated with the concept of air balloons. He developed his own ideas on the subject, and became obsessed with the idea of designing a flying balloon constructed of metal sheets and capable of carrying men. At this time he also began studying the theory of mechanics and Isaac Newton's Laws of Motion.

Tsiolkovsky realised that metallic airships, vehicles travelling on a cushion of air (hovercraft), and craft capable of travelling through the atmosphere (aircraft), would become important modes of transportation, and by developing a rocket-propelled vehicle, the age-old dream of human exploration of the vacuum of space could become feasible.

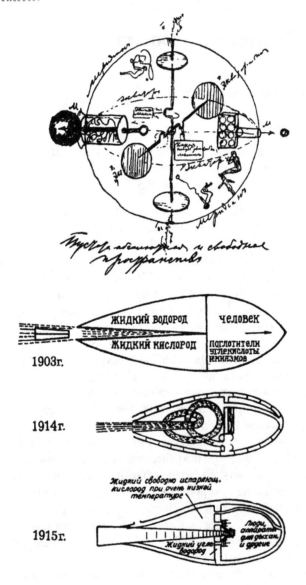

Some of Tsiolkovsky's design studies relating to manned spacecraft.

As a result of these studies, Tsiolkovsky became the first to suggest using liquid oxygen and liquid hydrogen as rocket fuels. In the 1903 edition of his paper, *The Exploration of Cosmic Space by means of Reaction Devices,* he drew what is widely thought to be the earliest concept of a liquid-fuelled rocket – a design which also featured a cockpit containing the word 'man'. In the 1911 edition of the same paper, the cockpit had been modified to include a restraint 'couch' for the passenger, to help withstand the expected forces of acceleration. This work also included details about achieving soft landings on celestial bodies without an atmosphere, where, Tsiolkovsky suggested, 'people will ascend into the expanse of the heavens and found settlements there.'

By 1924, almost half a century of experiments, research and precise mathematical calculations had led him to suggest the use of precisely fuelled cosmic rocket trains (multi-stage rockets) to attain the required speed (17,700 mph) for an object to enter orbit around the Earth (or 25,000 mph to achieve escape velocity).

Tsiolkovsky's writings were also featured in several science fiction novels. In the 1918 edition of *Beyond the Earth,* in which he recognised the importance of international cooperation for detailed and prolonged exploration of space, he continued to evaluate the use of a metallic envelope for his airships. This would

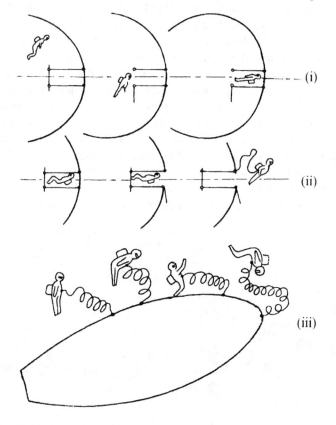

Tsiolkovsky's sketches of EVA techniques. (Astro Info Service Collection.)

allow a constant pressure to be sustained no matter what the altitude or temperature – his version of a pressurised cabin.

In another work, Tsiolkovsky explained a design of a special suit that covered the whole body; a flexible, lightweight and gas-tight garment giving the wearer full freedom of movement – a spacesuit. His design could withstand internal pressure, included provision for urine, and carried 'special cylinders' to provide an eight-hour oxygen supply. A special visor would be attached to the helmet to prevent sunlight from blinding the eyes.

Realising that both artificial satellites and manned spaceflight were theoretically possible, he also suggested the creation of orbital space stations that could develop into large cities whose shape resembled a huge wheel. But Tsiolkovsky was not just a theorist. He was also an inventor and skilled engineer, making models and experimental hardware to develop his theories and provide tangible results.

Tsiolkovsky also addressed the problems of the effects of spaceflight on the human organism – now commonly known as space medicine – including considerations on the results of prolonged exposure to the zero-g environment, and descriptions of what a human crew might see and experience away from Earth.

Fascinated by the forces of gravity, or the lack of it, Tsiolkovsky was restricted to ground-based experiments, but was still able to build several devices that simulated reduced gravity, as well as primitive centrifuges that he used to test acceleration and g-loads on chickens. He discovered that they could withstand up to 6 g, but invariably died at much higher levels – as indeed would humans.

From his many sketches and written descriptions, the Tsiolkovsky Museum in Kaluga later constructed a scale model of his proposed interplanetary spacecraft. This was planned to be 164 feet long, powered by liquid fuels, and capable of carrying a crew of three men. A separate crew compartment featured immersion tubs resembling baths, so that, as Tsiolkovsky theorised, the crew could withstand the forces of acceleration much better if they were immersed in a liquid of a similar density to themselves. Leading from the crew compartment was a connecting command room featuring an airlock, which the crew could use to venture outside, wearing spacesuits and tethers. The vehicle also featured a closed ecological system

Model of Tsiolkovsky's manned spaceship study at the Tsiolkovsky Museum in Kaluga.

capable of resupplying air from plant photosynthesis, using water and the radiation of the Sun.

On 19 September 1935 – just two days after his 78th birthday – this great Soviet space pioneer died at Kaluga, bequeathing all his work to the state. The following year his house became a museum, and in 1954 the Soviet Academy of Sciences inaugurated, in his honour, a Gold Medal which has since become one of the most prestigious astronautical achievement awards.

Following Tsiolkovsky's death, his legacy continued to inspire generations of engineering students across the Soviet Union. Some had been fortunate to visit the great man in his home town (including a young engineer named Sergei Korolyov), and through study of his writings, plans and models, Tsiolkovsky encouraged others to take up the challenge of attaining manned spaceflight. For many followers of Tsiolkovsky's ideas, it would become a personal quest to turn scientific theory into practical hardware. It would not be long before his followers realised his dream.

'Earth is but the cradle of reason. But one cannot remain in the cradle forever.'

Tsander and Kondratyuk

During the early years of the twentieth century, two other men emerged as pioneers in Russian rocketry: Fredrikh A. Tsander (1887–1933) and Yuri V. Kondratyuk (1897–1942).

Tsander had been inspired by the work of Tsiolkovsky, and in 1906, at the age of 21, he began work on rocketry. It was to become his passion. He designed rocket engines and boosters, as well as rocket-propelled aircraft that were early antecedents of the Space Shuttle. In addition, he analysed the problems of spaceflight, and the theoretical development, building and testing of rocket engines. He also predicted that manned flight to Mars would one day become a reality. Through his work in the 1920s, Tsander gained supporters in academic and political circles, and gave many lectures to those students who aspired to take up rocketry as a career, many of whom would become leading figures in the later Soviet space programme. But try as he might, Tsander was unable to obtain government funding for his studies on a full-time basis. Nevertheless, he advocated the use of multi-staged rockets to help alleviate the recognised problem of escaping the Earth's gravitational field. His rocket engine, Opytnyi Reaktivy No. 1 (OR-1), was eventually test-fired fifty times in the early 1930s, and led to the development of the OR-2, which was designed to be installed on the experimental R-1 glider.

Kondratyuk was more a mechanic than an engineer, and in 1929 published a popular book, *The Conquest of Interplanetary Space*. He studied the problems of rocket velocity in a vacuum, and concluded that a rocket flight would be dependent on the characteristics of the propellants chosen, and the initial and final mass. His researches also evaluated the effects of gravity and drag on an ascending rocket, as well as the acceleration, trajectory and guidance of the vehicle and the construction of the spacecraft itself. He was also one of the first to propose the use of smaller, lighter vehicles to achieve manned planetary landings, in order to save weight by not carrying unwanted propellant. Kondratyuk investigated methods of slowing down returning spacecraft in order to re-enter the atmosphere and land on Earth, and described the development of space stations and the harnessing of solar energy for spacecraft power.

Rocket societies

In Tsarist Russia, Tsiolkovsky's work remained largely unnoticed, having been dismissed by the scientific authorities of the day. It was only after the 1917 Bolshevik revolution that he found a new generation of interest from the developing scientific, technological and engineering communities of the young Soviet Union.

In the 1920s Tsiolkovsky inspired many engineers in a number of aviation institutes. The Moscow Higher Technical School (MVTU) and the Zhukovsky Air Force Academy invited him to lecture, and even organised trips to visit him in Kaluga. At a time of worldwide enthusiasm for the prospects of rocketry and space travel, enthusiasts formed societies whose purpose was to develop rocket technology in order to enable humans to achieve interplanetary space exploration.

On 1 March 1921 the Soviet military established a laboratory for the development of rocket research, based in a small two-storey house at 3 Tikhvinskaya Ulitsa, Moscow. But even with official backing, supplies and financing were irregular, although it achieved more than many of the other rocket societies, which received no official funding or support for rocket development. By the end of the decade, the First Five-Year Plan (1928–1932) was in force, and to support this the government's emphasis was aimed toward duplication of foreign technology rather than the development of radically new domestic hardware. A Moscow exhibition of interplanetary machines displayed models and designs from Europe and the USA, and in 1929 this was followed by the publication of a nine-volume encyclopaedia entitled *Space Travels*, which presented a summarised account of all known rocketry and space exploration concepts and ideas. However, government tolerance for such radical ideas was waning as the country followed a rapid pace of economic growth. Eventually, official support for rocket societies dwindled until, between 1929 and 1930, they were disbanded.

With Soviet strategy linked to military power, the only rocket development work that had been directly sponsored was that of the Red Army. Supported under the guise of rebuilding the defence industry as part of the First Five-Year Plan, the military rocket research team created in 1921 was, in July 1928, moved to Leningrad, where it would be closer to its artillery headquarters. Once in Leningrad, the group was renamed the Gas Dynamics Laboratory (GDL).

By 1930 the GDL had expanded its rocket research into solid fuel, electric rocket motors and, under the leadership of Valentin P. Glushko, the development of liquid-fuelled rocket engines.

During 1931, several civilian inventors, engineers and scientists, emerging once again from military efforts in rocket development, began re-establishing rocket groups across the Soviet Union. This time they sought the support of the Society for the Promotion of Defence, Aviation and Chemical Production (Osoaviakhim).

Formed in 1927, Osoaviakhim was a voluntary society of military and party organisations, formed to assist the development of the aviation industry, spread military knowledge among the population, and increase the defensive potential of the nation. This organisation was instrumental in sponsoring early Soviet aviation design bureaux, setting up aero-clubs, and sponsoring gliding clubs, and was the organisation sponsor of the Groups for the Study of Reactive Propulsion (GIRD), which supported the development of jet and rocket propulsion for aircraft.

One of the more successful of these groups was based in Moscow (MosGIRD) under the directorship of Tsander. MosGIRD was formed in 1931, by a group of engineers and scientists working on rocketry at the Tsentralyni Aero-Gidrodina-michescky Institute (TsAGI) – an organisation, formed between 1924 and 1927, that resembled the US National Advisory Committee on Aeronautics (NACA), the forerunner of NASA.

The membership of MosGIRD included a young engineer named Sergei Pavlovich Korolyov, who became very successful in securing funding from the military and official organisations. In addition to conducting work supporting the military effort, many of the members were also advocates of Tsiolkovsky and others, and worked to turn those theories into practical research and the creation of a manned rocket programme.

In April 1932 the first GIRD Rocket Research and Development Centre was established, in a basement at 19 Sadovo-Spasskays Street, near the centre of Moscow. Korolyov was in charge of the operation, the genesis of which would

Korolyov (*standing, left*) tests one of his early rockets, the GIRD-09, in 1933.

eventually become the Korolyov design bureau that guided Soviet space exploration for more than forty years.

In June 1931 the Director of GDL, B.N. Petropavlovskii, had complained to his superiors in the Artillery Administration that the GIRD groups were duplicating the efforts of his GDL. He argued that it was a waste of the limited funds to continue to support GIRD, and that they should merge their efforts with the official GDL programme. For three years, Korolyov argued in favour of the work of the GIRD at both the Moscow and Leningrad branches, winning financial support under the Directorate of Military Inventions. For Korolyov's group in Moscow, this allowed continued development of the first Soviet liquid-fuelled rockets. Choosing the simplest and most promising design of the GIRD engines to power the rocket – the GIRD 09, by M.K. Tikhonravov – the first launch occurred on 17 August 1933. The rocket reached a height of 120 feet, and although the flight ended in a crash it was deemed to be a success. It was the first modern-day liquid-fuelled rocket in the Soviet Union, and was a significant milestone for the GIRD group. With this success, Korolyov pushed for more funds to continue the programme.

The Commissar of Armaments, General Mikhail N. Tukhachevskii, therefore merged the MosGIRD and GDL to become the Reactive Scientific Research Institute (RNII). Most of the artillery researchers at GDL ignored this merger, and it was not helped by the appointment of Ivan T. Kleimenov as the Chief of the new research institute. He came from an aviation background and had no experience with the artillery. This caused considerable irritation at GDL, especially since Korolyov became Kleimenov's Deputy at RNII.

Korolyov's victory was short-lived, as, almost immediately, a Party Commission ordered that the RNII be placed under the command of the Commissariat of Heavy Industry. For Korolyov, this meant reassignment to a lower position – from Deputy Director of RNII to simply a Department Head. As a result, his two major GIRD projects (a liquid-fuelled rocket and manned rocket-powered aircraft) were abandoned.

During the rest of the decade, the RNII continued to make slow advances in rocketry development. But they suffered from poor co-ordination and restrictions in available hardware, with some elements from rocket flights having to be recovered and reused.

Having been put in charge of the RNII Rocket Section, Korolyov began to develop a series of designs for a rocket glider. In a paper entitled *Winged Rockets*, he described the work of 1936–1938, during which dozens of liquid propellant rockets were fired. In December 1937, the first ground-firing tests of a rocket glider (R 318-1) took place, and in January 1939, test article 212 made its maiden flight, with the assistance of a powder-type catapult.

On 28 February 1940, test pilot Vladimir Fyodorov went up in a rocket glider, RP 318, towed by a P-5 plane. At 8,500 feet, Fyodorov turned on the rocket engine, and reached a speed of 87 mph. This was the first free flight in the Soviet Union of a craft using a rocket engine. Korolyov designed the planes using Glushko's engines, establishing the early steps to the use of intercontinental rockets.

The 212 winged rocket with the ORM 65 engine (1937).

The RP-318-1 rocket glider with the ORM 65 engine (1938).

Manned rocketry during the 1930s

In 1932 the Soviet Academy of Sciences Institute of the History of Science hosted ceremonies marking the 75th anniversary of the birth of Tsiolkovsky. Even during this event, several speakers put forward suggestions that rockets were often proven unreliable, and that with this evidence, Tsiolkovsky's theories for using them to explore space were too impracticable to merit serious consideration. Frequent comments by the 'professional' researchers of such institutes as GDL were aimed at those 'amateur' inventors from the GIRD branches who continued to waste state expenses in 'crackpot' schemes.

Undeterred, Tsiolkovsky's followers continued to pursue their research, often in their own time. At the same time they tried to find other avenues to promote rocketry for advancing Soviet science and technology, to attain their real goal of manned interplanetary spaceflight.

During a 1934 conference on the exploration and study of the stratosphere, organised by the Soviet Academy of Sciences, Korolyov presented a paper that advocated the use of rockets to carry scientific instruments to the upper reaches of the atmosphere. Later that year, in his book *Raketnyi Polet v Stratofere,* Korolyov advocated the expansion of this idea with the use of rocket-powered aircraft to supplement the work undertaken by balloons and rockets.

However, government dissatisfaction with GIRD and the 'amateur' inventors led to the disbanding of such societies. This was part of a larger programme of changes that were influenced both by the five-year plans for rapid expansion (involving the creation of the collective farms and massive state industrialisation) and by political and social purges by the Stalin regime. There was also the growing threat of war with Germany.

Despite these darker years in Soviet history, and the slow pace for accepting rockets in the larger scientific plans, the Soviets had begun to establish a programme of exploration and discovery above the Earth, beginning with manned stratospheric balloons.

Stratospheric balloons

During the 1920s, the Soviet Army, the Air Force and the Academy of Sciences were engaged in a massive programme to develop a co-ordinated effort to explore the upper atmosphere. The objectives for such a programme were both scientific and military in origin, with plans to obtain data on the structure and composition of the upper layers of the atmosphere. This included the study of cosmic rays, air pressure and temperatures, wind speeds and direction, the ozone layer, and optical and stellar fields, as well as observation of cloud formations and ground topography.

In an era of frequent (and often very public) demonstrations of great advances and feats by the Russian people under the 'superior' communist regime, the Soviet leader Josef Stalin authorised the Air Force to stage demonstrations of Soviet military might. This included huge aerial fly-bys, the construction of large aircraft, and the conducting of well-publicised international aircraft flights and record-breaking events.

While this was clearly designed as a demonstration of the might of Soviet air power, it also spurred efforts in two areas of technological development that became important ground-work for the later manned space programme. Stalin's policy resulted in the first efforts in the Soviet Union to develop a pressurised cockpit (evolved into a spacecraft) and pressure suits (evolved into spacesuits).

In order for a crew to survive at altitudes of 10–25 miles, the Soviets pursued the development of pressurised compartments, to retain a breathable atmosphere under intense outside pressure, and suitable garments to protect a high-flying crew.

The golden age of human, balloon altitude record attempts occurred during the 1930s. Balloonists across the globe tried to surpass each other to attain ever higher and longer balloon ascents, both for scientific gain and for national pride. The Soviet Union also recognised that, with other nations developing a growing air power, air supremacy would be a major element in defeating the enemy in any future war. By ascending to the highest regions of the stratosphere, important information could be gained which would have an application in the design of aircraft. They could fly further and much higher than the enemy, or be beyond the reach of anti-aircraft batteries on the ground.

In the 1930s, the technology available to record this data was the stratospheric gondola and balloon envelope, with programme planning under the control of the Soviet Ministry of War. Funding came from the Osoaviakhim Society.

The high-altitude balloon USSR-1 being prepared for an ascent into the stratosphere in 1933. (Courtesy Astro Info Service Collection.)

The original programme was formulated by December 1932, and the following year, work began on the development of a life support system that could sustain a crew in a pressurised cabin. This work was co-ordinated by the Aviation Medicine Department of the Civil Air Institute of Scientific Research (AISR) in Leningrad, the Institute of Aviation Medicine (IAM) in Moscow, the Kirov Military Medical Academy in Leningrad, and the All Union Institute of Experimental Medicine. From this group of institutes came some of the earliest Soviet designs of pressurised crew compartments for altitude exploration, life support systems to sustain the crew, and preliminary designs of personal pressure garments. Thirty years later, each of these areas of technology would have direct application in the design of the first spacecraft to carry humans into space.

This era also generated new research in the science of material technology, in selecting the right materials from which to construct the gondolas. Special attention was given to the paint applied (to withstand expected high variances of heat and cold), and to the selection of glass that would be strong enough to withstand a near vacuum outside and the pressure from inside. Equally, the glass could not be so thick as to obscure the viewing qualities for visual and scientific observations, nor could it frost over.

The design of the gondolas featured weight-saving fabrication from aluminium tubes strengthened with steel brackets, over which was a skin of aluminium sheets. Inside this, a pressure vessel was inserted to support the human crew. This allowed for the design of an approximately 700,000–850,000 cubic foot hydrogen-filled envelope, enabling the balloons to lift a 650-lb gondola. The life support system provided up to 40 hours oxygen supply for a crew of three 'aeronauts', with internal temperatures ranging between 68° F and 86° F.

As with the development of any new technology, the technical difficulties and costs in designing and developing these craft led to opposition. It gave rise to arguments for instead flying unmanned sounding balloons. These would be far lighter, much cheaper, and ultimately safer. The complicated systems to support a crew were not needed, and in the event of accident, a crew could not be lost. Although unmanned balloons would be developed, supporters argued that the availability of an onboard crew would allow real-time adjustments to the flight programme, and would capture public interest

Thirty years later, the same argument was renewed concerning the benefits and disadvantages of flying manned or unmanned spacecraft. In fact, even after nearly half a century of sending a variety of spacecraft from Earth, the same argument continues.

Significant launches in the Soviet stratospheric balloon programme

Date	Balloon	Crew	Max. Alt.	Duration
1933 Sep 30	USSR-1	G. Prokofiev E. Birnbaum K. Godiunov	60,700 ft	8 h 15 m
1934 Jan 30	Osoaviakhim-1	P. Fedeseenko I. Usykin A. Vasenko	72,200 ft	7 h 08 m
1935 Jun 26	USSR-1-Bis	M. Christopzille M. Prilutski A. Varigo	52,400 ft	2 h 30 m
1939 Oct 12	Komsomol VR-60	A. Fomin A. Krikun M. Volkov	55,200 ft	6 h 15 m

Prior to the flight of USSR-1 in 1933, Soviet authorities eagerly preadvertised the event as a great scientific advancement in aeronautical research. It was only after several cancellations due to bad weather (all in the presence of specially invited guests and dignitaries) that the Soviets decided to conduct future launches in private and to announce them only after the event. This would continue to be the format used to announce space launches – after successfully placing them into orbit – thirty years later, and it continued well into the 1980s.

The ever-present risk of flying a human crew was tragically demonstrated in the events of 30 January 1934, with the deaths of the three-man crew of Osoaviakhim-1.

The stratospheric balloon USSR-1 prepares to ascend in 1933. (Astro Info Service Collection.)

As a direct result of trying to achieve the record altitude, too much ballast was dropped in order to lighten the gondola. This had fatal consequences in the attempt to perform a controlled descent at the end of the mission. Due to heat expansion, they lost a significant amount of hydrogen from the balloon canopy. As the Sun set, the gas cooled and the canopy lost its volume. Unable to sustain the weight of the gondola, the balloon descended rapidly, which resulted in violent buffeting and spinning from which three aeronauts could not escape. All three pioneers died as the gondola smashed into the ground.

Despite this tragedy, the programme continued, and it afforded the Soviets the opportunity to conduct some of their earliest research into the composition of the upper atmosphere and the phenomena of cosmic rays. The benefits of flying a crew became obvious. It enabled real-time adjustments to be made to the scientific programme, based on their visual observations and detailed recording of their findings, both on board in log books, via radio to the ground control stations, and during post-flight debriefing. Emphasis was placed on scientific observations, although many of these observations had direct military applications, such as the use of infrared cameras used to conduct area surveys of the terrain passing beneath them. The observations included measurements of the outside temperature and air pressure, and observations of the spectrum of the atmosphere from blue to deep violet.

The results of such research were to play an important part in the creation of the space programme and manned spaceflight three decades later. The need for flight planning and flight control was one such development, as was the demonstration of early air-to-ground communication links. Flight crew experience gave additional emphasis to the development of training and simulators, and in the art of relaying important information to future crews. Developments in in-flight tracking and post-flight recovery of a crew upon landing on Soviet soil were also established during this period.

Advances in the development of unmanned sounding balloons and pressurised aircraft eventually led to the demise of manned ballooning in the early 1940s. Several distance and duration balloon ascents were completed in the early 1940s, although it is not known if these were connected to the stratospheric programme or the developing sporting activity of ballooning and the new activity of parachuting – a technique that would feature in the recovery of the first cosmonauts. It would be more than a decade after the war before a new programme – Volga – would return Soviet aeronauts to stratospheric exploration by balloon. Volga would also have even closer application to the manned space programme.

Early pressure suits

Between 1933 and 1943, and complementary to the development of the stratospheric balloon programme, the Soviet Air Force High Command and the Pavlov Red Army Institute of Aviation Medicine (IAM) in Moscow pursued an extensive programme into the development of personal life support equipment to protect Soviet aircrews.

The Soviets quickly realised the value of providing pressure garments. They would facilitate a rapid domination of stratosphere air power, and would allow early reconnaissance over the territory of future adversaries while remaining virtually out of reach of enemy ground forces. They realised that without pressure garments, a pilot could be resistant to the factors of altitude flight – such as 'the bends', or greying (blacking out) – for only a short period. The question was: how high could a man fly safely with such a suit? To answer this and many other questions would require a simulation laboratory which duplicated the environment at high altitude, but without leaving the ground.

Simulated 'flights' to altitudes resembling the upper reaches of the atmosphere would need to be conducted, and the construction of three altitude chambers was therefore authorised. The IAM received its first chamber in August 1930 and its second in 1936. The smaller Civil Air Institute of Scientific Research (AISR), with a staff of just one and almost no support equipment, received its altitude chamber late in the decade.

It was also in 1936 that a task force from Union Aviation Medicine began the development of suitable protection for aircrews flying above 10 miles. Vladislav A. Spasskiy – a military physician attached to the IAM – determined the medical criteria for these protection devices. It soon became apparent that although completely pressurised cabins would be more favourable, there were considerable complications, with increased lead times and production costs beyond the highest predictions. Some of the problems encountered included ensuring a perfectly sealed cabin (the loss of cabin pressure would affect every member of the crew) and difficulties in devising effective crew escape methods, and it was therefore decided to proceed with the development of individual pressure garments.

Development of the first Soviet pressure suits was consigned to IAM, who were instructed to develop the garments for the Air Force. IAM's first director, Prof Dr Feodor Krotkov, had headed a five-year study of all known foreign pressure suits. Meanwhile, Spasskiy had completed his Candidate of Technical Sciences (CTS – the Soviet doctorate), and the resulting thesis, *Aircrew Protection in Stratospheric Flight,* became a standard reference source for all suits that were subsequently developed in the Soviet Union.

Spasskiy was detailed to head the development effort. The first prototype pressure garments were designed and made by A.A. Pereskokov and Rappaport during 1934–35, and featured an oxygen open ventilation system. By late 1936 the design requirements for a suit had been refined, and were devised into a prize-winning contest sponsored by the All-Union Society for Aeronautics and Chemistry. Organisations which participated in the design of a stratospheric suit included the Central Institute for Aerodynamics and Hydrodynamics, the Air Force of the Red Army and Peasants, the USSR Army Artillery Transportation, and the People's Commissariat for Heavy Machinery.

Each design submitted was tested under laboratory conditions (including altitude chamber runs) and in some cases also underwent actual flight tests in aircraft. The winning design was submitted by E.E. Chertovskoy of the Air Force Red Army Works organisation. By 1937 this suit had been tested in low pressure, low temperature altitude chamber runs, and again in actual flight tests where it was flown to a maximum altitude of 5.6 miles during a test programme that logged 70 hours of suit experience by the test subject.

The Soviets continued pressure suit research and development between 1937 and 1943, resulting in a number of prototype suits. These designs featured such details as electrical heating of a double-walled helmet visor, and a gas spray tube to prevent face-plate fogging. The suit fabrics and materials were examined for the effects of gases and the ozone layer. An important development was that of a closed-circuit fan, with an injector system for both the supply of oxygen (O_2) and the removal of carbon dioxide (CO_2) and water vapour. Designs also featured closed-cycle life support systems, a one-hour emergency bail-out capability, and electrically heated outer clothing.

In April 1943, as the effects of World War II began to tell on national resources, the Institute of Aviation Medicine was closed, and its work diverted to the war effort.

Design bureaux

The need for specialised design and development departments evolved in the 1920s and 1930s, as the Soviet Government developed a huge industrialisation programme. The creation of design bureaux (konstruktorskoe biuro – KB) evolved in all branches of Soviet industry and production – none more so than in weapons research and development. The primary objective of such KBs was to ensure that any newly introduced technology was quickly moved into full-scale production.

By the end of the 1930s the Soviets had established a unique three-tier system to compartmentalise Research and Development (R&D), which remained essentially unchanged until the demise of the Soviet Union in 1991.

From Central Aerohydrodynamics Institute (TsAGI) departments, the very first aviation design bureaux were formed, and from these evolved the separate organisations that became the strength of the Soviet aviation and space industries.

The first of these three-tier institutions; was the Nauchno-Issledovatel'skii Institute (NII) – more commonly known as a scientific research institute. A network of thousands of NIIs extended across the Soviet Union, featuring a significant number of military-based research institutes; and at the forefront of this network was the prestigious USSR Academy of Sciences.

These were the facilities involved in basic theoretical research, and the industrial ministries for specific fields. Access to specialist test and research equipment was the responsibility of these NIIs, which often dictated the type of technology and materials that designers should (and often had) to use. In practical terms, however, when new products were deemed necessary the resources went to the design bureaux rather than to the NIIs – a factor which was instrumental in creating a deep rift between the designers and scientists.

At the next level were the KBs that formed the link between the scientists in the laboratories and the workers on the factory floor. The role of the KB was to develop the scientific theory and basic research information (within the limitations and capabilities of the industry of the day), in line with the needs of the customer – which was invariably the military. A few larger KBs also had the capability to test the theory in practise, by producing their own experimental prototypes and mock-ups to prove the feasibility of the product, although they lacked significant resources for mass production. Major developments in defence R&D and design were handled by a few elite Experimental KBs, which were designated Opytno-Konstruktorskoe Biuro (OKB). These were specialists in their field, and often handled the more technologically challenging programmes such as defence, nuclear development, and the space programme. They also featured some of the prototype building and test facilities in the Soviet Union.

The head of OKB was the Chief Designer, who normally had to have a strong personality in order to achieve his own aims and those tasked to his bureau. The Chief Designer was a blend of inspirational leader, bureau organiser and manager, contract bidder, and project salesman, and was, invariably, a political pawn. A success ensured continued support and resources, but not general recognition. A failure to complete contracts (or even to win them) could lead to the closing of 'his' bureau. The Chief Designer was perhaps the closest a Soviet came to a Western-style entrepreneur, but without the financial rewards and international recognition.

The third level of the system was the factory production facilities, some of which were located near or integrated into the larger OKBs. Successful plant directors, (though institutionally lower in the system than the KB designers) were often more able to advance to higher political levels than were their counterparts.

This system was designed to help ensure a smooth transition from design to production. It was here, as with most areas of the Soviet system, that a combination of bureaucratic hurdles and personality clashes made interdepartmental communication much more time-consuming than it needed to be. The Soviet system rejected fast-paced technological change, but added political and power regime hurdles.

Conclusion

By the early 1940s the infrastructure was in place to support the expanse of defence related technological development in the Soviet Union. The onset of World War II and the ensuing bitter struggles against Nazi Germany led to the suspension of any further development in these fields in order to support the war effort and ensure that there was still a Soviet Union after the hostilities ended. For men like Korolyov, whose dream of space exploration remained the overriding long-term goal, their first priority in these difficult years was to ensure that they personally survived.

The era from Tsiolkovsky, through the Bolshevik October Revolution and the civil war to the brink of World War II, featured several notable milestones in military and scientific advances – technology that would be crucial to the dawn of the space age barely ten years after the end of the war.

From the theoretical ideas of Tsiolkovsky to the rocket designers of GDL, GIRD, and the work of the NII and KBs, there was accumulated a valuable resource of information on the characteristics of the atmosphere and how to travel through it. A wealth of data from the balloon, aircraft and rocket programmes became the pre-war legacy to post-war efforts to reach orbit.

Developments in the design, testing and construction of hardware were supplemented by operational experience, stratospheric balloons, high-speed aircraft, and the first rocket engines onboard early 'missiles'. Materials technology (in the construction of pressurised gondolas and aircrew cabins, plus developing crew restraint devices) gave rise to studies into the medical consequences of speed and altitude flight. Biomedical data were obtained on acceleration and deceleration, altitude flight, and the effects of pressure and g-forces on the human body. An infrastructure of command and control, crew training and mission simulation also provided important experience.

But of more significance, perhaps, was that many of the key figures in early Soviet manned space programmes learned their skills in the pioneering aeronautical programmes of the 1920s–1930s. Following World War II, these men formed the nucleus of designers, engineers, planners and administrators that led the Soviet Union into space. The period of the 1920s and 1930s had one other significant feature in a different area of the future space programme: they were the years in which most of the first cosmonauts were born!

WORLD WAR II TO THE 1950s

World War II

During the purges of the Stalin regime in the late 1930s, among the politically accepted professional researchers at the Academy of Sciences there still remained deep-rooted opposition to Tsiolkovsky's ideas. For members of the aviation industry (with their connections to the defence of the nation), association or involvement with a rocket research society was subject to close scrutiny and examination by the state authorities.

Further concerns were raised after the poor performances of Soviet-built aircraft used during the Spanish Civil War (1936–1939). This gave rise to the naming of several supporters of Tsiolkovsky's ideas in the RNII as members of counter-revolutionary Trotskyist organisations. As a result, the Government arms minister and the entire RNII staff, including Korolyov, was sentenced to a long term in a remote coastal Siberian gulag. By 1939, the RNII had been abolished, and had been redesignated NII-3.

Well before the German invasion of Soviet Union in June 1941, the Soviets had begun to evaluate the development of weapons that would produce an immediate result. With his experience in pre-war designs of winged rocket aircraft and gliders, Korolyov was released from the gulag and transferred back to Moscow. He began work at a special prison (under Andrei Tupolev), designing a rocket booster for a dive-bomber; but despite the limitations imposed during his 'deprivation of freedom' sentence, the dream of spaceflight was not diminished, as he gradually became involved in rocket research again, working with Glushko on rocket-assisted fighter–bomber designs.

During the latter stages of World War II, the German V2 rocket attacks on London clearly demonstrated the potential of the missile as a weapon of war. When the war ended in 1945, a significant change in Soviet R&D policy included a new economic plan to develop new technologies, including 'target rocket' development. The first fragments of German V2 missiles had been retrieved from artillery ranges during the Soviet advance through Poland in 1944. That same year, Korolyov was finally released from prison in recognition of his committed and successful work for the Soviet regime while interned.

The fragments of V2 had been shipped back to a remnant of NII (by then designated as NII-1), where for three months a small group of rocket specialists, headed by Glushko, worked on the pieces. Following a complete analysis of the hardware, they proposed the creation of a larger, longer-range Soviet version of the V2. Despite lack of official support for the project, the team continued to study the idea. The group included some of the key figures that were to become the early space programme leaders in Russia: N.A. Pilyugin, A. La. Berezniak, B.Y. Chertok, L.A. Voskrensky, V.P. Mishin, M.K. Tikhonravov and Iu. A. Pobedonostev. Korolyov was also a member of this group.

In May 1945 a special state commission was sent to Germany to assess the rocket technology that the Soviet Union had gained at the end of the war, from several Nazi rocket research facilities now under Soviet control in East Germany. Over the following year, several other teams – mostly former inmates of Soviet gulags – visited Eastern Germany to retrieve as much documentation of V2 operations as possible.

From this evaluation, the Council of Ministers passed decree No 1017-419ss, 'Special Committee on Reactive Technology', dated 13 May 1946. It set the structure of the Soviet rocket research programme back on course, with the design of a Soviet V2-based missile. As part of the decree, the control of rocket test ranges passed to the Ministry of Armed Forces. German specialists would also become an important part of Russia's post-war missile programme, with several hundred moving to the USSR with their families in October 1946. In August 1946, Korolyov had been assigned as Chief Designer of a department of NII-88, which was to produce Soviet versions of the V2.

Pressure garments

After the war, research into pressure garments also resumed, with the added advantage of the evaluation of captured German documentation and prototype pressure suits. Even though the war had severely curtailed work on pressure garments, its importance from a strategic point of view was well defined, and as a result, a basic research programme had continued.

A.I. Khromuskhin – a mechanical engineer, and author of several books – had continued to work on pressure garments during the period from 1944 to 1949. His fields of research were in regenerative jet circulation systems, the chemical absorption of carbon dioxide (using a silica gel for absorbing water vapour), and research into liquid and gaseous oxygen sources.

At the end of the war, Soviet pressure garment research began to feature stress analysis of suit fabric, quick-don helmets and gloves, and the beginnings of long-duration parachute jumps from high altitude. This latter research, to test the reliability of pressure suits during prolonged descents to Earth, reflected similar work being conducted in American programmes run by the United States Air Force and the United States Navy.

The Soviet research was conducted in conjunction with other Soviet Air Force studies in the development of ejection seats and emergency crew-escape capsules. It also featured compatibility (fit-and-function) evaluation of the suits and life support systems with these emergency procedures.

Following World War II, Soviet development of pressurised compartments reflected similar research around the world. One of the drawbacks of any pressurised compartment was the need to supply the atmosphere to sustain the crew during the flight, which added weight and volume to the design. The development of such a system also included complex leak tests, and provision for a second, back-up system, in the event of rapid decompression or primary system failure. As pressurised cabins became more of a feature in aircraft of the 1950s, the pressure garment research focused on the design of 'quick-don-and-doff' (put on–take off) helmets and gloves which became known in the West as 'get-me-down suits'. But as the reliability of pressurised crew-cabins improved, they became the primary life support system. As a result, the role of a pressure garment became more of a back-up to this life support system. This was reflected in the Soviet designs of aircraft pressure garments in the 1950s, which resembled that of their US counterparts (called 'capstan suits').

Aircraft pressure garment evolution moved away from true space pressure suits after 1955, but access to Western documentation allowed the Soviets who were developing high-altitude and partial-pressure garments to evaluate their standards against those in the West. The capture of U2 pilot Gary Powers in May 1960 also afforded the Soviets the chance to examine a flight-proven pressure helmet (MA2) and partial pressure suit (MC3A). From detailed examination of the construction and operation of this hardware, the Soviets conducted further testing and comparison with their own designs and, in some cases, adapted the most suitable features from the American suit into Soviet models.

Direct spin-off technology from these designs was applied to the development of the first pressure garments for spaceflight by the same facilities that developed the

Air Force suits. In the Soviet Union, the leading company pioneering this research was Semyon Alekseyev's Plant No. 918 (later named Zvezda) at Tomilino.

Testing the V2

One of the myths concerning the early stages of the space programme was that the German rocket scientists that went to Russia (rather than to the West) at the end of the war helped the Soviet Union beat America into orbit in 1957. The rumour at that time was that 'their Germans are better than our Germans'. This certainly helped fuel the Cold War fear of an era where any advances in technology could imply military supremacy. Spreading the idea that the best talent that created the V2 rockets was now behind the Iron Curtain also helped fuel the desire for America to remain one step ahead of the Soviets.

We now know that the opposite was true in that the West inherited the best German scientists from Peenemunde and other locations. Those who went to the Soviet Union were undoubtedly highly experienced in practical terms, with mass production and in launching rockets; but the Soviet German group was not nearly so experienced in developing theory as the Soviet scientists and engineers themselves had been over the preceding twenty years. In fact, when they came to work in the Soviet Union, the German teams often worked independently of their Soviet colleagues.

The Americans had secured the very best of the German rocket talent, including Werner von Braun and 400 V2 experts, plus hundreds of rockets and most of the production and testing hardware. In the Soviet zone, rocket research facilities had been mostly destroyed, but they had managed to retrieve detailed descriptions of the system, and enough parts to assemble about a dozen V2 missiles from the Nordhausen Institute, which were shipped by train back to Russia.

General Dmitry Ustinov (then Commissioner of the Ministry of Armaments) had recognised the importance of missile technology in the strength of future Soviet military forces. Josef Stalin had decreed that testing of a V2-derived vehicle would be completed before the development of the larger vehicle that Korolyov had really wanted. Realising that he would need to gain political and financial support for his long-term plans, Korolyov proceeded to launch a modified Soviet version of a V2 as soon as he was able. But he was in competition for limited funds, both with other Soviet departments developing a range of new-technology projects, and also with the 'Soviet' German rocket group.

Initially, Stalin favoured the use of proven German specialists for developing early Soviet missiles, but gradually, the Soviet team headed by Korolyov proved their skills. By October 1947, Korolyov had won the first lap in the race, by not only modifying and launching a V2 within six months, but also by repeating the feat two days later. The vehicles were launched from a new missile test range – Kapustin Yar, located south of Volgograd in the Ukraine, and set up in 1946. Facilities were primitive, with railway carriages housing both the laboratories and the launch control and accommodation provided by a 'village' of canvas tents.

The success of these flights prompted Korolyov to take a bold step. He invited five other leading Chief Designers, from each KB involved in aspects of the Soviet missile programme, for a series of informal meetings. The first meeting of what became known as the Council of Chief Designers – or 'The Big Six' – occurred in November 1947. The six were V.P. Glushko, V.P. Barmin. V.I. Kuznetsov, N.A. Pilyugin, M.S. Ryzanski and Korolyov. All had previously worked in Germany following World War II, and had kept in touch with each other during the development of their own research work. Korolyov had established a powerful group of key figures – an unofficial (and actually democratic) body of free-thinkers who influenced coordination, technical issues, flight planning, and long-range plans.

The group, working within the Ministry of Armaments, held no legal position, but had more of a moral agreement that resulted from long sessions in deep discussion, and so called 'chalk-board calculations'. Clearly, the origin of the operational Soviet space programme was to be found in this group of six men. Future progress was dominated by this closely knit group of individuals, who had been working on pioneering rocket technology for almost twenty years. The most prominent member of this group who was soon to become the leading force in the Soviet race for space was Sergei Pavlovich Korolyov.

Biological ballistic flights

After the flights using adapted German V2s, Korolyov and his design team spent the next year developing the first Soviet ballistic missile, the R-1. Similar in appearance

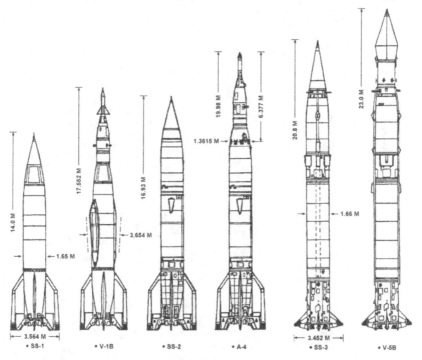

American artwork and military designations of three Soviet atmospheric rockets, based on military originals (six rockets). (*Left to right*) the R-1 (SS-1), R-2 (SS-2) and R-5 (SS-3). (© 1982 Charles Vick.)

(*Left*) The R-1 rocket – the Soviet version of the German V2 – and (*right*) the R-2.

The V-1A (*left*) and the V-2A (*right*) undergo tests.

to the German V2, the missile was of a more reliable and rugged design. Most importantly, it incorporated a more accurate delivery than did its German predecessor. The first flight of the R-1 over a distance of 180 miles took place on 2 October 1948. A test programme including thirteen R-1 launches was set up, although there were only a few hard-earned successes. Nevertheless, the R-1 had entered operational military service by 1950. It was soon joined by the R-2, which again was based on an earlier German design (designated the R-10). The R-2 featured a 35-ton thrust engine and, at 372 miles, had twice the range of the R-1. Its maiden flight occurred in October 1950. The next vehicle proposed was the R-3, which featured a 120-ton thrust capability and a predicted range of 1,865 miles.

Soviet missile experience was developing far more quickly than any work that the German group was achieving, and by 1950 only fifty of the 400 German group specialists remained working in rocket research in the Soviet Union. Enthusiasm, and experience gained by earlier rocket societies, had helped the Soviets demonstrate that future success in national missile defence rested with the domestic talents of Korolyov and his colleagues. However, to meet future demands, Korolyov knew he needed a 'new', larger vehicle. Although the role of the R-1 in developing larger missile programmes was ended, it was still to play a pioneering role in sending biological payloads towards space.

By 1948, M.K. Tikhonravov (one of the founding members of MosGIRD) was among the first former Tsiolkovsky advocates to suggest that manned spaceflight was within reach. He had already completed some work on the design of a manned capsule (designated VR-190) that could be launched by a V2 on a sub-orbital trajectory. During the presentation of a paper to the Academy of Artillery Sciences in July 1948, Tikhonravov went a stage further, stating that the technology already existed for the Soviet Union to launch an artificial satellite. This statement caught the attention of Korolyov (who was then a corresponding member of the Academy), who contacted Tikhonravov about the prospect of informally working together on the development of a satellite launcher. Officially, Tsiolkovsky's ideas were still discouraged, Tikhonravov receiving fairly negative reactions during the presentation of a paper at a conference in March 1950. By then, most of Tsiolkovsky's ideas for space exploration had begun to surface in print, but only in children's books. Information on the developing ballistic missile programme, on the other hand, remained highly classified, which obviously limited any public discussion about the possibility of placing an object in orbit.

However, progress was being made in other areas. In 1949 the USSR Academy of Sciences formed a Commission for the Investigation of the Upper Layers of the Atmosphere, headed by the General A. Blagonravov, assisted by N. Sisakyan. The R-1 would be adapted for high-altitude research as a 'sounding rocket', carrying a basic scientific payload developed by the Korolyov team, based at Kapustin Yar. However, not all the ascents would be purely scientific in nature, as many would feature applications for the military programme in both the aviation and ballistic missile fields.

Designated 'Akademik' tests (derived from the academic and scientific research experiments they carried), there would be five versions planned, each designated after the first five letters of the Russian Cyrillic alphabet: A, B, V, D, and E.

Despite claims of purely scientific objectives, these flights formed a direct branch of the military missile development programme, with the launch vehicles designed by the Ministry of Armaments, and the launch crews made up from military personnel, who used the experience as part of their missile training programme. The security of the launch and recovery sites was also handled by the military. Even some of the biological research experiments had direct application in the military programme, in pressure suit and crew compartment designs – technology initiated in the stratospheric balloon programme before the war.

Korolyov, of course, was extremely interested in the biological flights that were conducted during this programme, because of their long-term application to his own plans for human spaceflight. The Korolyov team initiated the first series of flights on 21 April 1949, with two flights that year to 63 miles altitude, carrying a scientific package located in two 143-lb payload canisters on each side of the rocket.

The programme was gradually expanded with an improved version of the R-1, to launch a series of flights in a three-phase operation between 1950 and 1960, with ever more complex payloads and research objectives.

Soviet high-altitude biological research flights

1949	Two rocket flights to altitude 63 miles
1952–1956	Six flights carried nine dogs to altitude 60 miles
1956–1957	Nine flights carried nine dogs
1956–1957	Four flights carried three dogs
1957 May 16	Two dogs reached 130 miles
1958 Aug 27	Two dogs reached 280 miles
1958 Sep 19	Two dogs reached 293 miles

Korolyov with one of the test dogs in the 1950s.

Choosing a suitable primary animal payload was the responsibility of the Pavlov Institute at Pavlov-Kuttishi, near Leningrad, where Ivan Pavlov had become a pioneer in animal psychology in the USSR.

Size and weight of the subject (between 13 and 16 lbs) were among the criteria that would determine what could fit inside the available capsules. It was also necessary to choose animals that would be best suited to such flights and which would provide useful data that could be used to predict how a human might survive in space. Scientists evaluated a wide range of specimens for the programme, including earthworms, flies, lizards, mice, rabbits, dogs and primates.

The Russians selected a group of canine crew-members as the primary candidates. Dr Nikolai Parin, of the Soviet Academy of Sciences, explained the reasons behind this decision: 'The Russian dog has long been a great friend of science. We have collected much information on our four-footed friends.' The Russians felt that dogs were highly organised, steady and easily trainable animals, unlike the primates chosen for the American programme, which the Russian saw as capacious, highly-strung and undisciplined. It was also reasoned that the blood circulation and respiratory systems of dogs were similar to those of humans. In addition, the dogs were found to be patient and durable during long experiments. This requirement was also instrumental in providing the dogs that would participate in preliminary tests of a manned spacecraft later in the programme.

Resembling the human space candidate selection process that would follow, a list of basic requirements was posted, to which suitable candidates would be nominated and evaluated before a 'class' selection was finalised. From here, the 'candidates' would be further examined and evaluated. These tests would result in a smaller 'training group' that would be prepared for a flight, or series of flights. Several dogs became veterans, going on to complete one or more 'missions'.

From a set of requirements sent to zoologists, canine vets, and specialists, a group of mongrels was selected. It was felt that they would be best suited to the rigours of ballistic flights, being strong, accustomed to hunger and cold, and already conditioned to a tough hard life. A search of the alleys of Moscow, plus pounds, and some owners, yielded a pack of hounds that all met basic requirements. As the specially designed pressure and anti-gravity suits and sanitation facilities fitted female dogs better than male dogs, the whole group consisted of female dogs, all with white coats so that their movements and behaviour could be observed clearly in the TV and motion picture coverage of each flight.

The group consisted of 24 dogs, which were put in isolation after a complete medical examination, including measurements of length, height, width, weight and so on. Each dog was given a nickname and placed in one of three 'training' groups that best suited its character.

The first group was even-tempered, showing moderate movements. This group would be trained for long-duration missions, and included the dogs Laika (Barker), Strelka (Little Arrow), Belka (Squirrel), Lisichka (Little Fox), Zhemchuzhnaya (Pearly), Chernuska (Blackie) and Zvezdochko (Little Star). The second group contained dogs that were more restless (producing more intense reactions), while those in the third group were sluggish.

As part of the preparations for their flights, the dogs' respiration and heart rates were recorded each day. They were X-rayed, and were flown on high-altitude aircraft to familiarise them with the sensations of flight. They were also placed in mock-up capsules and then subjected to vibration and centrifuge chamber runs, and long periods of isolation. Each dog was also subjected to extensive pre-flight and post-flight medical examinations, with veterans of more than one mission providing comparative data for subsequent ascents.

As the long programme evolved, so the emphasis changed from pure engineering and military applications to more biological research, as the prospects for manned flight became clearer. Between 1949 and 1960 there were 160 ascents, with gravitational forces of up to 5 g experienced in flights of 600 seconds, providing 370 seconds of weightlessness.

Phase 1: 1951–1952   The flight plan for this series consisted of six flights, each carrying a two-dog payload up to 100 miles and accelerations below 5.5 g. Nine dogs were chosen, resulting in more than one flight for three of them. The dogs used in this phase included Albinas (Whitey), Dymka, Modnista, Kozyavka (Gnat),

A biological canister attached to a V-2A rocket (*top*) in which dogs were carried in several sub-orbital flights during the 1950s (*bottom*).

Malyshko (Little One), Tsyganka (Gypsy), and possibly Laika. The dogs rode in hermetically sealed cabins mounted on the side of the carrier rocket. The cabin featured a 10 cubic foot volume pressurised to 680–760 Hg, with 70% air and 30% oxygen. The capsule separated from the carrier at 40 miles for parachute recovery, and the top, finned payload separated at 62 miles. During the flights, the dogs' pulse rate and temperature were measured and their behaviour monitored by onboard camera, using a reflective mirror in the dark capsule.

Phase 2: 1955–1956   This series was a programme of nine flights (each with two dogs on board), this time reaching 62–68 miles. From the first phase, veterans Albinas, Kozyavka, Malyshko and Tsyganka flew again, along with eight more dogs, each as part of a 4,000-lb payload. Each dog wore a custom-fitting pressure garment with a removable helmet, developed by Plant No. 918. The dogs each rode in a separate capsule within the nose cone of the launch vehicle. The capsules were separated at different altitudes. The left-hand capsule left first, at 20–30 miles altitude and a separation velocity of 3,300–4,500 ft/sec. Following separation, the capsule completed a free-fall to an altitude of 2.4 miles before the parachute opened. The right-hand capsule separated at 50–55 mph at 2,300 ft/sec. After 3 seconds, the parachute opened (at 47–53 miles), allowing a gentle decent lasting 50–65 minutes. Both dogs touched down at a velocity of 20 ft/sec, approximately 12–30 miles apart. This operation also provided early evaluation of ejection systems from a descending cabin that probably had application in the design of the system finally incorporated on the Vostok capsule.

Phase 3: 1957–1960   For this last phase, the special pressure garments were retained, but the 16 cubic foot compartment was hermetically sealed. Seven two-dog flights were accomplished, to altitudes of 130–300 miles. The payload weight ranged between 3,300 and 4,800 lbs. For this programme, a team of fourteen dogs was selected, with one dog becoming a 'veteran' of six launches. The group included Otvazhnaya (two flights), Snezhinka, Belanka and Pyostraya. From these flights, photographs of the Earth were taken, and instruments took measurements of the chemical composition of the atmosphere, pressure levels, and micrometeorites, as well as monitoring the behaviour of the onboard passengers. Just six months after the beginning of this phase, the dog Laika, onboard Sputnik 2, became the first living creature in orbit.

Design bureaux

In the early 1950s, rocket development began to take on a new pace that was both military (ICBM) and scientific (spaceflight) in application. The Soviet leadership realised that the Soviet Union had to match Western developments in nuclear technology by securing both defensive systems (surface-to-air missiles) and offensive systems (long-range bombers and missiles).

A three-tier system evolved. Initially, this was the long-range aviation force, a series of winged (cruise) missiles, and ballistic missiles. At the time, the use of rockets as nuclear carriers had not been proven, but in recognition of further developments the NII-88 was changed from a centre of artillery research to that of ballistic missile development.

As the structure of the Soviet armed forces began to reflect global changes in weapon systems and tactics, so this development was reflected across the whole aviation infrastructure of design bureaux and manufacturing plants used to create the hardware for these programmes.

Conclusion

From the end of the war to the late 1950s, the Soviet Union rapidly developed in areas of military technology in the scientific field. Responding to the growth of nuclear arms in the West, the Soviets accelerated their own programmes to develop ballistic and ICBM missiles. One of the results of these studies was a programme of scientific ballistic research flights, taking biological payloads high above the Earth. This was a prelude to creating an automated payload to enter orbit, and eventually to carry a man.

Support for such programmes was far from secure, or even constant, but research in areas of space pressure garments, plus advances in aerospace medicine and the growth of rocket technology, would soon also lend support to a programme of space exploration.

Rapid developments in both the nuclear and armaments fields required changes to the design bureaux and manufacturing infrastructure, in order to reflect these developments. Russian nuclear and missile development had progressed to a point where a large, long-range launch vehicle would be required to lift the weapons to their targets, and from that, a vehicle capable of placing payloads in orbit. In addition to providing the launch vehicle, this infrastructure needed to support a launch complex from which it could be dispatched.

First launcher, pad and satellites

A ROCKET FOR SPACE

The creation of an ICBM

Between 1945 and 1955, work progressed on the development of Soviet rocketry, based on the experiences of GIRD and GDL. It led to the R-1 to R-5 series of ballistic missiles used for both long-range weapon delivery and in the altitude biological flights up to 1960.

A system for designating weapon design had been adapted by the Soviets during the 1940s. Each weapon incorporated a letter and/or figure combination that consisted of two codes – one for the project definition and one for the department manufacturer. These projects were created when a customer (in this case the military) completed a technical proposal listing requirements and specifications that had to be met. Upon completion of all the technical specifications, the design received its appropriate code. Strategic and ballistic missiles were assigned a number and the letter R (Raketa – Rocket) such as R-1, R-2 and R-3.

From here, the development followed a sequence of: 1) technical proposals; 2) advanced designs; 3) the outline design, where the first working drawings were produced, and the development and manufacturer designation was assigned (which for ICBMs was generally 8A or 8K) and completed with a two digit number; and 4) the working design. For OKB-1, the designs assigned to the bureau were generally 8K or A 50s and 70s. The R-5 therefore became the 8A51, while Korolyov's new design of ICBM (the R-7) was designed 8K71.

Orders for creating new elements of defence and space hardware came from the top leadership of the Communist Party of the Soviet Union, with administration handled by a range of governmental bodies, such as Ministries, Chief Directorates, State Commissions and Committees. The chain of command was made easier, as the top leaders of these governmental bodies were also high party officials of the CPSU.

Chain of command originated from the Secretariat responsible for defence and space, in the Defence Council of the Presidium. The order was then issued to the Central Committee, where the Defence Industries Department handled issues

Chief Designer of OKB-1 S.P. Korolyov at the Kapustin Yar missile range in 1953.

concerning the space programme. From there, orders were passed to the top leaders in government to award the work to the appropriate OKBs.

Development of the flight operations element of the Soviet space programme originated in the Government Decree of 13 May 1946, to commence industrial and administrative development on a special sector of machine building. This was the normal euphemism for a defence-related project – in this case a ballistic missile programme.

Three months later, the NII-88 for Jet Armaments in Kaliningrad, near Moscow, was assigned early work on ballistic missile development, with overall supervision by Dmitry F. Ustinov of the Ministry of Armaments. Stalin was still fairly reserved concerning the idea of a Soviet-developed ICBM, believing that a Soviet ICBM would emerge from the German group. However, he ordered Ustinov to report to him directly, over and above the heads of ministers.

Three departments were created within NII-88. K. Tritko (as Chief Engineer of Plant) headed the experimental and testing facility, a Special Design Bureau (SKB). Within this was a group of departments working on the design of the missile, and the third section was a group of scientific subdivisions that undertook field research in material science, strength testing, aerodynamics, engines, fuels, flight and telemetry.

On 8 August 1946, Tritko delegated responsibly for long-range missiles to Korolyov, who was appointed Chief Designer for Ballistic Missile Development. Korolyov had V. Mishin as his deputy, and a team of 52 engineers.

Korolyov evolved a design that featured clustering the first and second stages together. This allowed the vehicle to lift the proposed nuclear warheads, with an additional capacity to incorporate upper stages for placing payloads in orbit at some future date.

In 1951 a group of mathematicians headed by M.V. Keldysh, at the V.A. Steklov Institute of Mechanics, USSR Academy of Sciences (MIAN), completed a study on the concept of clustering smaller rocket stages on the sides of a larger stage. Tikhonravov had first proposed this concept in his rejected 1948 satellite proposal, but from this, Korolyov asked the Keldysh group to investigate the mathematical principals of this idea.

At the time that the R-7 design was frozen, the Russians were working on far larger atomic weapons than were the Americans. Therefore a much more powerful missile was required – far larger than the American Atlas or Titan. A single or two-stage vehicle would have been extremely inefficient and difficult to transport, and although the larger UR-500 engines were in development they would not be available for the next decade.

With the R-7 manufactured in Moscow it was decided to use a rail transportation system to move the vehicle to Tyuratam. A core and strap-on design provided the advantage of a two-stage missile without the bulk and height of a 'stacked' two-stage design. The first stage split into four strap-ons that separated from the core after two minutes of flight. Following separation, the core stage continued to burn as the 'second stage'.

The following year – after experiencing development problems on his own engines – Glushko adapted the cluster principle, realising that it would be easier to build one engine with several smaller chambers, rather than one large chamber as he had previously attempted. The design he developed featured four combustion chambers that were fed by one set of pumps and associated equipment. From 1954 to 1957, Glushko incorporated these ideas into the development of his RD-107 and RD-108 engines, designed for use on the R-7.

By the early 1950s, Stalin had come to realise that a purely Soviet ICBM programme was the best option. A decree for work on the design of an ICBM by NII-88, and an engine with 500 tons of thrust to meet military requirements (to lift the warhead), was signed on 13 February 1953. For Tikhanrarov and his team this was at least an indication that there could soon be a launch vehicle capable of placing a small satellite into orbit, and they immediately began to work on designs for such a satellite.

Although the February decree initiated research into the feasibility of an ICBM, it was not an authorisation to begin any research and development on one specific design. Therefore, other design bureaux, including OKB-586 (Yangel) and OKB-52 (Chelomei), were working on their own ICBM proposals.

One of the major problems that faced Korolyov during the early 1950s was to develop his ICBM (designated R-7, and named Semyorka – 'Little Seven') towards a flight test programme as quickly as possible. It was a huge and important programme, and thus required the cooperation of a number of other design bureaux. To help him achieve this goal, Korolyov called upon the Council of Designers to provide a united front to government departments and also to coordinate all work across the bureaux. As a direct result of the work on ICBMs, OKB-1 grew in size and importance, with other smaller projects gradually being reallocated to other bureaux over the ensuing months and years.

The R-7 design originated from several sources. In 1947, M. Tikhonravov had proposed grouping more than one rocket together to provide a larger lift-off thrust, and discarding them as they emptied. His so-called 'rocket packet' was an early idea of staging. In the early 1940s, Korolyov had design ideas for ballistic missiles with solid fuel as a propellant, believing that liquid-fuelled rockets would be unable to provide sufficient thrust to carry the warhead to its target. After examining German V2 liquid-fuelled missiles, however, he began to change his mind; but although his work was originally based on V2 technology, he realised that he would need much larger rockets, with more powerful engines, advanced fuels and more effective launch techniques than had been experienced during the V2 trials.

The Kapustin Yar missile test site would be acceptable for the intermediate vehicles, but for the planned ICBM, a new and larger site would be required. Eleven Russian refurbished V2s were eventually launched (with mixed results), to be followed by the Russian-built R-1, which was more successful. Between the R-1 and the R-7 was a difficult development programme of three other missiles.

R-2 development began in June 1947, and featured a separable warhead. Weighing 20 tons, it was capable of delivering a payload up to 372 miles. It was first launched on 26 October 1950 from Kapustin Yar, and, powered by Glushko's RD-101 engine, did indeed fly 372 miles. The R-2 entered service in 1951, and was a technological step forward from earlier designs; but the missile that was planned to follow it was even bolder.

The R-3 was designed to deliver a warhead to 1,864 miles. Design studies between 1947 and 1949 examined the feasibility of achieving the distance with just one or two stages. This research concluded that it was not practicable to develop a single-stage vehicle beyond 900–1,200 miles. The first Soviet atomic bomb was detonated in August 1949, and Korolyov proposed that the R-3 would be the logical vehicle to deliver the operational version of the bomb. It was hoped that the R-3 would reach US air bases in England, and that it would be ready for flight tests in 1953. The problem with the missile lay in its demanding technology as a new vehicle. Tests on the engines for the vehicle – Glushko's RD-110 – revealed a 35% increase in thrust over the R-2, but also that the combustion chamber was not capable of withstanding the intense heat and vibrations due to the increased thrust. The R-3 was terminated in 1952, but the design team continued to work on the principle for a while, to increase their understanding of rocket technology. The RD-110 was of an earlier design than the RD-107/RD-108, and the added years of experience and experimentation probably helped to overcome the vibration problem on the R-3/RD-110 combination.

Success finally came on 15 March 1953, with the flight of the R-5. This missile featured a new engine delivering 44 tons of thrust. Glushko's RD-103 was capable of taking the missile 745 miles. At the time, Korolyov had study plans and proposals for a range of follow-on missiles and launch vehicles – one of them being the R-7.

During this period the case for developing nuclear weapon delivery by using cruise missiles was being proposed in both the Soviet Union and in America. The American Navaho missile had the capability to deliver a 15,000-lb payload 5,500 miles, at Mach 2.75, and with an accuracy of 0.25 miles. Korolyov, of course, knew of these plans, and set about developing two Russian cruise missiles – Sorokovka and Burya. Shortly afterwards, the Soviet espionage network revealed that America was developing the Atlas ICBM. Korolyov was instructed to pass the cruise missile designs to aircraft design bureaux, and to concentrate on long-range ballistic missiles.

In May 1953 a Spetskon No.2 (commission) meeting was convened to discuss the status of the missile programme. By then both the R-1 and R-2 were in series production, and the R-5 Intermediate Range Ballistic Missile (IRBM) and the R-11 Short Range Ballistic Missile (SRBM) were also ready for testing, although the R-3 ICBM was still behind schedule. Korolyov insisted that the R-3 should be scrapped in favour of an ICBM with a larger range (his R-7), although it was still in competition with the aircraft bureau cruise missile designs.

This comment certainly surprised the commission, and resulted in a heated debate concerning the adoption of Korolyov's reasoning against continued development of the R-3. Throughout the year, Korolyov took his case to those with influence in the decision. These included the Commander-in-Chief of the Artillery, Marshall M.I. Nedelin (whose branch of the military would be the primary customer for the missile), and the chief nuclear scientist, Kurchatov. It is unclear whether these arguments were to have any influence on the result, but the R-3 was finally abandoned in favour of the larger vehicle.

A significant and deciding factor in choosing the R-7 was in the progress of nuclear warheads after 1949. One of Korolyov's plans featured a design study of ICBM designs that could lift 3 tons, on a staged vehicle designated T-1. In October 1953, a successful test of a 5-ton thermonuclear warhead indicated that the T-1 would be redundant before it could be developed. In addition, the new warhead was well beyond the lift capability of the proposed R-3. Another reason was that the R-3's range did not extend to America. It was not necessary to reach Europe, as that was already covered by the R-5 and it seemed logical to Korolyov to produce a booster that could not only land on American soil but could, with little modification, also place a payload in orbit around the Earth.

Towards the end of 1953, V. Malyshev, Head of the Strategic Missile and Bomber Programme, visited the NII-88 facility in Podlipki and instructed a revision of the design of the R-7 to incorporate the new warhead. The original requirement was a lift-off mass of 170 tons in order to lift a 3-ton warhead. Now they had to develop a lift off mass of 300 tons, allowing the R-7 to lift a 5.4-ton payload.

A decree issued on 20 May 1954 initiated the development of the R-7, the design of which was already fairly advanced. Korolyov completed the Draft Project (EP) after the decree, and this was finally approved during July and August 1954. In support of the R-7 development, as soon as the project was approved, work on the support projects began, primarily on the selection of a suitable launch site.

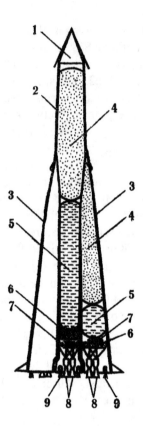

The R-7 satellite launch vehicle, showing the propellant tanks and stages. 1) The protective nose cone housing the first Sputnik inside a protective conical shroud; 2) the central core stage Blok A; 3) the strap-on first stage Blok B, V, G, D; 4) the liquid oxygen tank for the core booster; 5) the kerosene tank for the core booster; 6) the turbo-pump; 7) the RD-108 (core stage) and RD-107 (each of the four strap-on stages) engine assemblies; 8) twenty exhaust nozzles (four per stage); 9) twelve vernier engines (four on the core stage, and two on each strap-on stage).

Design features of the R-7

Soon after N. Khruschev took over the leadership of the USSR (after the death of Stalin in 1953), he visited Korolyov and was shown the design of the R-7. To the Soviet leader, it looked like nothing more than a huge cigar that would never launch. But Korolyov's confidence – the passion in his eyes, abundant energy and sheer determination – convinced Khruschev that the R-7 would indeed fly.

The design featured a central core (Blok A), with a four-chamber RD-108 engine providing 96 tons of thrust (all tonnages are approximate mass values) and 304 seconds burn time. The 92.65-ton propellant mass was a combination of liquid oxygen (LOX) and kerosene. The stage measured 94.3 feet in length and 9.6 feet in diameter, and weighed 6.51 tons empty. Strapped around this central core were four additional stages (in Russian, Blok B, V, G and D). Each of these featured a four-chamber RD-107 LOX/kerosene engine, providing 102 tons of thrust and 122 seconds burn time, from a propellant mass of 39.64 tons. These Bloks each measured 64.9 feet in length, and had a diameter of 8.7 feet.

On top of the vehicle was the provision for the warhead. It was here that Korolyov also allowed for the addition of upper stages in this design to take payloads into space. The first-stage thrust of the R-7 was, in fact, a combination synchronised firing of both the central core and the four strap-on boosters. When the four boosters had burned out, they would be explosively separated and allowed to fall back to Earth, while the central core continued to burn as Stage 2. Any upper stages would in reality be termed Stage 3.

R-7 engines: RD-107 and RD-108

The main engines for the R-7 had been developed at Glushko's OKB-456 design bureau over a three-year period. The combination of Korolyov's R-7 and Glushko's engines became one of the mainstays of the Soviet/Russian space programme over the next fifty years. The data presented by the Soviets indicated that the RD-107 used in the strap-on stages offered a thrust of 102 tons – about 30% greater than the H-1 engines used on America's Saturn 1B a decade later.

The R-107 design was simple but effective. It featured a four-chamber engine, with two smaller vernier chambers fed from a single turbo-pump. By adapting this design, it significantly reduced the length of the engine and also reduced the launch weight of the whole vehicle. This left a useful margin for payload capacity.

The turbo consisted of two main centrifugal pumps that fed oxidiser (LOX) and the fuel (kerosene) around the engine. Two auxiliary pumps powered a gear transmission that fed hydrogen peroxide to the gas generator, and liquid nitrogen to the tank pressurisation system. This arrangement delivered 114.6 lb of oxygen and 46.3 lb of kerosene to each engine chamber every second.

The oxygen was delivered directly to the propellant mixing head through a central pipe, while the kerosene was delivered to a ring-shaped collector, located near the chamber outlet. From the collector, the kerosene was distributed through cooling channels to the mixing head. The fuel was then vapourised using 337 injectors, arranged in ten concentric circles.

The combustion chamber design resembled a cylinder, with a flat injector head. The main chamber was 17 inches in diameter, while the nozzle throat was 6.5 inches in diameter. Each chamber was of a braze-welded construction.

A firewall of heat-resistant bronze was constructed, with brazed milled fins to attach it to the outer pressure jacket, which was subject to the highest thermal stress from the ignited engines. The less stressed areas were brazed directly to the jacket through a corrugated spacer that provided propellant ducts to cool the chamber. This design was extremely light, yet was still able to withstand 57.7 atmospheres and 3,250° C in the chamber, dropping to 0.038 atmospheres and 1,690° C when exiting the chamber. The chamber was cooled by a combination of the fuel passing through the corrugated spacer and an internal curtain formed by the rows of injectors.

For fine trajectory control, the vernier chambers could gimbal, on all of the engines, from ignition. Operation and shut-down was fully automatic, via commands

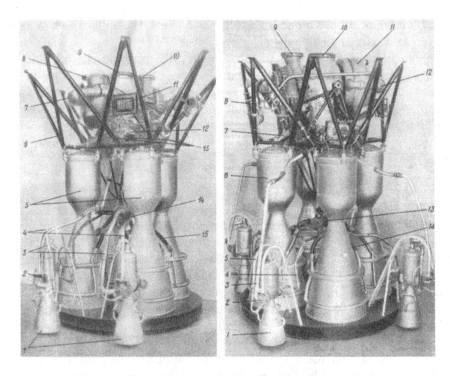

The RD-107 (*left*) and RD-108 (*right*).

An engine test firing as part of the R-7 development programme.

from the onboard flight control system. Pyrotechnics ignited the engine, which built up to full thrust while the restraint devices held the rocket on the launch pad. As propellant was fed into the combustion chambers under pressure, the levels of supply were maintained in the tanks by the pressurisation system, forcing the propellant towards the engine intakes. As the gas generator started, the transfer to full thrust was commanded. During powered flight, the thrust and the mix of fuel to oxidiser was controlled by a combination of commands from the flight control system and from the propellant feed sub-system.

The RD-108 was similar in design to the RD-107, and was used in the central stage. It differed from the RD-107 in that it featured slightly different performance parameters, four vernier chambers and modified control units. This was necessary, because the engine was ignited and shut down differently and had a longer burn time. It was lit first at launch, along with the four strap-on boosters, but continued to burn once they were discarded.

THE CREATION OF A COSMODROME

Finding a site

With the R-7, Korolyov had a vehicle capable of delivering an intercontinental nuclear warhead or placing a payload in space. The size of the vehicle meant that a new launch site had to be found, to cope with processing of the hardware for flight and for managing its launch.

The target zone for testing the R-7 was on the Kamchatka peninsula, between the Sea of Okhotsk and the Bering Sea, on the eastern edge of the USSR. Knowing the range capabilities of the rocket (not less than 4,375 miles), a backward calculation revealed three possible launch sites located in the south-central area of the USSR, in Kazakhstan.

The Kapustin Yar site near Volgograd (known as the State Central Trail Field) had been in use since 1946, but was too small to handle the R-7 and was also too close to Western listening posts, especially those on the American bases in Turkey.

General V.I. Vozniuk headed the State Commission that conducted a review of all proposed sites, including Kapustin Yar. Ground surveys were ordered, and over a period of several months the Commission heard testimony from numerous individuals and made personal visits to each site before making their final decision. Final approval for the construction of a new launch complex came from the Council of Ministers on 20 May 1954, and the actual selection of the primary site was decided upon towards the end of the year. The Commission chose the site near Tyuratam, in Kzyl Odra Raion in the Republic of Kazakhstan. At the time, Tyuratam featured nothing more than a rail stop. The local settlement consisted of a couple of two-storey houses for the railway men, and a couple of dozen mud-plastered wooden houses and tents for a group of prospectors and geologists. They had been looking for oil, but had found only salt water!

The site was a very harsh and remote location, and the topography was bleak and stony. In the depths of winter, several feet of snow covered the steppes, and in the height of summer the region endured scorching desert heat.

There were six primary reasons for choosing this site: 1) it was located 1,000 miles from the Afghan and Iranian borders – far enough for ground activities to go unobserved; 2) it enjoyed a relatively low annual rainfall of only 10 inches; 3) the vast and largely uninhabited desert surrounding the site ensured safety in the event of failed launches and dropped stages; 4) the sheer size of the area would enable the construction of two radio guidance posts 310 miles from the pad, to assist in guidance data during the R-7 ascent; 5) the small station of Tyuratam was already on a rail line (although admittedly primitive) from Moscow to Tashkent on the Syr-Darya River. This would make transportation of building materials (and later, of the huge rocket stages) from the industrial north much easier; 6) as the most southerly location of the three sites considered, it had the advantage of the highest possible initial launch speed (due to the Earth's rotation), which was a 4% increase over that from Kapustin Yar. This was a strong argument for trying to launch any satellite, and one that Korolyov frequently advocated.

There were, of course, disadvantages concerning any of the sites chosen, and Tyuratam was far from ideal due to its remoteness and its proximity to the potentially hostile southern border regions of the USSR. The harsh climate guaranteed extreme temperatures in summer (50° C and sandstorms) and winter (–30° C and icy, penetrating winds). There was no local surface water within easy reach, and although this ensured no natural flooding on the pads, it also implied restricted drinking water. The area was also known to be infested with rodents and snakes, and was prone to variants of the plague.

A harsh history

From ancient times, copper had been mined in the area of Dzhezkazgan (in Kazakh, copper mine). It also had an early connection with spaceflight. In the 1830s, Nikifor Nikitin, a citizen of the small township of Baikonur, was expelled by the Tsarist authorities to Kirghiz, on the grounds of his talking about flights to the Moon. The newspaper *Moskovskiye Gubernsiye Novosti* reported that the authorities had thought that hard work in a copper mine would return him to his senses.

In 1905 the railroad from Orenberg to Tashkent was completed, and in 1908 a consortium of joint stock companies, including the British Atbasar Copper Mines Ltd, discovered a larger copper field at Dzhezkazgan and coal fields at Baikonur. Near to these mineral locations, between the rivers of Karsakpi and Komula, a large factory was planned for processing the copper, powered by the coal reserves. During discussions, plans for a railway spur from the main line were discussed, but were deemed to be too expensive. With a shortage of labour in such a remote area, it was decided to lay a temporary narrow-gauge railway line for the transportation of men and equipment across the region. Work and facilities remained until 1919, when the British pulled out. The workers had to deal with disease and hardship, and living in such a primitive location, felt they were 'convicts without chains'.

Following the October 1917 Bolshevik Revolution, the whole area was designated as a primary source for copper and coal production for the USSR. Improved rail links with Moscow helped ease the transportation issues, but into the 1950s it was still only a settlement with large quarries. The land was unsuitable for pasture or for the plough.

Research and Trials Field No.5

In preparation for the construction of the cosmodrome, the first contingent of around thirty soldiers arrived on 12 January 1955. Preliminary surveys of the land

An army construction team begins the building of the Tyuratam cosmodrome in 1955.

had revealed the best locations for living quarters for the construction teams, and the responsibility for the building of both the civilian and military facilities was assigned to the Engineer Corps. In charge of this construction was Engineer-Colonel G.M. Chubnikov, while the supervision of pad facilities was under the control of Vladimir Barmin's design bureau. Korolyov argued that the complex needed to be completed in just two years, and in order to do this it would require 24-hour work.

The ministry decision to build the launch complex at Tyuratam came on 12 February 1955. The site received the official designation of Research and Trials Field No.5 of the Defense Ministry (NIIP-5). Lt-General Aleksy I. Nesterenko, known for his organisational skills, was appointed as the first Head of the Cosmodrome, with the Deputy Minister of Defence, Marshall M. Nedelin, helping to recruit the best specialists to staff the complex. In command of the site was Construction Engineer Georgi Shubnikov, and the Chief Engineer was Alexander Gruntman. The designer of the whole cosmodrome layout was Alexei Nitochkin.

Due to the remoteness of the site, the first major construction project at the complex was the creation of a main road from the railway station to the proposed pad area, known as Site Number 1. Construction was to include the actual pad; two radio guidance systems; nine tracking, remote sensing and orientation centres; and launch support installations. In just six months, the battalion of engineering troops had risen to 5,000, adding the problem of finding accommodation for the growing army of workers.

The problem was solved by the small village of Zarya (Dawn), which would grow to become what is now known as Leninsk. The area is also called Zvezdograd (Star

Living quarters at the Tyuratam cosmodrome in the late 1950s. The village settlement of Zarya (Dawn) became the town Leninsk on 28 January 1959, and is also known as Zvezdograd (Star City).

City). Originally intended to be a temporary military hostel for the duration of construction, the first stone was laid on 5 May 1955, and for a while, the 'town' consisted only of tents and wooden houses on the banks of the Syr-Darya, about two miles from Tyuratam. In order to maintain as much secrecy as possible, the postal location was changed to Number 10, Tashkent 50. The town was not without its problems. Apart from the harsh weather, there were frequent fires in the canvas tents and wooden buildings. At night, ground squirrels (whose bite carried diseases) dug holes that had to be refilled during each day, and avoidance of snakes was an additional problem during construction.

It took until the end of the summer of 1956 for the first two-storey brick building – the barracks – to be erected, but this was soon followed by streets, houses and offices. Most of the workers had originated from the forestry regions of central Russia, and in the barren desert they soon missed the sight of thousands of trees. They therefore decided to plant trees over a vast area, which they then called Soldalski Park.

Digging the pit
Due to the urgency of the work, the construction team could not wait for the completion of the railway and road network (which would not be completed until 1 November 1955), so a fleet of hundreds of trucks, earth-movers and excavators was employed to shift material to and from each location. This mini-army created giant clouds of dust from the unsurfaced roads that the vehicles themselves had carved out of the ground. From sunrise to sunset, the dust became so dense that headlamps had

Construction begins at the Tyuratam launch facilities.

to be used in broad daylight. To support this army of vehicles, an on-site sawmill, wood-conditioning plant, concrete plant, and a vehicle repair shop and garage were constructed. These last locations were in frequent demand, as the harsh conditions took their toll on the vehicles. But the workers liked to boast that 'out here, only machines break down, not men.'

Work started on the excavation of Pad 1 in July 1955, and used the best topography available to assist construction. The ground consisted of clay topsoil on a layer of coarse sand, supported by hard clay. To begin with, no machine could reach the area, and teams of workers resorted to using picks and shovels (and, reportedly, even bare hands) to dig. As they dug deeper, the soil was so compact that explosives were required to loosen it. The huge vent hole, which would expel the rocket's exhaust gases while sitting on the pad, was constructed by two teams. With one team drilling and digging, the second crew used more than thirty bulldozers, pushing sand and clay up the slope as far as the huge excavators could reach to extract the fill. The sand was kept for the concrete needed to fabricate the pad slab.

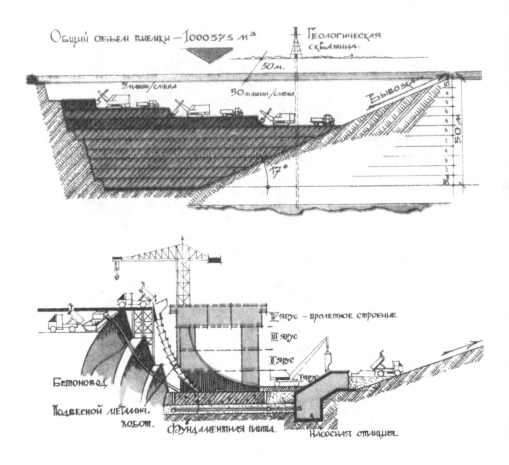

Soviet drawings illustrating how the pit and stadium were constructed, showing the huge evacuation of soil and laying of concrete to form the launch complex that is still used almost fifty years later.

In three months, 1.3 million cubic yards of earth was moved – 13,000 cubic yards per day (and night, under floodlights provided by a steam generator housed on a special train; the electricity supply from Leninsk would not be completed until 1956). The 'Pit', as it was termed, was claimed to be the biggest man-made hole in the world.

Hot summers and cold winters

Actual launch pad construction was ready to begin in August 1955. On the 19 August, the first pad construction teams arrived at the remote train 'station' of Tyuratam. They soon called it 'one house, one yard, one fence'. Many had formerly worked at the Semipalantinsk atomic test site, located near the Irtish River in north-east Kazakhstan. This was another remote site between the towns of Omsk, Karaganda and Novibirsk, and was under the leadership of leading Soviet atomic physicist Igor Kurchatov.

Many had worked close to the site of test explosions, and had suffered burns and injuries from explosives, as well as receiving dangerous levels of radiation. They soon developed serious nose and throat bleeding, and constant noise in the ears. Known as the 'doomed inmates of Semiplatinsk', many thought they would leave the site in a box, or with terminal illnesses. Upon learning of their move to the new launch site, there had been much joy and relief – until they arrived.

Initially arriving at the pad area by truck, they had to sleep on reed mats in the open air, and were informed that they had to build their own shelters before winter set in. The larger shelters housed up to a hundred men, while smaller dugouts housed three, four or five workers. Doctors informed the group that the area was not safe, as the rodent population carried cholera and plague. Many workers also became affected by inhaling poison, as they shovelled buckets-full into scores of rodent holes. The water supply was also not completed, and the frequent fires in the tents burned quickly, trapping many inside.

In winter it was so cold that the men dressed under the blankets, and during the winter of 1955–1956 they had to lay 25 miles of water pipes to allow work on the mains to start in the spring. The hurricane-strength winds reduced temperatures to a bitter –35°C to –40°C, and it was difficult to dig just a few hundred feet of ditch over a couple of days. To overcome the problem, teams worked day and night, laying half a mile of pipe every 24 hours. Concreting in winter also proved very difficult, because it froze when it was poured. To alleviate this difficulty, warming sheets were used to prevent premature hardening of the mix.

At the other extreme, during the heat of the summer, five fire engines were on constant state of readiness, with firemen hosing down the tinder-dry wooden framework. At night, the men sought respite from temperatures of 40° C by pouring water on the floor and wrapping themselves in wet sheets in order to sleep. Unfortunately, drawn by the water, scorpions and tarantulas swarmed in from the desert. The conditions often led to mental or physical breakdown.

With the construction site spread over several miles, it was proving to be a difficult site to work. This was not helped by the authorities operating the 'Kurchatove Rule' by which a project had to be put into service at all costs. When ordered equipment did not arrive, the workers had to sign papers declaring that they did not actually need the equipment! Drawings were often incorrect, and several surveys had failed to reveal uncharted underground lakes. These resulted in frequent wall collapses during digging, after which water began seeping in, flooding the area in a few hours.

A move to a new site would prove costly in time and materials, and while the situation was being reviewed a solution was found by removing the water in the rock by explosion and by laying concrete before the water seeped back. The authorities, after consultation, decided to try this method, and built a sliding bridge to support the tipper lorries, which poured tons of concrete from the edge of the 'bridge' onto the framework of the concrete slab. Using traditional building methods it would have taken months to build a slab of this size, but at Pad 1 it was achieved in just a week! However, the deep ruts created in the soil by the vehicles not only damaged their chassis, but also generated the huge cloud of dust that found its way into everything: engines, petrol, food – and lungs.

The construction of the stadium above the pit at Tyuratam – one of the largest civil engineering construction projects in recent history.

Work on the building of the structure that supported the pad slab had begun on the evening of 19 April, and by June the desired height had been achieved. However, geological cross-sections revealed that the load-bearing capacity would be 20% less than previously estimated. In order to reduce overall pad weight and avoid the more difficult and costly option of expanding the pad area, hollow support columns were introduced to compensate for the difference. A further setback occurred on 19 August, when part of the supporting framework sagged as the embankment gave way. Work progressed at a crawl, as additional support was needed to restrain the construction before progress could be made. Between April and September 1956, 39,000 cubic yards of concrete was poured into the site to construct the enormous flame vent that measured 820 feet long, 330 feet wide and 147 feet deep.

Facilities at the stadium

The cost of creating the launching system amounted to more than half the cost of the entire cosmodrome. Selecting a suitable designer was straightforward, in that there was only one person qualified for the task.

Vladimir P. Barmin was the First Chief Designer of the Spetsmash State Specialised Design Bureau, which had specialised in mobile and static rocket launch platforms for ten years. It was the only domestic enterprise for building launch platforms for both the Army and the Navy.

For this new project, Barmin had to take into account the features of both the R-7 and the location. Transportation was made easier by the rail link, which was to provide horizontal movement of the large multi-staged rocket from the assembly building to the pad. The next challenge was to design the hardware to raise the rocket to the vertical, ensure it was pointing at the correct angle (azimuth) for launch, load the vehicle with propellants, and install the required connections for a comprehensive check of systems and payload during pre-launch.

Theoretical and laboratory studies were conducted to determine the thermal, dynamic, hydraulic, gas, acoustic and radiation effects of launching the R-7 from this pad design. Studies were also made into the effects of vibration, shockwave, pressure and exhaust gases on ground support equipment, and the location of the range-safety 'fallout' area for the spent or failed rocket stages.

As work continued to construct the launch site, the fabrication of the launcher erection system and gantry commenced in a Leningrad engineering plant, again under the direction of Barmin. He had been instructed to complete the work within a year, but actually achieved it in just six months, using time-saving refinements in the design, including the fuelling system, which reduced rocket fuelling time from 90 to 40 minutes.

As there were no heavy lifting cranes installed at Baikonur, all static and dynamic test stands had to be constructed at the Leningrad plant before being shipped out to the new cosmodrome.

By October 1956 – just a year before the launch of Sputnik – the system passed a simulated launch test and was shipped for installation at Baikonur. The previous month, building operations at the launch pad had been completed. Here, workers had begun to refer to the pad towering over the pit as the 'Stadium'. When Korolyov heard this, he informed them that in time it would become a stadium that would have all of mankind as spectators. On 5 October 1956, the road from Tyuratam to Baikonur was completed. In January 1957 a model of the R-7 was installed on the pad at Baikonur, where service lines were laid and launch personnel were trained. During March and April 1957, a full systems test programme of vehicle transport, launch erection, and fuel and control systems was completed. The State Commission officially accepted the launch site after the completion of these tests.

It had been two years and three months since the official approval of the project, and it was now time to test the R-7 that it was designed to launch.

Cosmodrome facilities

The cosmodrome was located in the Kzyl-Orda Oblast of the Republic of Kazakhstan, and had a total area of 2,593 square miles. In addition, 40,262 square miles was isolated from habitation for the fall of spent rocket stages. Of this huge area, 17,760 square miles was located in the Republic of Kazakhstan, 17,231 square miles in Russia, 4,594 square miles in Turkmenistan, and 656 square miles in Uzbekstan.

Launch Pad   Site 2 Pad 1 at the cosmodrome was the first to be built, and was the most famous. In 1961 a second pad was constructed to supplement manned and unmanned launches of the R-7, and was designated Site 31.

Launch Control Blockhouse Located deep underground near the pad. From here the R-7 launch was remotely controlled. Everything from erection on the pad to the filling of tanks and full systems checks were automatically controlled. Communication links were established across the cosmodrome, and when the pad was cleared the launch team could view the pad area via periscopes.

MIK (Site B) The large rectangular assembly building located on the rail link, where elements of the launch vehicle and payload were assembled.

Cryogenic Plant (Site 3) Housed the liquid nitrogen and liquid oxygen supply and other propellants. Construction supervised by Gen-Major A. Berezin.

Experimental Testing Service The department and laboratories for the assembly of the launch vehicle and payload. Technicians maintained close contact with the designers and engineers of the OKBs and manufacturing plants.

Operations Control Service Responsible for telemetry from the rocket and payload during ground tests, launch preparations and powered flight. Sensors around the vehicle sent data to ground stations, and supplied information on the operations of onboard systems. There was also a network of automatic optical devices that recorded the ascent and transmitted data to the Control Centre computers.

Time Service: Calculated and executed a sequence of launch preparation steps through processing, countdown and launch.

Tracking A network of ground-based tracking stations across the ground track of the launch profile to the eastern edge of the launch complex Site 4 Tracking Station Vega; Site 9 Tracking Station Saturn.

Ballistic Support Calculated basic data on the time of lift-off, and the temperature of the engines when filled with cryogenics. This department also guided the ground tracking station aerials during powered flight.

Analysis Department Assessed measurements from the flight trajectory, summarised the data, and submitted a report to the State Commission and Cosmodrome Chiefs. This department provided objective data on the performance of onboard systems of the launch vehicle.

Geodetic Department Established the geodetic fixation of radio direction for each flight, and then provided forecasts for future flight profiles and mission requirements, based on actual flight data.

Weather Service Provided temperature on the ground at launch time, and wind strengths from aircraft and balloons at altitude, gained from weather reports across the USSR and the world's oceans and linked to air and naval support fleets.

Military Facilities In addition to the main R-7 facilities there were several military missile silos and launch facilities, and an infrastructure which formed part of the Strategic Rocket Forces.

Launch preparations

The system of flight check-outs and tests was devised by Leonid Voskrensky, one of Korolyov's deputies based at the MIK, which Korolyov called one of his OKB sub-departments. MIK received components that had passed factory tests, which then underwent full systems tests, both as separate elements and as assembled components.

Each stage of the R-7 was assembled horizontally on a rail-mounted trolley, after which the payloads were attached. Autonomous horizontal testing of the launch vehicle and payload provided system data, recorded on telemetric tape. The data were then analysed, and corrections were made as required. No rocket was moved to the pad until all data were verified at the MIK.

The Vostok R-7 (8K72) carrier rocket in the preparation building at Tyuratam, with the Top Six training group standing nearby.

A Vostok carrier rocket in the preparation building a Tyuratam.

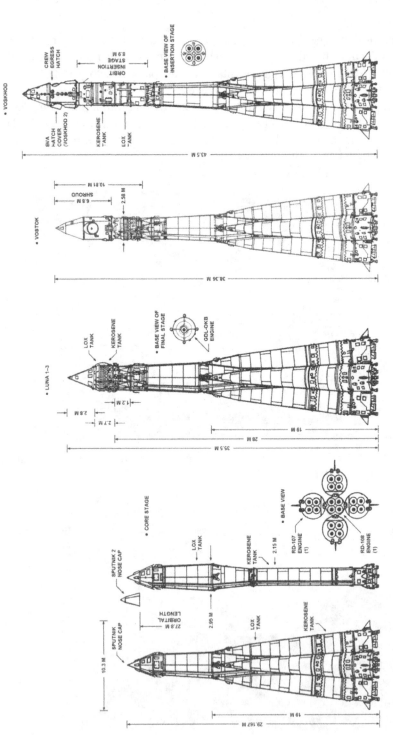

Comparison of variants of the R-7 launch vehicle, showing the evolution during Sputnik 1–3, Luna 1–3, Vostok and Voskhod. (© 1982 Charles Vick.)

The vehicle was rolled backwards to the pad via the railtrack network, the boosters were hoisted horizontally in the MIK onto a transporter adjacent to the assembly bay. Huge doors were opened, and the boosters were then transported towards the launch pad 1.6 miles away. There were no clean-room facilities during this stage of assembly.

The move to the pad took place about two days before the launch, and was a 50-minute journey. Hydraulic jacks stabilised the erector arm and anchored it to the pad structure. Once secured, it took the R-7 from the horizontal to the vertical in 70 minutes. The rail carriage was then withdrawn from the pad area. Vertical lattice support structures were then raised, along with the four support arms that held the launch vehicle at mid-body and at three feet above the pad.

The missile was then lowered several feet below the level of the pad, to direct the exhausts down the flame trench and out into the huge pit. This prevented damage to the pad, or a shockwave running back up to the rocket. The launch table was then moved to the correct azimuth, while support towers and two light masts were brought in to provide access for ground crews making launch preparations.

The R-7, carrying a Vostok payload, is raised from the back of the locomotive transporter at the No.1 pad at Site 2 Tyuratam.

Support vehicles (including a decontamination vehicle) were brought up to the pad during pre-launch operations. Fuelling of the spacecraft and upper stages was accomplished in the MIK, as they would be sealed within the launch shroud. The R-7 was fuelled on the launch pad, as an empty launch vehicle weighing only 20–25 tons was much easier to manoeuvre than was the loaded mass of approximately 300 tons. The LOX evaporation was also topped off at the pad. With all pre-launch operations completed, the launch access arms were retracted one minute before lift-off.

The command to ignite the five first-stage engines comprising of twenty main nozzles was then given, and while thrust built up, the support arms restrained the

The R-7 is raised to the vertical position over the exhaust vent in the 'stadium' construction.

The R-7 (8K71) ICBM on the pad at Tyuratam in 1957.

vehicle. The thrust exhausts passed down below the vehicle and out into the pit. The design of the flame trench was deliberately simple, and enabled the pad to be used repeatedly over many years, even in such harsh conditions.

Upon reaching the required launch thrust, the support arms were released and the R-7 left the pad, its progress tracked by the support facilities across the cosmodrome.

Tyuratam or Baikonur?

The cosmodrome launching station was officially created on 2 June 1955, and from 1961 the complex was publicly identified as the Baikonur Cosmodrome. It was not until 1991 that the Russians used the official name – Tyuratam.

For several years the name of the complex was reported in the West to be either Baikonur or Tyuratam, or both! The actual launch site of Tyuratam is located 1,300 miles south-east of Moscow, 100 miles east of the Aral Sea, 500 miles west of Tashkent and 230 miles south-west of a little town with the name Baikonur, although following the first announcements of unmanned launches from the site, its actual name was withheld for military reasons. The site was described as being in

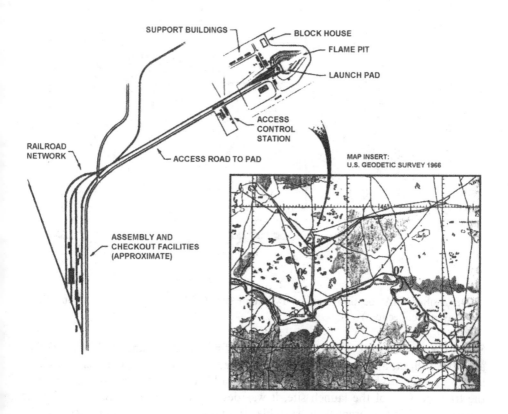

Map showing rail links of the Tyuratam area. (© 1982 David Woods.)

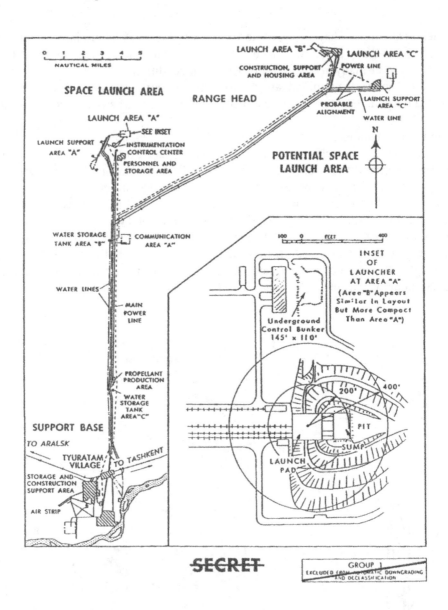

Declassified CIA location map of Tyuratam site. (CIA declassified 1997, from the Pesavento collection.)

Soviet Central Asia – a vast area. For the later first manned launch to be classified as an official flight, and to register the achievement with the FAI, a launch site and a landing site had to be named. In order to mislead Western intelligence agencies as to the true location of the launch site, it was described, in 1961, as Baikonur by the Soviets, who even erected a suitable signpost outside the entrance to the cosmodrome.

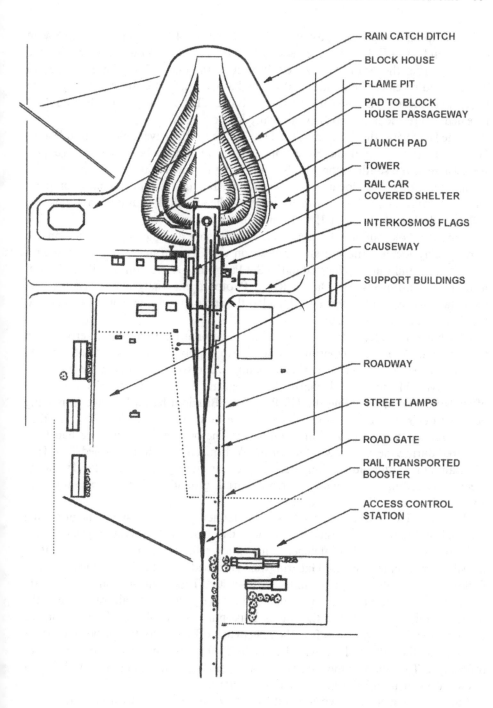

RAIN CATCH DITCH

BLOCK HOUSE

FLAME PIT

PAD TO BLOCK
HOUSE PASSAGEWAY

LAUNCH PAD

TOWER

RAIL CAR
COVERED SHELTER

INTERKOSMOS FLAGS

CAUSEWAY

SUPPORT BUILDINGS

ROADWAY

STREET LAMPS

ROAD GATE

RAIL TRANSPORTED
BOOSTER

ACCESS CONTROL
STATION

Detail of Tyuratam launch pad area. (© 1983 Dave Woods.)

The map references were posted as 47°.4 N and 63°.4 E, but in reality they were 45°.6 N and 63°.4 E. It was some years before the real location was admitted by the Soviets, during the Apolo-Soyuz Test Project in the mid-1970s. However, the US military had revealed the existence of the pad as early as 1957, during U2 spy-plane flights over the area. By the early 1970s Landsat/ERTS satellite pictures had publicly revealed where the launch pads were really located.

In the late 1950s, the Americans used the best maps then available – World War II German maps – to find the Soviet missile site, known to be under construction somewhere to the south of the Aral Sea. Locating the site became a high priority for U2 missions. From the documents, it was found that an old British mining company had run a railroad north-east from a 'sleepy rail head' called Tyuratam, and this was still recorded on World War II German military maps.

Using these old rail lines for guidance, the new spur and the pad construction site were quickly located and cross-referenced on the German maps. These revealed a small hamlet and rail stop located in the Bet Pak Dala Desert, south of the Aral Sea, near the north-flowing Syr Dar'ya River. Photographic analysis revealed the new construction area, and a huge launch pad at the end of a rail spur, some 15 miles into the desert from the main line.

As Chief CIA Information Officer for the National Photographic Interpretation Center, Dino A. Brugioni's primary role was to supply all known facts of the area that was to be flown over during forthcoming U2 missions, including the provision of any available maps and charts. In preparing the briefing notes for military officials up to and including the US President, Brugioni had a key role in naming new features revealed by the spy-planes. Once these names had been selected they were rarely changed, to avoid any confusion. The best maps of the area had been prepared during the war by Mil-Geo of the Wehrmacht, and were labelled Tjur-Tam Bf (Bahnhof – railroad station) near a pre-war quarry. Interestingly, Tyuratam means (according to Brugioni) 'arrow burial ground' – which was not very appropriate for a military test centre.

When the first ICBMs and Sputniks were launched, with the site name not being revealed by the Soviets, President Eisenhower instructed that Western leaders were to be briefed on what US intelligence had known for some time as the Tyuratam Missile Test Center. When Gagarin was launched, the Soviets named the site Baikonur – actually a town 200 miles north-east of Tyuratam. Although well aware of the U2 photography (as it was also one of Gary Powers' main objectives), the Soviets continued to refer to the site as Baikonur, in fear of any US attack in the event of war. This attempt to fool the Americans officially continued beyond ASTP in 1975, to the fall of the Soviet Union in 1991. Only then was the site known officially as Tyuratam. If the Americans had been fooled, it could only have been for a few weeks or months at most, in the late 1950s.

Future civilian engineer, cosmonaut Vladislav Volkov, was once asked why such a remote place was chosen for the site of a cosmodrome. He replied (half jokingly) that it was achieved by maximising the coincidence of inconveniences. First, find a place which is uninhabited, and extremely difficult to reach by any mode of transport, including camel or donkey. There should be no water but plenty of sand, preferably in excess, and the wind should stir up enough dust to block all vision, with the added bonus of tasting it in your mouth as well as in the food – allowing you to recall the benefits of your wife's cooking. Volkov reasoned: 'If you find such a place, you can build a cosmodrome there!' This was exactly what the Soviets did!

R-7 test flights

Between 1954 and 1956, Korolyov and his team progressed with the development of the R-7 launcher at a (self-inflicted) brisk pace. It was only after a personal visit to OKB-1, by Premier Khruschev and other State leaders in January 1956, that political pressure was added to Korolyov's desire to achieve a quick flight of the R-7. Khruschev's party was shown round by Korolyov, and saw a full-scale mock up of the R-7. Korolyov explained that it would not only be able to fly 8,000 miles to strike the United States, but that any air defences would not be able to stop it. On hearing this, Khruschev reportedly told Korolyov that his work would not be rushed to bypass a full check-out, because the success of this rocket was crucial for the Soviet Union.

For the Soviet Premier, stating the power of 'his' rockets was to become one of his favourite political tools in future years. He hoped, believed, and stated, that the R-7 would solve the problem of creating an intercontinental nuclear delivery system that was capable of defending the country without bankrupting it in construction and maintenance costs. Without seeing an R-7 fly, Khruschev had boasted that the nuclear missiles had rendered all other weapons obsolete.

The construction of a second operational launch site, north-west of Tyuratam, would also be authorised for R-7 launches. This was the Plesetsk complex, on which construction began in July 1957. For Korolyov, political support from the Soviet leader would be beneficial to the designer in gaining support for bolder ventures. Korolyov was granted almost unlimited direct access to Khruschev, allowing the Chief Designer to appeal personally to the Soviet leader for his needs, and bypassing the bureaucracy of ministerial levels and oversight committees. A powerful but risky liaison had been achieved, but it was dependent upon the success of the R-7. Political and military pressure to achieve early success with the R-7 test programme was so important that Korolyov was asked to telephone Khruschev directly if there were any setbacks.

By the spring of 1956, General of Engineers V.M. Riabikov had been named as the Head of the State Commission for the R-7 test programme. The first missile actually capable of being launched arrived at the cosmodrome in March 1957.

The first launch of an R-7 (8K71 – project designator for the ICBM) occurred on 15 May 1957, and ended with an explosion just 100 seconds after leaving the pad as the Blok D strap-on broke away from the ascending rocket. Subsequent investigations revealed that fuel had leaked from the engine pump, and had been ignited by the heat from the surrounding engines to create a huge fireball. Before the R-7 flew again, several modifications were required, including the addition of steel heat-shields on each strap-on, facing the core unit that reduced the heat flux in the aft compartments.

A second attempt ended in three pad aborts on as many days, due to valves being left open in the wrong position. Following these setbacks, a Special State Commission met to discuss the test failures. It had been an inauspicious beginning.

The Commission found a wide range of problems with the R-7 design, and argued both politically and technically against further pursuit of the R-7 programmes. Meanwhile, Khruschev's own political problems, and his victory over the so-called 'Anti-Party Group' during June 1957, narrowly managed to secure continued support for Korolyov's ICBM programme.

The third test, on 12 July, also failed only 33 seconds into the flight as all four strap-ons suddenly separated as the rocket rotated about its longitudinal axis; but on 21 August 1957 a fourth test was deemed a success. The missile flew as programmed, but the dummy warhead disintegrated upon re-entry. The announcement of this flight was made public by the Soviets one week after the event, with the aim of advertising the capability of the missile to the West. However, the event generated more attention within the USSR than outside of it.

Following further ground tests and test flights, engine performance was modified, with new valves and reductions in propellant pressure flow. Operationally, 20% less fuel consumption was required for coolant jacket feeding.

On 7 September 1957 a fifth test was performed, repeating the 21 August flight. The missile flew, but once again the warhead did not survive. Despite this, the State Commission decreed that the R-7 was ready for immediate deployment to the military, although this was postponed for several months until a new warhead became available in January 1958. By then, resistance against incorporating the R-7 in the military was growing.

By end of 1958, 97 changes had been incorporated into all areas of design, structure, engines, control systems, warhead, and launch equipment. This was not surprising, as it was Soviet practise to carry out more system flight-testing than trouble-shooting before flight tests. This also meant that significantly more design modifications would be needed once the flight tests began than would be expected in Western programmes. Most of these changes were only small adjustments rather than major design revisions. As the size and complexity of the R-7 was relatively new technology in the 1950s it would be expected that some modifications would be needed after in-flight data were analysed.

A launch failure of an 8K71 test vehicle occurred on 24 December 1958, followed on 17 February 1959 by the firing of the first 8K71 ICBM from the initial production batch – which also failed. Between 28 March and 18 June 1959, six 8K71 test launches were flown successfully before the first series production ICBM 8K71 flew without incident on 30 July 1959. On 22 October this was followed by a fully loaded and operational 8K71 with peak range capability, its dummy warhead splashing down in the Pacific Ocean target area.

By January 1960, even as the R-7 was put into the operational service of the Soviet Rocket Troops of Strategic Destination, its future as a carrier for nuclear weapons still remained in doubt, but its future as a space carrier rocket was about to begin.

As an ICBM, the R-7 could not be stored 'on the pad' or in launch silos, and was deployed only in limited numbers. The authorisation to begin the development of the

R-7A (8K74) was given in July 1958, and it made its first flight in December 1959. It was this variant that was deployed as an ICBM at both Tyuratam and Plesetsk.

Korolyov's missile had the added problems of using LOX and kerosene, which are not storable, and it was therefore not a practical 'instant response' missile. A better design was Yangel's smaller R-16, which used storable propellants, while Korolyov's R-9 design, while still using LOX/kerosene, was also smaller than the R-7 but could be stored in silos, and needed only propellant loading before being launched.

THE ARTIFICIAL SATELLITE PROGRAMME

The International Geophysical Year

In 1952 the American International Council of Scientific Unions (ICSU) appointed a committee to expand upon a US proposal for a third International Polar Year (IPY). The plan suggested the period of 1957–1958, which would be at a time of maximum solar activity, to study the whole Earth over an 18-month period. The plan also included the suggestion of renaming the IPY as the International Geophysical Year (IGY).

The first IPY, in 1882, established a precedent for international scientific cooperation. Scientists from twenty nations agreed to combine their efforts to study polar conditions. Fifty years later this was followed by the second IPY of 1932. Lloyd Berkner, head of the Brookhaven National Laboratory in New York, was a member of a group of American scientists who, since 1950, had been evaluating the acquisition of measurements and observations of the Earth and upper atmosphere from a 'distance above the Earth'. Berkner's suggestion was to obtain support for such a large project from a third International Polar Year. This idea was endorsed, and was put forward to the ICSU.

The ICSU acted as the headquarters of a non-governmental organisation of scientific groups. In 1953 it adopted a list of scientific fields to be covered by the 64 countries that were to participate in the IGY, and in September 1954 these were submitted to the Comité Spéciale de l'Année Géophysique Internationale. The list proposed the inclusion of the scientific use of a satellite. The following month, the CSAGI endorsed the proposal. The date of the endorsement was 4 October 1954.

Military proposals

As with the Soviets, American prospects for orbiting a satellite originated from military studies of captured German V2s at the end of World War II. Their ideas also included a USN study proposal for orbiting a 2,000-lb unmanned instrumented 'spaceship' (launched by a single-stage liquid-propellant rocket) that could remain in orbit for several days.

By March 1946 the 'Earth satellite vehicle' was abandoned as being too difficult to pursue, although the principle appeared to justify a major programme. Over the next decade, each branch of the armed forces conducted further studies, but these often failed to balance a military or scientific advantage against the expected cost of the project.

During 1952, articles in three issues of *Collier's*, by leading American figures interested in space exploration, advocated the prospect of scientific space exploration and a possible threat to the nation's security. Dismayed by the apparent lack of progress to put America in space, the editorials asked 'What are we waiting for?', and expressed alarm that a Communist nation might beat the United States to space, as by doing so they could perhaps control Earth by placing atomic bombs on manned space platforms.

In May 1955 a military evaluation prepared by the US Department of Defense for the White House, the CIA, the Joint Chiefs of Staff, and the National Security Agency, reviewed a proposal for adapting existing rocket technology to launch a small satellite before the end of 1958. The report also stated that the Soviets had recently announced the creation of a high-level commission for interplanetary communications, and that a group of top Russian scientists was working on a satellite programme. The report also highlighted an 'unmistakable relationship to intercontinental ballistic missile technology,' but also implied that the USSR was unlikely to outstrip the US in a satellite 'race'.

In commenting on the evaluation, President Eisenhower's specialist assistant, Nelson A. Rockefeller, was concerned with the 'costly consequences of allowing the Russians to outrun [the US] through an achievement that will symbolise scientific and technical advancement to people everywhere. The stake of prestige that is involved makes this a race that we cannot afford to lose.' He proposed that the US government should announce a civilian satellite project, and stress that military missions were not involved. This would pre-empt any Soviet claim that such a satellite was a threat to world peace.

On 28 July 1955, Eisenhower's Press Secretary, James Hagerty, announced that the President had approved plans for America to launch small Earth-orbiting satellites as part of the IGY. The satellite would be scientific in nature – but a press release issued by the Pentagon stated that plans would include technical contributions from the three armed services. Though not stated clearly, there was real concern that Russia might beat America to the 'high ground'.

As early as 1947, an Army Air Force feasibility study by Project Rand (a division of the Douglas Aircraft Company) had suggested the capability of launching a spaceship to orbit the Earth, and had foretold of the advantages of putting the first artificial satellite in orbit. 'To visualise the impact on the world, one can imagine the consternation and admiration that would be felt here if the United States were to discover suddenly that some other nation had already put up a successful satellite.'

A Soviet satellite?

During the early 1950s a number of satellite projects were already under development by military and scientific organisations in the Soviet Union, but all of them were low priority – until 1956, when Korolyov's ICBM programme neared completion and offered the opportunity to place a payload into orbit. Then the idea of a scientific satellite 'piggybacked' on an R-7 became a top priority.

In November 1953, A.N. Nesmeyanov, of the USSR Academy of Sciences, stated that satellites and Moon-shots were already feasible, and with Tsiolkovsky's work being recognised, the Americans had reason to believe in the capabilities of Soviet science and technology. This belief was further supported by a Radio Moscow announcement in March 1954 (at the time of commitment to the IGY) that Soviet youth should prepare for space exploration. The very next month, Moscow Air Club announced that it was beginning studies in interplanetary flight. This public show of a country dedicated to exploring space was not supported by hard evidence, however, and although Russian delegates had attended many of the meetings establishing the IGY, they had listened to the discussions but had offered no comments, voiced no objections, nor asked a single question.

By 1954, Tikhonravov – then a scientist at NII-4 – had been working on his own designs for some time. He presented his calculations and files (including some foreign press cuttings on a proposed US military reconsat programme) to an official of the Council of Ministers, who, convinced of the seriousness of the US plans, decided to support Tikhonravov's research at NII-4, to create a Soviet satellite that had a dual role for later variants that encompassed civilian and military objectives.

Meanwhile, Korolyov began pushing for approval of a scientific satellite programme during the same period. In a memo dated 26 May 1954, he explained that the Soviets would be capable of creating an artificial satellite 'in the next few years,' and also insisted that Tikhonravov's satellite design team be transferred from NII-4 to OKB-1. But this received no immediate reply.

In the annual report to the USSR Academy of Sciences, submitted on 25 June 1955, Korolyov advocated a major Soviet space effort. In the report, he stated that launches of both satellites and *piloted* spacecraft had been proven possible, and that a satellite design project could be completed by the end of 1956. He also proposed a single commission to coordinate all work on space, instead of the existing two commissions – the Blagonravov Commission for the Investigation of the Upper Layers of the Atmosphere, and Sedov's Commission on Interplanetary Communications. The following day he sent a memo, with the same satellite and human spaceflight proposals, to the Central Committee. The memo proposed a unified method for the control and management of space programmes and the creation of a special government sub-division (podrazdelenie) on the development of Earth satellites, and also addressed the problem of piloted flights on rockets. However, it is unclear whether these reports had any affect on the final decision to proceed with a satellite.

On 2 August a Moscow press announcement stated that the Soviet Union planned to place a satellite in Earth orbit during the IGY. At that time, Academician Leonid I. Sedov was attending the 6th International Astronautical Federation Congress in Copenhagen, and added fuel to the story by stating that the Soviets would launch their satellite in 1957. He also said that it would be much bigger than anything that the Americans were planning.

For the Americans this seemed absurd – just another inflated claim of Soviet technical superiority that had become common during the 1950s. For other observers, it was disappointing that both nations were announcing nothing more than a space race. President Eisenhower did not want any sort of race with the Russians, particularly in what looked like being an expensive race to orbit.

Authorisation to proceed with a Soviet satellite project came on 30 August 1955, with a meeting of the Presidium of the Academy of Sciences, during which the new project was discussed with both Korolyov and Glushko in attendance. The proposal was not a major weapon project and thus received a lower priority, not requiring approval from the party leadership.

Korolyov stated that his R-7 would be ready in 18 months, and as such, the Academy should draw up a consolidated scientific programme of space research. To facilitate this, Keldysh was named chairman of the Third Commission on Spaceflight (commonly known as an Interdepartmental Commission), along with Korolyov and Tikhonravov as his deputies. This commission eventually became the powerful coordinating body that Korolyov had wanted, and was similar to the Council of Chief Designers.

Object D

By 1956, Tikhonravov's design of satellite had envisaged a 2,645-lb laboratory known as Object D (Orientrivanny D – Orientated D). It resembled an instrument module and a cone-shaped re-entry capsule, and would be launched by Korolyov's R-7. The D designation followed earlier letter designations that were assigned to warhead development payloads of the R-7. In addition to being an unmanned scientific satellite, Object D also implied future application as a military reconnaissance satellite.

In January 1956, Khruschev visited OKB-1 to view the R-7. During the visit, no mention was made of anything larger in the bureau's space ambitions, until, almost at the door, Korolyov was asked if there were other applications for his rockets. Korolyov took Khruschev's group to a room full of exhibits from rocket research programmes of the 1940s and 1950s, including some ideas for satellites. However, no enthusiasm was shown concerning an Object D model, and even Korolyov's enthusiasm for Tsiolkovsky and the prospects for Object D did nothing to impress Khruschev. Korolyov immediately changed his approach, and stated that the US was developing a satellite programme but did not have a launcher large enough to put a satellite into orbit. Almost bursting with pride, Korolyov stated that with the R-7, the Soviets could launch a satellite many times heavier, and well before the Americans. Khruschev liked this idea, and indicated his support for such a project, but not at the expense of the missile programme.

By the end of January 1956 the satellite had become a high priority and an integral part of the R-7 programme, with Khruschev an influential supporter of the idea. Official authority to proceed with Object D was finally given by the Central Committee and the Council of Ministers on 30 January 1956. Korolyov was pronounced Chief Designer and was mandated to launch Object D by the end of 1957 in a programme structured to beat the US into space. He signed the completed Draft Project (EP) for Object D on 25 September.

Object PS-1

By the end of 1956, matters were not proceeding well for Korolyov, despite OKB-1 being given independence from NII-88 during the summer of that year. Adapting to a new Five-Year Plan, and a weakening of industrial management and control processes, led to serious delays for the satellite project – so much so, that at the meeting of the Academy of Sciences on 14 September 1956, Keldysh asked for urgent help, blaming the lack of cooperation between various ministries and research institutes.

The causes of frustration included the solar cells that could not be completed, because the appropriate ministry refused to supply the silicon to make them. The Department of Chemistry was not helpful in supplying fuel for the upper stage of the rocket, and the radiotechnology industry was behind schedule. In addition, the Institute of the Academy of Sciences was slow in delivering engineering models of instruments that were to be installed in the satellite. Keldysh concluded that if these 'vendors' were not pressurised by higher authority, the US would launch first and steal priority from the USSR. By the end of the year, some improvements had been made; but overall, Object D was still far behind schedule.

Korolyov received authorisation to concentrate his OKB team on satellite and launcher design. At the same time, Tikhonravov's satellite design group finally moved to OKB-1. But as coordination and economic problems still plagued the project in late 1956, Korolyov realised that to meet the deadline launch late of 1957 and beat the Americans, he would need to launch a less complex payload to orbit. At the end of 1956 he began work on a simpler satellite – prosteshii sputnik (PS) – to replace Object D. Korolyov faced stiff opposition from Keldysh, who did not agree to the change, but by January 1957 Korolyov had begun to convince his colleagues that this was their only hope if they wished to beat the Americans.

On 5 January 1957 a memo was sent to the leadership proposing that two new smaller satellites (PS-1 and PS-2) be constructed between April and June. In the memo, Korolyov also requested that a new variant of the R-7 be developed, specifically to carry a smaller and simpler satellite. He further suggested that a single, powerful, government committee should be appointed to supervise all work, and that the whole project should receive the highest government priority, regardless of cost. On 25 January 1957 his endorsements were approved.

Upon receiving the authority to proceed, Korolyov completed the design specification for PS-1 by the end of the very same day. His design was a simple, highly reliable device, transmitting a radio signal that could be received by the maximum number of listeners around the world. With an eye towards the future and new spacecraft designs, Korolyov's satellite incorporated two important features: a round shape, from which the effects of atmospheric drag could be calculated; and a pressurised interior, to ensure even cooling of instruments and to provide a test of the ability to maintain pressure and temperature in space. This would be an important requirement for future piloted missions. One of the young engineers who worked on the fabrication of PS-1 was apprentice coppersmith Gennedy Strekalov, who was later to become a cosmonaut.

By March 1957 Keldysh had procured the most powerful computers available to conduct trajectory calculations for the satellite project. By July, plans had been drafted to modify the R-7 to carry the satellite payload. This was the 8A91 variant, capable of lifting a payload of 3,300–3,750 lbs. Object D fell by the wayside as efforts shifted to prepare PS-1 for ground tests in May. Upon completion of these, it was shipped to Tyuratam on 24 June, but had to wait until after the first successful test of the R-7, which took place on 21 August.

On 20 September the satellite project was at the forefront of launch operations at Tyuratam. With the R-7 now flight proven, the State Commission (under the Chairmanship of Riabikov) met to approve the launch of the First Scientific Satellite of the Earth. The plan was to launch on 6 October, but this was moved forward two days. Korolyov had heard that on the 6 October the Americans were to present, in Washington, an IGY paper entitled 'Satellite In Orbit' and thought that this might be an indication of the imminent launch of their satellite.

However, KGB sources were certain that no US orbital launch was planned for that year, even though the Americans had indicated a sub-orbital launch of their Vanguard rocket before the end of 1957. Korolyov has been described as being a no-risk, cautious man who did everything step by step; but there is evidence that, in addition to the official line from the Kremlin that space spectaculars were important propaganda tools for military purposes, the Chief Designer was envisioning his own spectacular feats to steal the thunder from the Americans and to achieve his own space records. It was therefore decided to launch the first space satellite on 4 October 1957, for which the State Commission prepared a press statement for Tass to release after the first orbit had been confirmed. The date was exactly three years after America had indicated that it was to launch a satellite during IGY.

On 15 September, in a speech marking the centenary of Tsiolkovsky's birth, Korolyov told his audience that future plans would soon include a second satellite in Earth orbit, a parcel on the Moon, a probe around it, and then piloted spaceflight.

Sputnik

The PS-1 (and its back-up, PS-2) was of a spherical shape, 23 inches in diameter, with four whip aerials 7.8–9.5 feet in length. The sealed sphere (weighing 185 lbs) was constructed of aluminium alloys, in two halves, and supported both the 'scientific' equipment and the power supply from batteries. Its radio transmitter operated at frequencies of 20.005 and 40.002 MHz, with wavelengths of 15 and 17.5 m, providing a signal of around 0.3 sec resembling a telegraphic message – a 'beep-beep-beep'. Onboard power supply ensured a three-week lifetime.

The objectives for the satellite were stated to include the determination of the characteristics of the upper layers of the atmosphere, and in particular the density recorded from the deceleration of the satellite at its lowest point in orbit. This was of the 'utmost value in determining the life duration of artificial satellite and spaceships yet to be launched'.

In addition, the ionosphere was to be studied by observing the propagation of radio waves emitted by the satellite. This had applications in the development of reliable systems of communication with space vehicles. Special studies were to be made of temperature patterns, in order to evaluate the tolerances required in the design of future equipment such as the temperature regulation system being prepared for the biological experiment on the next satellite.

On 4 October 1957 at 10.28 pm MT, with most of the 'Top Six' designers present,

(*Left*) the R-7 (8K71PS) used to launch Sputnik, and (*right*) the launch on 4 October 1957.

Boris Chekanov depressed the lift-off button in the blockhouse, upon command from Alexander Nosov (chief of the launch control team), who was watching through the periscope. The R-7 ignition command began the ascent into history for PS-1, which entered orbit five minutes later. The orbit was inclined at 65°.1 to the equator (the same as that which the Vostok missions would fly). Orbital altitude was 588 miles, providing a lifetime of 92 days. Sputnik would re-enter Earth's atmosphere and be destroyed on 4 January 1958.

After the 'all clear' was given at the pad, the launch team waited nervously in a bus for 90 minutes for the satellite to circle the globe. Only when they heard the 'beep-beep-beep' did they take out the champagne and vodka to celebrate.

Standing on the launch pad, Korolyov gave a speech to his co-workers: 'Today the dreams of the best sons of mankind, including our outstanding scientist Konstantin Eduardovich Tsiolkovsky, have come true. The storming of space has begun.'

American reactions: a technological Pearl Harbor

Sergei M. Poloskov was a member of the Soviet delegation assembled at the National Academy of Sciences in Washington DC. On 30 September 1957 he had presented a paper at the opening sessions of a CSAGI conference on rocket and satellite activities for IGY. Entitled 'Sputnik', Poloskov stated that this Russian word, meaning 'travelling companion', was the name chosen for their satellite, 'now on the eve of launch'. Many at the conference had been involved in the IGY since its inception, and though surprised by the implications of an imminent launch, when the launch actually took place their surprise paled in comparison to that of the general American public. They were taken aback by the thought that a Soviet-made object was orbiting over their heads.

Across America news of the Soviet success caused a period of intensive soul-searching and revelation in the military and in the academic and scientific communities. It shocked the general public that a nation thought by the United States to be second-rate in science and technology had clearly demonstrated their ability to put a new instrument in such a place as to seriously threaten the very heartland of the American nation.

Both psychologically and patriotically, America was unprepared for such an event, and it gave rise to many questions about what had happened to prevent the US achieving this feat first. Were they lagging behind in their new technology? Did the fault lie with Congress, the military, the scientists – or even in the very heart of the American education programme? All eyes looked towards the expected launch of Vanguard to bring America back to a level of equality.

Lt-General James M. Gavin, who had a long association with the American ICBM programme, called the launch 'a technological Pearl Harbor'. Some tried to state that it was nothing more than a neat scientific trick, while others suggested that perhaps the Americans had captured the wrong Germans at the end of Word War II!

Soviet leader Khruschev was not even at the launch, but was returning from a holiday in the Crimea. In an almost casual attitude, he indicated that when the satellite was launched he was telephoned and told of the event, and that everything was going to plan. 'I congratulated the entire group of engineers and technicians on this outstanding achievement, and calmly went to bed.'

Sputnik 2
After achieving success with Sputnik, Korolyov had the authority to carry out at least one more of his other satellite projects. The January 1956 decree of the Central Committee and Council of Ministers (calling for the launch of Object D) still remained in effect, and it was expected that this would follow Sputnik into orbit. However, Korolyov had other ideas.

There was a second R-7 that was assigned to the PS-2 satellite (the back-up to PS-1), but which was not required, due to the success of the first article. There does not appear to have been a clear schedule for launching PS-2, and most of the leadership of OKB does not appear to have expected a launch. Immediately after the first launch, Korolyov's top deputies went on holiday. It was a short holiday, and they were not happy about having to return early for Korolyov's new intensive programme to launch a second satellite so soon after the first. Apparently, it was done in such a hurry that they worked from the Chief Designer's sketches and not from a full design plan.

Sputnik 2 was Korolyov's idea to capitalise on the success of Sputnik. There was no strategic advantage in launching a second R-7 missile with a heavier satellite over

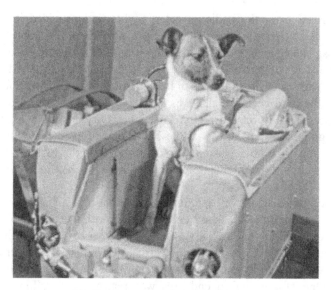

Laika – passenger on Sputnik 2.

large distances, as the launch of Sputnik had very dramatically proved the capabilities of the missile to the Americans. But by flying this new satellite, and incorporating a significant biological payload, Korolyov could imply that the Soviet Union had a large, robust space programme, orientated towards human spaceflight, and that Sputnik was not an isolated stunt but was part of a much more structured programme.

Less than one month after Sputnik opened the Space Age, a second R-7 carried Sputnik 2 off the pad at Tyuratam on 3 November. The satellite weighed more than 1,000 lbs, and remained attached to the final stage of the launch vehicle. It demonstrated Soviet capability for placing 7.5 tons in low Earth orbit (although most of this mass was the still-attached core of the launch vehicle at approximately 6.5–7 tons). The biological passenger was a dog, called Laika. However, even though the Soviets had the capability of putting a living creature into orbit, they had not yet determined how to retrieve it alive. Despite this, the flight of Sputnik 2 proved that a living being could survive in space, and indicated that manned flight was the ultimate objective, for the near future.

As with Sputnik, Sputnik 2 was enclosed, for launch, in a nose cone that was spring-ejected upon reaching orbit. On the front of the satellite, instruments were installed to study ultraviolet and X-ray radiation from the Sun. Behind this was a spherical airtight container (the original PS-2), which housed support equipment and transmitters for gathering a second set of data complementary to Sputnik, and a support structure for the detectors.

The cylindrical pressurised cabin containing Laika was at the rear. This also contained the necessary life support system and food supplies (via a feeding device) to sustain her during the flight for up to seven days. The life support system featured a regenerative plant and temperature regulation sub-system.

Results derived from the ballistic rocket flights had provided important data concerning acceleration and deceleration, but not about prolonged periods of weightlessness. The chance to use an artificial satellite for this purpose was one not to be missed for the research teams. Results from rocket flights had proved that animals could withstand short periods of high gravitational loads without any visible disorders. All that had been revealed was an increase in arterial pressure, pulse and respiration rates directly related to the force of gravity encountered. In the short period of weightlessness at the top of the ballistic arc (up to three minutes), these rates reduced. What the Soviets wanted to determine was the variation of rates over several hours or days. Laika – one of ten dogs trained for the flight – would provide that first set of figures.

The capsule was designed with an automatic method of air circulation, whereby active chemical compounds released oxygen for breathing, and absorbed carbon dioxide and excess water vapour. This was also a further test of the technology for a prototype life support system planned for the manned spacecraft then in development and evolved from the earlier design used during the 'sounding rocket' programme. One of the conical end-caps was removable to allow access to the capsule, and was also fitted with a perspex porthole to illuminate the cabin and allow observation of the passenger by the onboard camera.

In a few brief statements, the Soviets reported that Laika had stood up well to the flight to orbit, and after a brief and insignificant increase in heart rate, had settled down well to her new environment. Data revealed that although some change was noted in the functional state of the dog's circulation and respiratory systems, it soon returned to normal and thereby indicated the absence of any harmful influences on the body. The Soviets also stated that the selected design of the life support system, to ensure as near 'normal' conditions as possible, was correct, suggesting that they were to continue to develop this flight-proven system for their manned spacecraft.

Laika became the first space celebrity and 'the most famous dog in history'. She also became a heroine of the dog lovers and animal welfare groups around the world who, upon learning that she was not to be recovered, began to urge an international daily minute of silence in tribute to her. They even went as far as sending a telegram to Moscow, suggesting that the next Soviet spaceship included a 'Russian Hero' instead of a dumb and defenceless animal. The Soviet Union was accused of a lack of compassion in using live animals for such a cruel experiment.

After ten days in orbit, it was reported that the oxygen had run out and the space heroine had slipped into unconsciousness – but the truth emerged several years later. Apparently, Laika was in great distress very early in the flight, as vital insulation was either stripped from the capsule as it failed to separate correctly from the carrier rocket or as the insulation was ripped away during ascent (similar to the fate of the Skylab orbital workshop sixteen years later). There have been conflicting reports as to whether the satellite should have separated. The original plan was to have the satellite separate, but the separation system was excluded so that the mass limits would not be exceeded. Other reports indicate that the system was flown but failed to operate, or that a more stable orbit would be achieved with the core still attached. Whatever the reason it caused a sharp increase in internal temperature, and Laika began barking and moving about the tight cabin in an 'agitated way'. She was apparently frightened, but otherwise unhurt, as the temperature rose to 40° C. According to Dr Oleg Gazenko, a pioneer of Soviet bioastronautics, she died from the unbearable heat inside the capsule, well before the oxygen was depleted.

Sputnik 2 re-entered Earth's atmosphere and incinerated during the night of 13–14 April, after 2,370 orbits. Soviet scientist I. Ivanov recalled that the flight of Laika made it possible to talk more boldly about manned spaceflights. But Laika's fate also demonstrated the dangers of such flights.

The flight of the second Sputnik continued to surprise, and in some cases alarm, Westerners. Not only were the Soviets able to orbit a satellite – they could also put animals in orbit. The Soviet lead in space technology was dramatically underlined when the USN Vanguard rocket exploded on the launch pad, in full view of TV cameras, on 3 December 1957. The following day, banner headlines printed what most of America was thinking, 'Oh what a Flopnik!', above a picture of an exploding Vanguard. Eight weeks later, on 31 January 1958, von Braun's US Army team finally succeeded in launching America's first satellite, Explorer 1, into orbit. The satellite was the first to reveal the belts of radiation around the Earth that were named after the American scientist whose instrument discovered them – James van Allen. At last, America had something to shout about.

Sputnik 3

The long delay between Sputnik 2 and Sputnik 3 was a result of technical difficulties rather than specific leadership decisions. Object D, authorised two years earlier, remained on the launch schedule, but fell behind Sputnik 1 and Sputnik 2, and was not ready before April 1958. The first launch attempt, on 27 April, failed, and the vehicle was lost. Finally, on 15 May 1958 the back-up Object D was successfully launched as Sputnik 3.

Onboard operations were apparently not a complete success, as the tape recorder system, which had been designed to store and transmit measurements from onboard instrumentation, malfunctioned. This restricted data downlink to real-time transmissions, as the spacecraft flew over ground stations.

The Soviets called this satellite a scientific space station, enabling research in all the subjects that had been assigned to them as part of the IGY. What was more impressive to the West was the sheer size of the object orbited, weighing 2,900 lbs and measuring nearly 12 feet in length. In an orbit of 140 × 1,168 statute miles, the satellite remained in orbit for 692 days and completed 10,037 orbits before re-entry on 6 April 1960.

A scientific package of 2,134 lbs contained twelve experiments. These included a magnetometer for recording solar radiation; instruments for registering photons in cosmic rays; magnetic and ionisation magnetometers; ion traps; electrostatic fluxmeters; mass spectrometer tubes; measurements of heavy nuclei in cosmic rays; intensity of primary cosmic radiation; sensors for recording micrometeorites and data collection on the ozone layer; upper atmosphere density; pressure measurements; and the function of solar cells to gather electrical power from the Sun.

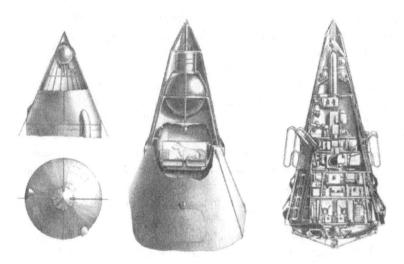

Comparison diagram of the first (*left*), second (*centre*) and third (*right*) Sputniks.

American response

After Sputnik 3, Khruschev boasted that America was sleeping under a Soviet Moon and, although not mentioning the R-7 failures, reminded the West that America had had six rockets blow up before they reached orbit. A missile gap was becoming very obvious to the citizens of America.

The Eisenhower administration was accused of not being strong enough in pursuing a defined space exploration programme, for losing the lead in nuclear arms and ballistic missiles, and for allowing the competition between the US Army, Navy and Air Force to hold back progress in placing America ahead in orbit.

Eisenhower considered that the military could be given their own control of space, with a joint space command, but due to the demands of secrecy and national security it would not wish to discuss any results and new developments. He therefore adopted a proposal by Vice-President Richard M. Nixon to convert the National Advisory Committee for Aeronautics into the civilian (and space science orientated) National Aeronautics and Space Administration. Congress supported the idea, as did the important scientific community and the members of the NACA that would be absorbed in the new agency.

On 29 July 1958 the National Aeronautics and Space Act was signed, creating the civilian agency NASA with effect from 1 October 1958. In August, Eisenhower assigned the task of developing a man-in-space project (previously under development with the military) to NASA. On 7 October, NASA Administrator Dr T.K. Glennan, upon reviewing plans for a manned satellite, approved the project by saying 'Let's get on with it'.

Korolyov's future plans

In a letter to the Kremlin, dated 27 May 1959, Korolyov outlined plans to hand over the bulk of his OKB work to civilian institutes as the space programme grew in complexity and diversity (the letter was also signed by Keldysh). He would continue to work on the development of the R-7 and of his next projects – the first probes to the Moon, Mars and Venus. He suggested that overall development of artificial satellites, and planetary exploration in the long term, should be handled by an organisation that combined scientific research institutes; and development of manned spaceflight should be handled by an institute of medical and biological specialists. Unfortunately, no such 'Soviet NASA' was forthcoming, and Korolyov was to abandon any further unmanned Earth orbital satellites in favour of the first lunar and planetary space programmes.

At the same time, OKB-1 would head the development of a major project to put the first humans in orbit before the new American NASA could do so.

American intelligence estimates, 1959

An intelligence report dated 21 July 1959 NE 11-6-59 – prepared by the CIA for the White House (President Eisenhower), the Army, the Navy, the Air Force, the Joint Chiefs of Staff, the Atomic Energy Commission, the National Security Agency, the Department of State and the FBI – presented the current level of Soviet science and technology and its prospects for future development.

The report considered that the Soviets possessed the capacity to move forward with accomplishments in space at a considerable pace. It was believed that there was a major effort to achieve manned spaceflight ahead of the US, with a highly developed research and development programme. High priority was given to missile technology, electronics, meteorology, space medicine, astrobiology, astrophysics and geophysics.

A second report (3 November 1959 – NE 11-5-59) forecast that the main objective of Soviet scientific research was the development of military applications to attain manned spaceflight in support of Soviet propaganda and politics. The report stated that over the following twelve months (to the end of 1960), the Soviets would complete a 'programme of vertical down-range (sub-orbital) flights of a manned capsule, and the orbiting and recovery of a capsule carrying instruments, animals, or a man, from ICBM missile ranges at Kapustin Yar and Tyuratam.'

Future objectives were to include the development of scientific research to gather information to support 'moral and military objectives'. The report reviewed past Soviet accomplishments, and stated that the capacity to place biological specimens in orbit (Sputnik 2) revealed a programme to develop the technology for specimen recovery that was 'highly desirable, if not necessary, prior to manned capsule recovery.' These tests were to be expected in the next 6–8 months, and with the nature of Soviet secrecy, any failure would not necessarily result in adverse publicity, as any setback could simply not be mentioned.

The report predicted that the first Soviet manned flight would use a capsule-type recovery (not a winged vehicle), and that current Soviet missile technology demonstrated the ability to meet this requirement, although reliability and safe recovery still needed to be demonstrated.

The Soviets' long-term goals were thought to be in manned interplanetary travel, including glide tests of a winged vehicle by 1962–63, followed by demonstrations of manned manoeuvring vehicles, rendezvous and docking, and the creation of a space station with 'military functions' by 1965. They also expected manned lunar flights (with large boosters) to begin with circumlunar missions in 1964–66, lunar orbital flights in 1965–66, and a Soviet manned lunar landing by 1970. That year could also see advances in unconventional (nuclear) propulsion systems and large 'super space stations' with closed cycle ecological systems.

The report concluded that a Russian manned flight could be expected between mid-1960 and mid-1961, and with past experience and data, would stand every chance of success, with significant propaganda purposes and an acceptable risk of failure. The document also indicated that the Russians possessed a 'current' capability for sub-orbital manned spaceflight at the end of 1959.

First spacecraft and first cosmonauts

VOSTOK SPACECRAFT DEVELOPMENT

Origins

The development of a vehicle to carry the first Soviet citizens into space began to gather pace in the late 1940s, when Tikhonravov, a member of OKB-1, began work on Project RD-90, in which he designed a cabin to carry two men on a sub-orbital trajectory.

Then, in September 1955, during the 125th anniversary celebration for the Bauman Higher Technical School (MVTU), Korolyov announced that five different manned spacecraft designs for vertical flight were currently under consideration. Korolyov also made clear that the purpose of the programme was 'to ensure that Soviet rockets fly higher and farther than has been accomplished elsewhere to ensure a Soviet man be the *first* to fly a rocket [and that] Soviet rockets and Soviet spaceships are the *first* to master the limitless space of the cosmos.'

The following year, at the height of the unmanned high-altitude ballistic rocket programme, a series of manned sub-orbital flights from Kasputin Yar, using a modified R-5 missile, were considered. Three medical specialists (Abram Genin, V. Sheryapin and Ye Yuganov) had volunteered as candidates, but with the development of the R-7 ICBM taking priority and consuming most of the resources of OKB-1, the sub-orbital proposal was abandoned.

During the spring of 1956, Korolyov had indicated that human flights on an R-7 were possible, but the idea had been temporarily shelved because of the requirement to develop the ICBM programme first. Korolyov's long-range plans appeared to be based directly on Tsiolkovsky's work on the establishment of manned spaceflight capabilities. These plans included the development of nuclear and liquid hydrogen rocket engines, and the assembly of manned orbital space stations to be used to explore regions around the Moon. He also predicted that large boosters and spacecraft would be required to carry out lunar surface exploration and manned flights to Mars.

Designers at OKB-1 continued working on the requirements for manned spacecraft, and by November 1956 these designs were far enough advanced for the

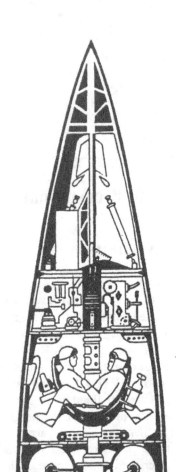

The VR-190 – the first Soviet concept of a piloted spacecraft after World War II. Two 'stratonauts' would complete a sub-orbital trajectory. (Courtesy A. Siddiqi and NASA.)

Council of Chief Designers to begin serious discussions on the potential for launching piloted spacecraft.

Project Section 9 was created by Korolyov, as a department of OKB-1, on 8 March 1957, and was headed by Tikhonravov. The first research projects included designs for the first artificial satellites (with both civilian and military applications), and probes to the Moon, Mars and Venus. By April, the department had completed preliminary research for the creation of a manned satellite, using the R-7 as a launch vehicle. These studies indicated that the addition of a third stage to the R-7 could lift a 5-ton payload into low Earth orbit.

As plans solidified, some argued that a step-by-step programme should be run,

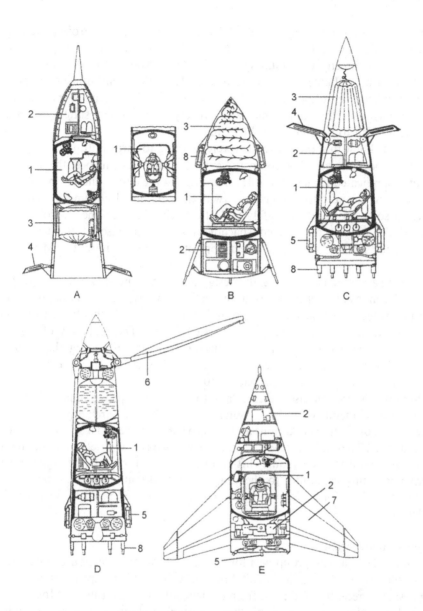

Five concepts for sub-orbital manned spaceflight begun in 1955. A) a design using parachute and stabilising aerobrakes for return; B) a design using parachute and solid propellant rockets for recovery; C) this concept incorporated all three methods of return by parachute, air brakes and retro-rockets, and braking by aerobrakes and rockets; D) the inclusion of helicopter rotor blades with rocket engines at the tips of the blades; E) the use of aircraft-like wings for a glide return aided by stabilising engines. Key: 1) capsule; 2) equipment; 3) parachute system; 4) braking and stabilising surfaces; 5) position stabilising nozzles; 6) rotor; 7) wings; 8) braking engine. (Courtesy A. Siddiqi and NASA, reproduced from a 1987 East German original diagram in *Sowjetischer Raketen* by P. Stache.)

beginning with sub-orbital flights (as NASA later proposed with Project Mercury). Korolyov decided that this was unnecessary, and proposed a vigorous programme that would lead directly to orbital spaceflight. On 1 May 1958, all proposals for manned sub-orbital flights were abandoned, as Korolyov's enthusiasm for orbital flights had spilled over into his team. There was an overriding desire to try to address all the problems at the same time, and Korolyov had often stated that there remained so much to be done in so little time.

The Soviets knew that the Americans were working towards putting the first man into space, and this led to a competitive rivalry that also helped the designers overcome complicated technical hurdles. Throughout the design and testing phase of the programme, the need to be consistent (with the minimum lead-time) added to the pressure to solve any problem safely and effectively. At this time, Korolyov also advocated the pursuit of manned spaceflight at the expense of the unmanned reconnaissance satellite, and drew direct opposition from Ustinov, the Minister of Defence.

In creating a manned spacecraft, the design team was venturing into new fields of research. The areas of re-entry, thermal protection, and hypersonic aerodynamics of a returning vehicle from space, were groundbreaking fields at this time. Computer technology was also a new avenue that was embraced. The Academy of Sciences mathematicians provided a supply of trajectory calculations, using the BESM-1 electromechanical computer. The data revealed that a ballistic re-entry would produce a maximum gravitational load of 10 g.

Tikhonravov's section used these data to conduct a programme of thermal research from September 1957 to January 1958. This research investigated the expected range of heating conditions and surface temperatures on a selection of heat-shield materials. They also reviewed maximum payloads on a variety of aerodynamic forms that featured hypersonic lift-to-drag (l/d) ratios of the spacecraft's aerodynamic lifting capability to the drag force of the atmosphere. Once results were obtained, the information was again fed through the BESM-1 to further refine the calculations.

Korolyov's kindergarten

In December 1957 a new group was formed in the Planning Section of OKB-1, to begin detailed studies of manned orbital flight. Most of this group consisted of very young designers recently graduated from the technological institutes in Moscow and Leningrad. Korolyov called them his 'kindergarten', and many of them – including Konstantin Feoktistov – would later become cosmonauts. The drive and passion of Korolyov and the developing skills of the 'kindergarten' designers often led to conflicts between them.

Feoktistov (then in his early 30s, and holder of a degree in rocket design) held considerable authority in his group. Although he never seemed to raise his voice, he was very obstinate, which caused Korolyov to flare up on more than one occasion. Despite this there was mutual respect, and the young team was assigned the task of developing the manned spacecraft.

Plans for the requirements for the manned spacecraft and the required third stage for R-7 to lift it began to be drafted on 1 January 1958. Over the next ten months, the 'kindergarten' approached the work at fever pitch, solving a host of problems in quick succession. The principles of rocket flight and orbital mechanics had long been evaluated during the development of the R-7 and Sputnik, but returning an object safely back to Earth was one challenge still to be met, even without the additional complexity of ensuring that a passenger would come back alive! As work progressed, reports were passed to Korolyov for approval, and so the work moved on to the next stage.

Studies by Tikhonravov's section had found that for winged spacecraft (that had the highest lift/drag ratio and the lowest net payload weights), the temperatures would far exceed the capabilities of the then current heat-resistant alloys. It was determined that the ratio of lift/drag should be greater than zero, and ideally between 0.0 g and 0.5 g, to provide the necessary lift and reduce the g forces on human passengers during re-entry flight.

It was soon decided that a cone would be a suitable shape for the spacecraft, with a rounded nose and spherical base, having a maximum diameter of six feet – in effect, the headlight shape later adopted for Soyuz. The landing system for the cosmonauts would be designed around ejection at a few miles altitude after re-entry, to land by parachute into a land area of the Soviet Union. This would obviate the need to develop a capsule soft-landing system, so the capsule need not be recovered.

Refinement of the lifting design was another major challenge for the designers, but in April 1958 the Aviation Medicine Research unit found an answer. Testing human subjects in centrifuges, it was found that pilots could endure up to 10 g without ill-effects. This changed the flight profile to a pure ballistic flight (removing the lift/drag requirement) and allowed the design to move quickly to advance project stage. By the spring of 1958, Konstantin Bushyev's department in OKB-1 was developing the idea of a spherical capsule – the simplest design shape – launched by a variant of the R-7.

In May 1958, Tikhonravov proposed development work on a heavy satellite for manned spaceflight. Konstantin Feoktistov and his team were already working on the design of Object OD-2 – intended for manned spaceflight – derived from the unmanned reconnaissance satellite project and featuring a large cylindrical instrument module and a small conical re-entry capsule.

During June 1958, Feoktistov presented these ideas to Korolyov. The design of the descent capsule had been enlarged to 7.5 feet diameter, and also featured an unmanned version of the same capsule, in which the crew-member was replaced by photoreconnaissance equipment. Korolyov preferred this design, and abandoned all other ideas, concentrating all efforts to develop this version. The layout, structure, equipment and materials were each designed by different sections of the team at same time, and each required at least two or three redesigns.

The project name, Vostok (East), was selected from suggestions by members of the design team. One other name that was rejected was Sharik ('Little Sphere', referring to the shape of the re-entry vehicle). Later that month, Korolyov presented the Vostok design to the Committee on Science and Technology, as a two-programme proposal. One was configured as a manned orbital spacecraft, while the

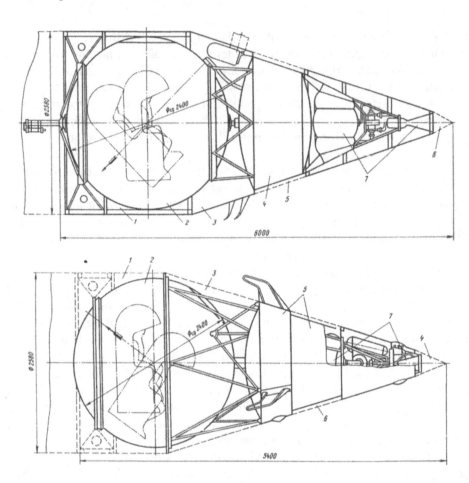

Two concepts for the Object OD-2 spacecraft, the first design for an Earth-orbital piloted spacecraft in the Soviet Union. Originating in August 1958, the designs featured the spherical return capsule later adopted into the craft named Vostok. The conical instrument section on the right of both designs resembles the Sputnik 3 satellite. (Courtesy A. Siddiqi and NASA, reproduced from a 1991 Soviet original in 'Materialy po istorii kosmicheskogo korablya 'Vostok'', Ed. B. Raushenbakh, Moscow, 1991.)

other would be an unmanned orbital reconnaissance platform. Building upon the success and the propaganda coup of Sputnik, Korolyov also sent a letter to the politburo on 1 July 1958, explaining that by August 1958 the advanced project concept would be completed, and telling of the advantages of the project, both politically and scientifically.

Finally, in November 1958 the Council of Chief Designers took the decision to develop the manned version of OD-2, with the unmanned reconnaissance version taking second place. A decree dated 22 May 1959 saw OD-2 re-named Vostok, with a programme that envisaged several variants and authorised the development of the necessary launch vehicles:

Vostok-1 (Object 1K) The unmanned research and development experiment version, intended to test hardware applicable to both the manned and unmanned versions.

Vostok-2 (Object 2K) The unmanned military reconnaissance satellite.

Vostok-3 (Object 3K) The spacecraft for manned spaceflights.

Vostok-4 (Object 4K) A future design concept of high-resolution photoreconnaissance.

Work on the construction of the first Vostok spacecraft hardware began in December 1958, but despite this, final authority for the programme appears not to have been awarded immediately. This was in response to a government call for alternative spacecraft proposals from other OKBs. Korolyov was building an authorised manned spacecraft, but apparently the government had not agreed to actually use it! This was a time of decline in Korolyov's influence. He still had strong support from Ustinov, the Academy of Sciences, and much of the missile industry, but this did not guarantee approval for his projects.

After the creation of a special design section devoted to the project on 15 August 1958, Korolyov, who was personally managing the project, decided to take one final look at a headlight-shaped lifting capsule; but it was decided that there would not be enough time to carry out research on the characteristics of such a design.

On 15 September 1958, Korolyov signed the final project document allowing full production drawings to be sent to the factory and systems testing to begin. A bitter fight with the military over the nature and priority of the manned programme over that of the photoreconnaissance programmes meant that the final decree was not issued until 22 May 1959, when authorisation was awarded for a single design to be used for both programmes.

Only after the programme was finally approved did the search for crew-members commence. The programme to select suitable candidates to fly on Vostok began more than a month after America had selected her first seven Mercury astronauts. On 10 December 1959 the Soviet leadership issued a Comprehensive Decree on Space, in a final agreement ratifying the concurrent planetary probe and manned spaceflight programmes, consisting solely of Korolyov's Tsiolkovsky-inspired plans.

By April 1960 the draft project, defining the numerous variants (and the designations of these variants), had finally been completed. Vostok-2 was renamed Zenit-2, and Vostok-4 became Zenit-4, to distinguish them from the manned programme. On 4 June 1960 the first decree was issued with an amended flight schedule:

- May 1960: completion of two 1KP prototype spacecraft (no heat shield or life support system).
- August 1960: three 1K systems completed for the photoreconnaissance and radio reconnaissance missions.
- September–December 1960: three 3K systems for manned flight.
- 11 October–December 1960: first manned spaceflights attempted.
- The 1K and 3K versions would employ a 2.4-ton re-entry capsule, a service module and a braking engine.

The plans called for the use of a three-stage R-7 that could place a mass of approximately 4.6 tons in a circular Earth orbit at 155 miles altitude, and could contain sufficient payload and life support supplies to support one human passenger, together with a limited amount of scientific equipment. A spherical design for the ballistic entry capsule would withstand 2,500–3,500° C surface temperature during entry, and 8–9 g maximum load. This would require a heat-shield mass of 2,866–3,300 lbs.

To recover the vehicle a re-entry burn would be required, using a –2° entry angle at 62 miles. This would yield a landing accuracy of +109 miles to –62 miles from the target area. The pilot would need to eject at 5–6 miles altitude.

To reduce acoustics and vibration to tolerable levels, suitable insulation was required. The spacecraft would be orientated via a control system, using cold gas jets and flywheels. On the first manned flight, the pilot would not be required to control the vehicle. Sub-systems would include a limited avionics and orientation control system, a guidance command processor, and redundant voice radio. All orbital flight control requirements and the deorbit-braking rocket were to be housed in a module separate from the re-entry vehicle containing the passenger.

The design would feature a functional redundancy in onboard systems. The cabin life support system would include a spacesuit that would also need to feature a four-hour independent supply, in case of cabin depressuration or failure of main systems, and which would still allow re-entry over the next two full orbits, with a further margin of 60 minutes to align the vehicle.

Orientation would be provided by infrared vertical sensors and manual orientation by the pilot. Re-entry would be controlled by a command timer, heat sensor or radio command. For final recovery the pilot would use a parachute ejection system activated by both inertial and barometric sensors. The only system that could not be provided with a back-up facility would be the TDU deorbit-braking engine, as there would be no additional margin in the restricted overall mass. To compensate for this, the orbital flight path chosen would naturally decay in no more than ten days, and adequate provisions would be included to sustain the crew-member in this eventuality.

Upgrading the R-7 to manned flight

Across the Soviet Union, a total of 123 organisations and 36 factories participated in the Vostok project, and 7,000 engineers, designers and scientists worked on the development of the capsule.

By the end of 1958, OKB-1 had designed a new faring and cylindrical adapter section to be placed on the forward end of Blok E, in order to lift Luna probe payloads to the Moon and to test unmanned versions of Object 1K.

In January 1959, OKB-1 began design work on the launch vehicle, while Kosberg's OKB-154 was tasked to develop a new uprated version of the RO-5 (8D719) engine for Blok E stages (later designated RO-7). Over next twelve months, the new three-stage version of R-7 (designated 8K72K) was developed, based on the modified ICBM. This new booster also featured uprated RD-107 and RD-108 engines (8D74 and 8D75). The new Stage 3 also incorporated an improved control

A Soviet representation of the Vostok capsule in orbit. This heavily disguised design gave rise to countless interpretations of the actual spacecraft before it was finally revealed some years later. (Still photograph from the 1961 Soviet documentary *To the Stars Again.*)

system and uprated payload capacity, from 5.04 to 5.56 tons. The engines were an in-house design based on OKB-154's RO-5 engine, and had taken fifteen months to develop. They were finished in December 1960.

Stage 3 (Blok E) featured a single RO-7 engine, burning LOX/kerosene propellants, with a total thrust of 5.6 tons and 430 seconds burn time. The stage measured 10.17 feet in length and 8.53 feet in diameter. Typical data included a dry mass of 1.44 tons, a propellant mass of 7.39 tons, and payload mass of 4.72 tons. The launch shroud, designed to cover the spacecraft during ascent through the lower regions of the atmosphere, weighed 2.5 tons, and was 8.53 feet in diameter and 7.87 feet long, producing a total launch mass of 287.03 tons. The Vostok launcher stood 125.85 feet tall.

When the R-7 was first used to launch the Sputniks and Vostoks, the design was still a State secret. In order to disguise the design, the Soviets provided a range of highly stylised drawings and paintings of what was proposed as the launch vehicle and spacecraft – and they even added tail fins to the spacecraft models displayed. The public first saw a full-scale replica Vostok spacecraft at the Exhibition of Economic Achievement in Moscow in April 1965, and at the 26th Paris Air Show the following June. It was at the next Paris Air Show in 1967 that the first replica R-7 was on display to the public.

Vostok design

One of the first decisions taken in the design of the Vostok was the method of recovery (either under a parachute or by using a glide landing technique), including whether the craft could be steered to a desired point. It was discovered that, in order to effectively slow down the vehicle, a ballistic re-entry would solve the problem of how to steer the craft in the dense layers of the atmosphere, although it would complicate the provision of a lifting capability. To achieve any entry, a braking engine would have to be incorporated into the design and be fired against the direction of flight, with the exact amount of thrust to allow descent into the atmosphere.

Added to these difficulties, the need to sustain the man in the capsule meant that the vehicle would need a life support system, as well as support systems to keep the occupant warm and capable of working. All of these added both to the complexity of the design and to the weight. Another concern was the provision of an effective recovery and soft-landing system to ensure that the passenger would survive at the end of the flight.

This raised other questions. An electrical power supply had to be designed and incorporated into the vehicle. And what about a steering system, or food and water for the passenger? How would the pilot communicate with the ground – by radio, or could the developing TV systems be used? All of this would increase the payload weight, already limited by the capability and dimensions of the R-7, and set further barriers before Feoktistov and his team.

A design emerged, featuring a modular spacecraft with a separate crew recovery section – the Descent Module, or SA – and a compartment that contained most of the support equipment – the Instrument Module, or PO. At the base of the PO would be the retro-rocket package, to initiate the descent from orbit. The maximum weight allowable was 4.5 tons, to be within the lift capability of the R-7. By designing the braking rocket as a non-recoverable unit, it allowed for a larger weight limit on the recovery parachute, and although the pilot ejection system added weight, some of this weight was compensated for by not using soft-landing rockets to cushion the landing.

A suitable soft-landing system (with shock absorbers) would protect the crew-member against the effects of hitting the ground, which, in the event of landing on a rocky surface, could be severe. But it was decided that the development and testing of such a complex system was too time-consuming, and the designers therefore turned to proven aviation ejection seats, which were adapted to fit into the capsule. These could be used both for emergency ejection during ascent or landing, and also as a primary method of crew recovery after re-entry and prior to spacecraft landing.

Despite all these alterations, there was no allowance in the lift capability of the R-7, or in the defined dimensions in the nose cone. The designers were told that they could build the spacecraft in any shape they wished – even a corkscrew if they really had to – provided they did not infringe the maximum dimensions by even a fraction of an inch.

The Descent Module
With no previous designs as a guide, several shapes were evaluated, including cones, cylinders, hemispheres and spheres. Of these, the sphere appeared to be the most promising, with several advantages over other configurations. The heat flux would be less, and its aerodynamic properties were easily calculable. It was also found to remain stable at any speed to which it would be subjected, and provided the

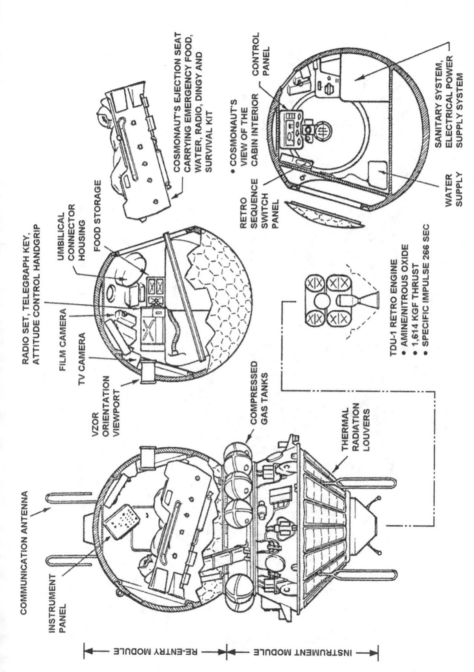

COMMUNICATION ANTENNA

INSTRUMENT PANEL

RADIO SET, TELEGRAPH KEY, ATTITUDE CONTROL HANDGRIP

UMBILICAL CONNECTOR HOUSING

FOOD STORAGE

FILM CAMERA

TV CAMERA

VZOR ORIENTATION VIEWPORT

COMPRESSED GAS TANKS

THERMAL RADIATION LOUVERS

COSMONAUT'S EJECTION SEAT CARRYING EMERGENCY FOOD, WATER, RADIO, DINGY AND SURVIVAL KIT

• COSMONAUT'S VIEW OF THE CABIN INTERIOR

RETRO SEQUENCE SWITCH PANEL

CONTROL PANEL

SANITARY SYSTEM, ELECTRICAL POWER SUPPLY SYSTEM

WATER SUPPLY

TDU-1 RETRO ENGINE
• AMINE/NITROUS OXIDE
• 1,614 KGF THRUST
• SPECIFIC IMPULSE 266 SEC

|← RE-ENTRY MODULE →|←— INSTRUMENT MODULE —→|

An artist's impressions of the Vostok capsule. (© 1983 Dave woods.)

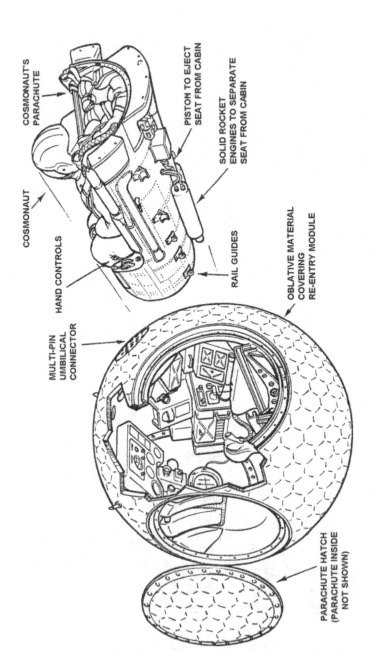

The Vostok cosmonaut ejection system. (© 1983 Dave Woods.)

maximum internal volume for the space available. By placing the centre of mass aft of the centre of the sphere, it would naturally assume the correct orientation during its re-entry trajectory.

In a very early practical test of the shape, Feoktistov had members of his team stand on different landings of a long staircase. Each of them would then drop a plasticine-weighted tennis ball from different heights, into the stair well, and each time the projectile 'landed' in a stable manner. Though crude, the test demonstrated that the real spacecraft could be suitably weighted to achieve self-stability during entry. This meant that orientation and stabilisation devices would not be required for a lifting re-entry, and once slowed down by the braking rocket, the spacecraft would follow a simple ballistic entry. This saved weight on the inclusion of control surfaces, and eliminated the need for the passenger to do anything during the re-entry sequence. It also meant that the structure of the module would require fewer surface breaks in the thermal protection system, which was designed to completely cover the module.

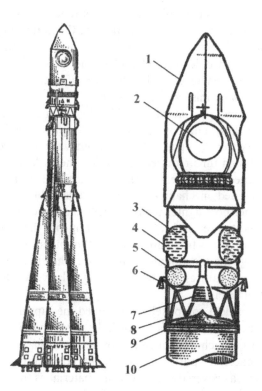

The Vostok payload in launch configuration on the R-7 (8K72K). 1) The launch shroud; 2) the Vostok spacecraft; 3) the torus tank containing liquid oxygen; 4) the upper stage (Blok Ye); 5) the torus tank containing liquid kerosene; 6) two of four vernier engines; 7) the R-7 engine assembly; 8) open truss network connecting upper stage with the central core stage; 9) the attachment ring; 10) the upper core stage of the R-7.

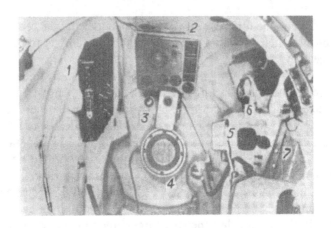

The interior of the Vostok crew compartment.

A programme of computer calculations was followed by wind-tunnel tests, which generated further mathematical data to provide the theoretical proof that the design was the correct choice. Years later, Feoktistov commented: 'We had [but] one aim and a firm belief that we would succeed, and a belief in Korolyov. We knew that he could get the necessary machines and funds and, if need be, political support to ensure that our work was finished successfully.'

The size of the Descent Module was fixed by the volume inside the launch shroud (calculated at 90.5 inches), and covering the whole sphere was the heat protection system. Several ideas were proposed for this, but were rejected due to the cost, complexity, or the design time required to develop them. A subliming material that was developed to protect the nuclear warheads in the ICBM programme was finally selected. This maintained mechanical stability while subjected to extremely high temperature. The innermost lighter pressure shell had structural integrity, but poor heat conductivity, and to prevent heat from radiating into the crew module, an outer thermal protection shield was fitted around it. A thin outermost layer was applied to the finished capsule, giving the module a shiny look when new. This was designed to reflect the extreme temperatures in orbit, but being ablative, would blacken and char in the intense heat of re-entry.

Although the heat shield completely surrounded the capsule, there still had to be provision for hatches, access panels and portholes through the material. The hatches were for the recovery parachutes, entry/exit by the cosmonaut, a ground inspection cover, and the connection port for equipment connections to the Instrument Module. Each hatch was covered with the same material as the capsule to withstand re-entry heating. Any break in the protection system was a weak spot, and although it would be sealed, this added complications to the design and a risk to operations. For the thermal protection engineers, the fewer hatches the designers included in the capsule, the better.

Each of the three portholes was made from refractory glass, and was deeply recessed into the outer covering. There had been discussions about attempting to attach outer covers to the windows for re-entry, as the excessive heat might fracture the glass. This proved impractical, but the concerns remained, and so Korolyov personally intervened and scoured the Soviet Union for a suitable manufacturer of heat-resistant glass for the spacecraft.

The Instrument Module

Once the external design of the Descent Module had been finalised, the team had to decide on the best shape for the Instrument Module that would carry most of the equipment to support the cosmonaut in orbit. As this module was not intended to be recovered, there was no need to include the additional weight of a thermal protection system.

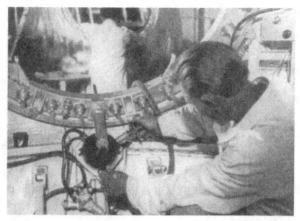

A technician works on the inside (*foreground*) of a Vostok capsule while a colleague works on a second sharik (*background*).

A mated sharik and instrument unit during construction at OKB-1.

This module had to house the braking rocket, support equipment, and the re-entry module, and still fit into the nose cone of the R-7. In an already restricted area, its maximum diameter was also defined by the inner diameter of the faring, while its length was restricted by the gap between the top of the upper-stage tanks and the bottom of the spherical Descent Module housed at the front of the nose cone. The designers settled for two flattened truncated cones attached by bolts and airtight seals, and fitted base to base. At the uppermost point would sit the re-entry module, and at the lower end the braking rocket system.

The next problem was how to firmly attach the Descent Module to the Instrument Module so that it would not separate during launch and orbital flight, but could be separated for re-entry. This was resolved by using four metal straps. Triggered by a separation command, the straps would be severed at the Instrument Module to allow the Descent Module, now trailing four steel streamers, to separate for re-entry. It was originally planned to use several locks that would simultaneously open upon

A pressure-suited cosmonaut is helped inside the Vostok crew module for systems tests during the construction programme.

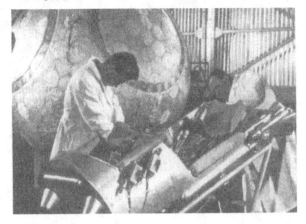

A technician adjusts the Vostok ejection seat during construction. Behind him is a sharik capsule.

command. However, if one of these did not fire, the capsule could not be separated. It was therefore decided to use only one lock to release all four straps. It is unclear what could be done if this failed as the spacecraft headed for re-entry, although the heat would eventually melt the straps, causing separation. Although this would certainly be dangerous, partial separation of the Instrument Module on several missions clearly demonstrated that this was a possibility .

Inserting all the equipment inside the small Instrument Module was difficult, and often resulted in arguments between the designers of the retro-rocket system, parts of the temperature regulation systems, and of the telemetry systems. All argued for more space inside the module.

To help resolve the disputes, the spherical gas containers for the attitude control system were relocated outside the modules, arranged in a ring on the Instrument Module's upper cone. Joining the two modules required a dozen connections, operated electrically with hydraulics or compressed air. These were fixed to the outside of the Descent Module. Clusters of cables ran around the outside of the Instrument Module to the various systems and components.

Once in orbit, the spacecraft would deploy a number of aerials for radio

At Tyuratam, a Vostok spacecraft undergoes a pre-mating systems check prior to certification for flight.

A Vostok sharik is winched across the assembly room at OKB-1. The hatches and thermal protection system can be seen in this photograph.

communications between the cosmonauts, spacecraft and ground control. Four antennae for telemetry transmission extended from the rear.

As the configuration evolved, regular meetings between the designers of different components reviewed the progress of each system and how they fitted into the spacecraft. Each team argued how important it was for a little more room for their hardware, or how another team's system occupied too much room. Disputes were frequent as each team tried to gain space and not lose it. At the end of the meeting the teams would go to their laboratories and work further on the latest outcome, and the following week would reassemble to argue their point once more and to refuse or accept the others' suggestions and requirements. And so it went on.

Gradually, each problem was resolved and moved the design towards completion. The next stage was the ground and airborne test phase, prior to completion of a series of unmanned test flights in orbit.

The first manned spacecraft
Vostok measured 13.12 feet long, and weighed, on average, 10,430 lbs. The spherical Descent Module measured 7.54 feet in diameter, and averaged 5,292 lbs. In total, Vostok contained 1,764 lbs of instrumentation, consisting of 6,000 transistors, 56 electrical motors and 800 relays and switches. The overall length of the Instrument Module was approximately 8.53 feet, with a maximum diameter of 8.53 feet and a volume of 52.97 cubic feet.

Mass breakdown of the Vostok-3 (Object 3K) variant

Systems	21.5%
Structure	20.0%
Heat shield	17.7%
Electrical system	12.5%
Cables	8.6%
TDU braking engine	8.4%
Ejection seat and cosmonaut	7.1%
Landing systems	3.2%
Orientation and ECLSS gases	1.0%

Pre-launch preparations

Vostok was the first spacecraft to be manufactured on a series production line. This was an important decision in the overall Soviet approach to space vehicle construction. The R-7 is (in 2001) still being used (in a modified form) for manned

The Vostok assembly line.

At the foot of the pad, a cosmonaut reports to officials that he is ready to complete his mission, as photographers (*left*) record the event.

launches. Proven design used frequently is very cost effective and was the basis of the Soviets' success, and they applied this to almost all spacecraft used from the 1960s onwards. Construction was completed at the OKB-1 factory, close to the design bureau, and after acceptance testing they were shipped to the launch site at Baikonur to be tested further, prior to mating to the R-7 launch vehicle in a horizontal position. The combination was then 'rolled out' to the launch pad on a railway transporter, where the 124-foot tall stack was erected vertically for the final check-out and rocket fuelling.

The prime and back-up cosmonauts, fully suited, arrived at the base of the pad in a bus, and received the greetings and farewell from leading officials, fellow cosmonauts and pad workers. Having reported their readiness to carry out the flight programme, they then climbed the short flight of steps and, with a wave, took the elevator to the top of the vehicle and climbed into the spacecraft.

The pad crew ensured that all was ready before the hatches were sealed and the pad area vacated. When the count reached T–20 seconds, the engines in the central core and the four strap-ons ignited simultaneously, the vehicle being held down as thrust built up until the point where the combined thrust reached slightly more than the weight of the R-7. Once the required thrust was attained, balance supports were released and the rocket climbed from the pad.

As there was no launch escape tower on Vostok, the personal ejection-seat system would be employed for emergency pad evacuation. Each cosmonaut rode in the semi-reclined position, leaning at 65° to the horizontal. This ensured proper attitude in the event of ejection, and minimised the forces of acceleration during launch. In the event of a deviation from the prescribed trajectory, the escape sequence started with the explosive detachment of the outer hatch by pyrotechnic devices. This was followed by the firing of two (powder) solid-fuelled rockets under the seat to launch the seat and cosmonaut away from the capsule to sufficient height for safe parachute deployment. Cosmonaut and seat would then separate, and a personal parachute would deploy to bring him down safely.

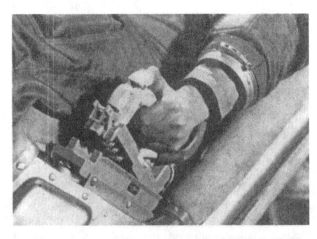

A cosmonaut grips the Vostok ejector handle.

During tests of this system, it was found necessary to adjust the angle of ejection from the rocket from sideways to slightly forward to take it clear of the slipstream of the ascending launch vehicle and to ensure a safe distance if the launcher was about to explode. In order to speed up parachute inflation at low altitude, the seat was provided with a special 'gun', the 'shell' of which was a tether connected to the apex of the parachute.

The ejection seat also had a special restraint system, to withstand forces of acceleration and buffeting during normal powered ascent of the launch vehicle. There was also a cut-out in the launch faring so that the cosmonaut, propelled by the ejector seat, would be unhindered by additional aerodynamic covers once the spacecraft hatch was jettisoned in the event of a launch abort.

The descent system incorporated a number of survival devices in the event of emergency ejection during either launch phase or landing. These included parachutes and pressure sensors for the landing system, temporary oxygen and life support for high-altitude ejections, an inflatable dinghy in the event of a water landing, emergency food and drinking water supplies, a radio, and standard pilot survival equipment (knife, flares, desalination tablets, and other items).

As the rocket climbed it began to pitch over to approach a horizontal flight trajectory over the steppes. At about 120 seconds into the flight, the fuel in the four strap-on boosters would be depleted and the engines shut down. Explosive bolts separated the four spent stages, which fell back to Earth. This was followed by the separation in two halves of the nose shroud 60 seconds later. During powered ascent, the cosmonaut would be subjected to a force of 5 g.

The central core would continue to burn for 300 seconds into the flight, until it too was depleted, separated and allowed to fall towards Earth. The third stage burned for 430 seconds, and upon reaching orbit was ejected after a total flight duration of 730 seconds, placing Vostok in a 112–150 miles orbit at 65°.

The Vostok pressure suit

The need for a suitable pressure garment for a 'passenger' for spaceflight stemmed from design studies in the early 1950s and the provision for life support systems for canine passengers of the ballistic rocket programme and Sputnik 2. Factory 918's research, design and production enterprise at Tomolino, Moscow, was the leading facility for these 'suits' and escape systems, as a spin-off from work conducted on military aircraft programmes since 1952. The design of the pressure garment for Vostok began in 1959, and the work contract included the design and development of personal escape, rescue, survival and life support systems, including the ejector seat.

In the early days of capsule development, immersion of the cosmonaut in liquid was proposed, as opposed to the creation of a pressure suit to be worn solely inside a spacecraft (an Intravehicular Activity (IVA) garment). The designers of the spacecraft thought that the use of such suits was a waste of resources, as cabin depressurisation was highly unlikely, and was an acceptable risk. If depressurisation did occur it would be either so small that there would be time to recover the cosmonaut, or else death would be instantaneous! The addition of a suit would only add weight and complexity to an already restricted payload envelope. Korolyov (supported by leading aerospace medical specialists and Air Force authorities) intervened, and insisted that a full space pressure garment would be incorporated as an additional redundancy to the integrity of the spacecraft. In the summer of 1960, Korolyov allocated 1,100 lbs for the suit system, but he also added that it must be ready by the end of the year.

The objective of the suit was 'to protect the cosmonaut from the hazardous conditions of space and to isolate him from the cabin environment in the event of decompression or harmful gases entering the system loop.'

The design relied on many years of evaluation of stratospheric and aircrew suits. The Vostok pressure garment was never intended to support EVA, and remained the equivalent of a US 'get-me-down' pressure garment, as developed for Mercury from the US Navy suits.

The Vostok suit was designated SK-1 'Sokol' ('Falcon') and featured several exacting design specifications. The design needed to be strong yet flexible, lightweight and durable. It was required to function unpressurised in the spacecraft cabin, provide full safety in the event of cabin depressurisation, and be able to withstand an unforeseen water landing, sustaining the cosmonaut even if the landing was face down.

The design featured an inner comfort garment and a multilayered pressurised garment, with integral boots, gloves and pressure helmet. Special joints at the shoulder, elbow, knee and ankle retained a limited range of mobility under positive pressure. Each glove also featured minute joints to allow limited bending of the fingers.

A constant-wear biomedical garment, containing fifteen sensors attached directly to the skin, was used to measure and monitor medical parameters during the flight, including pulse, temperature and blood pressure. These readings were relayed, via a transmitting device, to the ground. This garment also incorporated the waste management system.

On top of this was a woollen comfort layer, which featured a heat-insulation layer that also included the heat ventilation system. Constructed of a lightweight porous material resistant to temperature change, this vented the suit by means of a system of ventilation pipes, removing excess body heat. Weighing no more than 4.5 lbs, it protected the cosmonaut from overheating or freezing.

The next layer was a thin elasticised rubber layer, which was the primary pressure layer of the suit design. Over this was a layer of dacron, the outermost layer of the pressurised garment. The layer also featured a system of ropes and hinges that allowed each suit to be moulded to the contours of the wearer.

An orange overlayer completed the suit, in the recognised international rescue colour which is distinguishable in water, snow or on land in the event of emergency landing.

Weighing between 22 and 26 lbs, the suit remained fairly mobile while unpressurised, as witnessed in film of cosmonauts wearing them. This was achieved by the use of 'special fabrics that would stretch in only one direction'. This outer

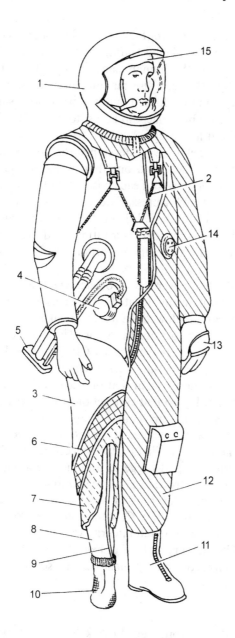

Vostok pressure garment layers. 1) hard shell helmet (white with red CCCP lettering); 2) system of sizing cords and pulleys to allow perfect fitting; 3) outer layer of Dacron; 4) connectors between the suit and spacecraft life support supply; 5) ECLSS umbilical for spacecraft–suit connections; 6) second thin elastic layer; 7) blue rubber pressure layer; 8) third heat insulation layer with ventilation system; 9) constant wear biomedical garment; 10) integrated booties; 11) black lace-up over boots; 12) orange coverall layer; 13) pressure glove; 14) suit regulation gauge; 15) soft communications cap. (Astro Info Service collection.)

layer was also fabricated from a fireproof material, which also served to protect the suit from mechanical damage, accidental tearing, and excessive wear.

The shape and dexterity of the human hand is one of the most challenging elements of space-suit design that any suit manufacturer faces, so the gloves were custom fitted for each man, and were locked onto the arms of the suit near the wrists. The cosmonaut also wore large black lace-up boots, to protect the lower part of the suit and the wearer's feet and lower legs.

Several different designs of helmet were developed for Vostok, including Apollo-type, dome-shaped full-view helmets, and one that turned on its axis as the head turned (similar to the Gemini design). But each cosmonaut flew with the same design on Vostok missions. This featured a pressurised helmet and visor that had a seal connection along the 'face cut'. In normal flight, the visor was raised up and under the helmet shell. Only during cabin depresurisation (and ejection) was the visor automatically sealed down. The visor also incorporated light filters to block out direct sunlight.

The helmet had a hard white cover that protected the head from the rigours of launch and entry. It locked onto a ring located on the neck of the cosmonaut's suit, and also featured a feed port that allowed the wearer to take food and water through a one-way valve without having to raise the visor. The only identification marks carried on the helmet were the large red letters 'CCCP'. No Vostok cosmonaut wore any 'mission emblems' or the hammer-and-sickle flag on their suits.

The suit featured a regeneration system that purified exhaled air and enriched it with oxygen to be used again. It also had connections to the spacecraft life support system. A communications soft cap (snoopy-cap) was also worn for radio communications. The suit was designed to protect the cosmonaut in most environments, with ventilation controlled by the spacecraft ECLSS, which 'sensed' any slight drop in cabin pressure and triggered automatic closing of the visor. The pressure level in the suit also controlled the air regulation.

In the event of a landing on water, the suit was designed to keep the cosmonaut afloat on his back for up to 12 hours in 0° C, with protection from the cold afforded by the several layers of suit insulation. Suit integrity was maintained with a neck dam (similar to those on the US Mercury suits) that held the head high above the water, even without the helmet. However, during most training sessions, most cosmonauts tended to leave their helmets on.

Vostok in orbit
Once successfully placed in orbit, the Vostok spacecraft performed mainly automatically, with very little input from the cosmonaut. In the capsule, space was at a premium, and with movement restricted everything had to be within reach of the cosmonaut on the couch. Instruments were usually grouped in their functional role. Several elements of the landing and telemetry system and elements of the air conditioning system were located beneath the couch and ejector seat, taking advantage of the additional space.

To the left were the control panels for regulation of the temperature and air humidity, the radio equipment and orientation controls, a tape recorder activated by the sound of the cosmonaut's voice, the landing system direction finder, the emergency heat regulation system, and the water supply.

To the right was the control handle for manual orientation, further radio equipment, the food container, a second TV camera, the waste management system, and the electrical storage batteries. A porthole on the right was used for clear observations and photography, and a second porthole was located behind the head of the crew-member, in the exit hatch.

Directly in front of the occupant was the main instrument panel, which was very sparse in terms of controls and displays. There was an orbit counter, and a clock that was activated at the moment of launch (Mission Elapsed Time) and used in relationship to ground control's instructions. An Earth globe rotated in time with the spacecraft's movement around the planet to allow the cosmonaut to determine his geographical location at any time and to predict the area of descent should an emergency landing be required from any point in the orbit.

A bank of indicator lights on the right of the panel also recorded the operating condition of several onboard systems. Other dials on the lower part of the panel indicated the status of temperature, humidity and pressure (bottom, second left), levels of cabin oxygen and carbon dioxide (bottom, third from left), and the levels of attitude-control system propellant (bottom, far right).

Beneath the panels was a TV camera, and under that, a third porthole which incorporated the 'Vzor' optical orientation device mounted in the 'floor' of the cabin. Designed to allow the cosmonaut to manually line up the spacecraft with Earth's horizon, this system featured an optical device located in the view port. This consisted of a central viewer and eight ports arranged in a circle around the centre of two annular mirror reflectors, a light filter and a lattice glass.

The light rays from the horizon struck the first reflector before passing through the pane of glass and the light filter in the porthole, and reaching the second reflector which directed the rays through the lattice glass to the eye of the cosmonaut. In the correct attitude, the horizon appeared as a circular pattern, with all eight ports lit up. An incorrect attitude would require the use of a control handle to orientate the spacecraft until the concentric circle had been achieved on the glass. This could be worked manually only over daylight portions of the orbit, although the automatic system could operate either by day or by night.

The air generation and conditioning system was based on stores of oxygen and absorbents that could handle the water vapour and carbon dioxide at small values of partial pressure. As these were pioneering excursions into outer space, each flight of Vostok was planned to be relatively short in duration. During the planning phase, however, it was anticipated that longer flights would be required – possibly flown by improved designs of Vostok. This additional capability was incorporated into the basic Vostok design, and combined with emergency requirements. In theory, flights of 7–12 days could be attempted, with a gradual increase in air temperature and the temperatures of cabin equipment up to $+35°$ C. Onboard systems were designed to support a single cosmonaut for twelve days in an hermetically sealed cabin with a temperature of 35° C. Sudden decompression enforced a mission rule to return at the earliest opportunity. Sealing and

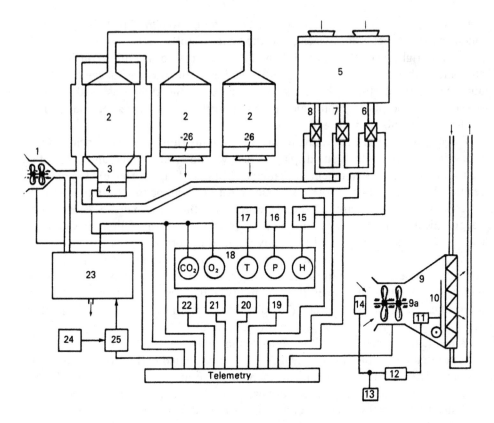

Soviet diagram of the Vostok life support system for regeneration and conditioning of the atmosphere for the hermetically sealed Vostok cabin. 1) ventilator with electric motor; 2, 3, 4) regenerator with regulator device; 5) moisture absorber; 6) automatic valve; 7, 8) two valves with a manual actuator; 9) liquid-air temperature exchanger; 9a) ventilator with electric motor; 10, 11, 12, 13, 14) automatic temperature regulator; 15) humidity measuring device; 16) pressure sensor; 17) thermometer; 18) control panel indicator dials in crew cabin (carbon dioxide (CO_2), oxygen (O_2), temperature (T), pressure (P), humidity (H)); 19, 20) humidity measuring device; 21) thermometer; 22) pressure sensor; 23, 24, 25) automatic oxygen and carbon gas analysers; 26) filters for dust and harmful impurities. (Courtesy NASA.)

pressurising the spacesuit could sustain the occupant long enough to select a favourable site and land the spacecraft.

The system worked by cabin air being forced across an alkali metal superoxide, which absorbed carbon dioxide and released oxygen. The air was then purged of any noxious odours and excess moisture. The temperature was regulated by a heat exchanger inside the capsule and the heat radiators located on the outside of the instrument compartment.

During normal operations the cosmonaut lay in the couch with his visor open (allowing the suit to be ventilated), and breathing via the cabin oxygen supply. The Vostok life support system ensured that the internal pressure was maintained at between 755 and 775 mm Hg, in a temperature range of 13–26° C. The humidity in the cabin remained at 51–57%, the oxygen content at 21–25%, and carbon dioxide content at not more than 1%. The system was automatically controlled, and the whole life support sub-system, including food and the waste management systems, had provision for up to ten days in the event of a braking-rocket failure.

Under emergency conditions, such as cabin decompression, the visor closed automatically and the suit functioned using the spacecraft's systems through the suit connections. A reserve supply of oxygen and air allowed the cosmonaut time to contact the ground and make a decision on when and where to land. Should the levels of oxygen drop, or carbon dioxide increase, a warning sensor gave a signal to adjust the levels. This also applied when too much oxygen was supplied or when the humidity rose or fell beyond prescribed levels. Filters were used to prevent harmful contamination of the atmosphere, and excess heat was transferred to the radiators on the outside of the Instrument Module.

Communications were maintained with a signal radio transmitter on a frequency of 19.995 Mc/s for ground tracking. The cosmonaut maintained a two-way radio communications link with Soviet ground stations located at the Central Control Room in Moscow, at the cosmodrome, and at remote sites located at Novosibirsk, Kolpashevo, Khabarovsk amd Yelizovo. The radio communications link used shortwave on 9.019 and 20.006 Mc/s, and ultrashortwave on 143.625 Mc/s, while the FM channel was used for reliable contact up to distances of 930–1,240 miles from the tracking station. This allowed the cosmonaut to talk directly to ground control for most of the orbit while in line of sight with Soviet ground stations as they came over the horizon. For the period out of this range, an onboard tape recorder was used, for later downlink on the next pass of ground stations on the following orbit.

The spacecraft's attitude was controlled by the orientation system with the aid of two solar sensors, one automatic and one manual. Vostok was unable to change its orbit, but the cosmonaut could manoeuvre the vehicle by firing compressed gas from the set of micro-jets around the yaw, pitch and roll axes using the Vzor optical device, the manual orientation system and the control stick.

Telemetric medical data were supplemented by the two TV cameras, which provided a frontal and a profile view of the actions and reactions of the occupant in orbit. Sensors attached to the cosmonaut's body under the suit recorded pulse, respiration, blood pressure and electrocardiograms. Radiation levels were monitored through a dosimeter reading in the capsule via telemetry. Vostok itself shielded the occupant from most of the harmful rays, but in the event of a solar outburst during the mission there were chemical serums on board if required. Doctors also analysed voice communications, and listened for signs of fatigue, hyperactivity, stress and tension.

Food rations on Vostok consisted of two sets of provisions. Upon confirmation of a flight's duration, a daily calorific intake of 2,500–2,700 kcal/day, average protein of 4 ounces per day, fats 3 ounces per day and carbohydrates of 10.5 ounces per day, was provided. These figures were determined from research by the Institute of Nutrition, USSR Academy of Medical Sciences, resulting in a set of standards (1951). The second requirement was for a supplement, including the emergency rations for increased flight duration, of a calories content of 1,450 kcal/day. During the evolution of the Vostok programme, the composition of the food rations changed with each flight, and was gradually improved from purée-like canned food to more solid foods on later missions.

The puréed food was packaged in 5.6-ounce aluminium tubes, with metal screw-caps sealed with an edible resin. The tubes were lined with canning lacquer. Stored food was sterilised using high-pressure steam, while solid food products were vacuum-packed in polymer sheets. Later flights also included an assortment of foodstuffs (including meat products) to increase nutrient value, and the food was provided in bite-size pieces for ease and convenience of consumption. A daily ration was four meals, with approximately 4–5 hours between each meal, which was determined to be the best balance for digestion.

For the water supply, the designers had to create a system to allow the cosmonaut to drink under weightless conditions, select the material for manufacturing such a system, and devise a reliable method of water preservation. They then had to develop a test procedure to evaluate whether the preserved water could be stored in containers made of the selected material.

The drinking water supply consisted of a container to hold water for one crew-member for twelve days, calculated at a consumption rate of 3.5 pints per day. The container was made of two layers of high-strength polythene film (each hermetically sealed) within a metal container. On the outside was a valve nipple, connected by a pipe to the mouthpiece. The only way to extract the water was by placing the mouth over the closing device and applying pressure. The mouthpiece also served to sterilise and deodorise the water. The cosmonaut grasped the mouthpiece, opened the closing device by pressing a button, and sucked in the water. The vacuum created in the oral cavity was sufficient to induce water flow from the storage container, as well as restricting the seepage of gas bubbles into the mouth.

For sanitation, with the cosmonauts wearing their suit for the duration of the flight (which could be as much as ten days under emergency conditions), a waste management system was required. On Vostok, the system chosen was a method of air stream. Faeces receptacles were inserted into the suit and then removed when used and stored in a special compartment. The collector for solid waste also had a cotton wiper.

Liquid waste was transported by means of a rapid airflow. Urine passed through a hose into a container by using the airflow generated by a fan device. Inside this container was a moisture-absorbing material (polyvinyl formol in the form of ¼-inch cubes), offering minimal resistance to the air flow. The transporting air – now completely cleared of any liquid – then encountered an absorbent filter, and was cleaned of harmful gases before being returned to the spacecraft's atmospheric system by the fan.

Other hygiene facilities were very rudimentary, with no provision for washing, shaving, or cleaning of teeth. The cosmonaut slept (when able to do so) in the seat.

Each cosmonaut had to carry out an individual programme of visual observations of the Earth, stellar field and solar bodies. In addition, they carried cameras, and recorded the performance and function of themselves and their spacecraft during the flight. The flight programme was not crowded with experiments.

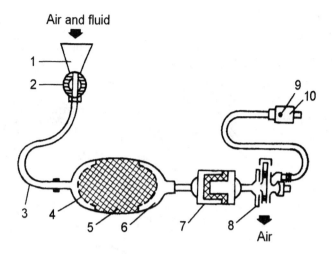

Vostok system of pneumatic transport of liquid and phase-separation of moisture absorbing material. 1) urine receiver; 2) shutoff valve; 3) feed tubing; 4) perforated wall; 5) moisture absorbing material (polyvinyl formol); 6) collector; 7) adsorptive filter; 8) fan device; 9) signal lamp; 10) switch-on panel. (Courtesy NASA.)

One of the most critical onboard sub-systems was the guidance system. There were actually two guidance systems: an automatic solar orientation system and the manual visual orientation system. Each was capable of operating the cold nitrogen gas thrusters' sub-systems, and each offered redundancy to the other system. Research conducted by a team headed by B.E. Chertok found that the greatest risk to a flight on Vostok was incorrect orientation for retrofire – hence the need for system redundancy.

The automatic solar orientation system consisted of solar sensors, angle of flight sensors, and an analogue computer. This system would operate only if the angle of flight sensors (a slit over three photocells) indicated correct orientation. The TDU then fired, and the computer used the input derived from the solar sensors (and a two-step double gyroscope with mechanically opposed directions) to generate an impulse to complete the required burns.

The sequence required to re-enter the atmosphere began with a manual check of the clockwork globe showing the position over the Earth. Pressing a button to the side of the globe showed a projected re-entry position, assuming that the spacecraft followed a standard re-entry from that moment.

At the end of the mission, the cosmonaut was required to orientate the Vzor assembly to view the Earth horizon and make an alignment. The spacecraft was then turned around, allowing the braking rockets to ignite. Normally, such manoeuvres (aimed directly against direction of flight and fired for a predetermined period) slowed the spacecraft's orbital velocity sufficiently to allow gravity to draw the spacecraft back towards Earth.

Vostok used an infrared radiation system (combined with sophisticated gyroscopic devices) and a Sun-seeking system (since the Sun was the largest and

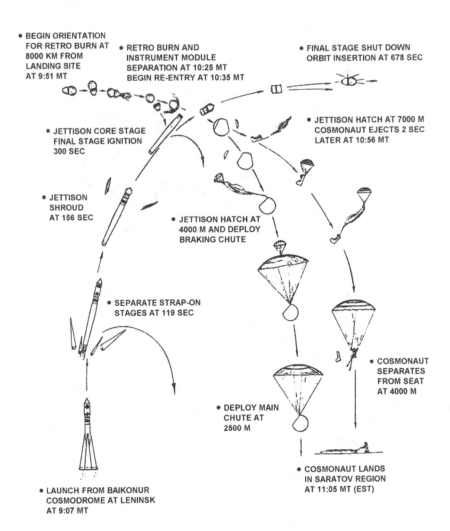

The Vostok mission profile. (© 1983 Ralph Gibbons.)

brightest object in the field of view) to command the engine to fire in the right direction. The attitude control system directed the nozzle of the retrorocket towards the Sun, which would be 'ahead' of the spacecraft, which resulted in thrust away from the Sun, allowing a downward movement towards Earth.

The retro-rocket braking engine was designed by the OKB-2 bureau headed by A.M. Isayev, and was designated TDU-1 (Braking Engine Installation). By using a nitrous oxide oxidiser and an amine-based fuel this produced a self-igniting propellant that resulted in a thrust of 1.58 tons and a specific impulse of 8,281 ft/sec. The fuel load was approximately 600 lbs with a total ignition time of 45 seconds that was sufficient to reduce the velocity of the Vostok by about 508 ft/sec to begin entry into the atmosphere.

The landing site of a sharik.

Calculations indicated that even with the Sun high above the horizon, re-entry would still be feasible. A direct re-entry into the atmosphere from such a high Sun angle was possible by a vertical pulse that 'pressed' the spacecraft towards Earth. But since the amount of fuel needed was four times the amount required for deceleration, this was not adopted for Vostok. In reality, Vostok flight plans were determined by the need to have the Sun in the correct position at time of entry (just ahead and not too close to the zenith).

On re-entry, the straps joining the Descent Module and Instrument Module were severed, as were all connections, and the Instrument Module was allowed to burn up in the atmosphere. Meanwhile, the Descent Module (being weighted forward of the geometric centre) reorientated to point the thickest part of the shield forwards to protect the cosmonaut from the high temperatures of re-entry. The spherical design (with the offset centre of gravity) ensured the correct orientation was attained for entry. The ballistic entry profile flown by Vostok endured up to 8 g from Earth orbit. (The American Mercury also flew a ballistic entry, but could refine its trajectory using thrusters. On the later Gemini, Apollo and Soyuz vehicles, an offset gravity reduced entry g loads to 3 g for Earth orbit and, in the case of Apollo, 8 g from the Moon).

Deceleration forces for Vostok levelled at 8–9 g (as compared with up to 5 g for launch), while the outside temperature rose to 10,000° C. As a plasma 'sheath' built up around the descending spacecraft, all radio communications with the spacecraft and cosmonaut were blocked. As the capsule broke through 4.3 miles, the side hatch was blown off, followed two seconds later by the ejection of the cosmonaut. The cosmonaut separated from the seat at 2.4 miles, and both fell to Earth under their own parachutes.

With no soft-landing rockets, the spacecraft continued to fall to 2.4 miles altitude, at which point its parachute was activated and detached from 'the boot' container. The drogue and main parachutes deployed at 1.5 miles. The landing was generally in the area of Kazakhstan, and was often witnessed by local people. Shortly afterwards, the landing support team arrived to perform medical checks on the cosmonaut, and the spacecraft was airlifted back to OKB-1 for post-flight evaluation. The cosmonaut usually stayed overnight within the vicinity of the landing area before returning to Moscow for the post-flight celebrations and awards, followed by the debriefing and the planning of the next mission.

The Soviet authorities deployed a recovery team in the designated area. Helicopters ferried doctors, support personnel and engineers to the site, some concerned with the recovery of the crew and others with the craft. A special group of doctors was trained to parachute, in emergencies, to treat the cosmonauts if they were injured. Over a period of time, a special unit was formed to recover the crew and the spacecraft.

SELECTING THE COSMONAUTS

When America and the Soviet Union began the process to select their first astronauts and cosmonauts, no one understood, or could even imagine, the 'hero' status that these people would attain. Every consideration was given to their qualifications, their fitness, and their personal (and, in the Soviet Union, their political) backgrounds. No thought was given to the pressure they would be submitted to from politicians, from their public relations duties, or in their personal lives.

In the Soviet Union, those who flew became heroes, appearing at political rallies and being the modern icons of the success of Soviet citizens. They were pawns in a political game between the superpowers. For those who did not fly, the reverse was the case, and their identities remained secret. It was only in the mid-1980s that their names, the reasons that they did not fly, the training accidents they suffered (in one case, a fatal accident) were revealed. No one prepared them for that either.

Who would fly?

Once approval for the manned satellite project had been acquired, some thought had to be given as to who would be best qualified to fly the missions. In February 1959, Keldysh chaired a meeting that discussed the plans for manned spaceflight in some detail, including who should fly and how they should be selected. At that time, NASA was completing the selection of America's first astronauts for Project Mercury, and had undergone a similar phase of deciding who would fit the criteria as space explorers.

Like the Americans, the Soviets had no precedents on which to base their decisions. Therefore, as well as the engineers who would build the vehicles, the list of possible candidates included submariners, as they were accustomed to long periods of isolation. Another suggestion was to use mountain climbers – who were familiar with living and working at high altitudes, reduced pressure and harsh environments – and pilots, who worked with high-performance machinery.

Support for a fighter pilot selection gradually grew, as it was realised that they would be experienced in a variety of disciplines. They were skilled in the exploration of the stratosphere in high-performance aircraft, and as such were not only pilots but also navigators and radio operators, and normally had a background in aeronautical engineering. Military pilots would already be medically fit, would have been screened by the security services and would be qualified parachutists – a requirement that would be called upon during the Vostok recovery procedure.

Criteria

Once it was decided that all applicants chosen would be serving military pilots, additional criteria were defined. All candidates would need to be qualified to fly jets with a 3rd Class Certificate (the basic level for operational flying in the Soviet Union). All applicants would be aged under 30 years, with a maximum height of 5 feet 7 inches (to fit inside the capsule) and weighing less than 154 lbs (to fall inside payload limits set by the R-7's launch capacity). These criteria were similar to those used to select the Mercury astronauts, apart from the higher academic qualifications or the requirement to be an experienced test pilot.

The selection encompassed mainly younger pilots from the Air Force, Navy, and Air Defence regiments (PVO), who were physically fit but had limited flying experience. Those considered by the selection committee had only 200–1,000 hours flying time, and not all of this in high-performance jets. By comparison, the American Mercury astronauts averaged more than 1,500 hours in high-performance aircraft, and most had test-pilot experience. Some also had combat experience from World War II and Korea. One reason for this difference was that Soviet rocketry and spacecraft at that time allowed for more automated systems, to reduce the cosmonaut's involvement.

Responsibility for recruiting the first cosmonauts was given to the Air Force, commanded by Marshall Konstantin A. Vershinin. He appointed Professor Vladimir I. Yazdovsky to take charge of the actual selection process by heading a special Medical Selection. His deputies included Oleg G. Gazenko, Abram M. Genin and Nikolai N. Gurovsky. Colonel Yevgeny A. Karpov of the Central Aviation Research Hospital was also involved, and Colonel-General Fillip A. Ogol'tsov was responsible for the 'political side' of the selection board, although he was a medical man.

When asked how many candidates he needed, Korolyov replied that he would require at least three times the number selected by America (seven), which could create a first cosmonaut class of around two dozen. However, the results of the first medical screening produced an insufficient number of suitable applicants, and the height and age restrictions were therefore relaxed slightly. The new, older applicants had a greater depth of experience in flying skills (although not as test pilots) and in engineering.

The selection process

In August 1959 a recruiting team of military medical specialists (headed by Karpov) began visiting air bases from the Urals to the Far East. The selection was restricted to European Russia, as it was assumed that here there existed a suitable pool of pilots that would result in the number of required candidates. All the pilots interviewed were told that they were being considered for a programme involving 'aircraft of a completely new type.' Many thought that this meant helicopters, as there had been a number of transfers of jet pilot to these new vehicles. This was not a very popular move, as the pilots preferred flying fast jets to the slow, cumbersome helicopters. G. Shonin recalled that at his interview, he was quickly reassured: 'No, you don't understand. What we are talking about are long-distance flights; flights on rockets; flights around the Earth.'

Around 3,000 suitable candidates were considered, and between October 1959 and January 1960, Karpov (who was considered as a potential candidate) chaired the selection committee to select a suitable number of candidates for further consideration. From the 3,000, a group of 102 were nominated for a rigorous screening programme of medical and physical tests at the Air Force's Central Aviation Science Research Hospital, in the suburb of Sokolniki, Moscow. All applicants 'demonstrated a fundamental desire' to fly into space.

Medical screening

Data gathered from extensive animal research programmes helped define the conditions likely to be experienced in spaceflight. The closest proximity to spaceflight at that time was in the aviation field, from which a well-established and approved system of medical screening for flight personnel provided a baseline for the selection of flight crews for spaceflight. Flights into space are accomplished by experiencing varying degrees of exposure to acceleration, vibration, noise, weightlessness, long-term isolation, and disturbance to Earth-based daily routines and body rhythms. The medical specialists therefore devised a range of facilities to reproduce these conditions, and used them to select candidates who displayed resilience in each condition.

Today, more than forty years of direct participation in space has provided a wealth of data on human responses to spaceflight. At the beginning of the programme, there were no such databases, and the selection of the first humans to fly in space was therefore reliant on wide-ranging and intensive investigations covering the clinical, psychological and physical capabilities of every candidate. In the late 1950s, even the medical specialists were unsure of exactly what faced the first explorers of outer space. The range of tests and examinations that were devised has often been compared to mediaeval torture chambers, in which nothing was left to chance and no orifice in the body left unprobed!

Each candidate had to exhibit good general health, a strong will, quick reactions, and the capacity to make and execute rapid decisions. They would be tested under flight conditions and environments similar to those that might be experienced during actual spaceflight. From data provided by their medical records, and examinations by clinicians, physicians, physiologists and psychologists, those who were best prepared and most resilient to the tests were put forward for further consideration.

Even then, the whole process – from initial selection to spaceflight – was one of continual evolution, and passing one element was no guarantee of progression, even for those who were physically fit and fully trained. Each candidate was monitored throughout the selection, during work and rest, and even during sleep. At any point, a candidate might be withdrawn from the process for the slightest reason. Some of

An Air Force volunteer undergoes early space medicine tests in the 1950s.

the largest, fittest, brightest and most skilled of the applicants were rejected during the process, as further results were obtained and analysed.

Selection was completed in three phases: first, under ambulatory conditions, to reveal clear pathological and functional disorders (which prevented any further participation); then under clinical conditions; and finally during actual spaceflight training programmes. The clinical tests were designed to reveal concealed pathological problems, to evaluate changes in the functioning of organs and systems of the body, and to determine body reserves such as stamina and strength. The physiological tests were designed to reveal personality disorders and behavioural and emotional reactions, such as determination or fear under stressful conditions. These were completed both individually and as part of a group, to determine which candidates would be ideally suited for high-risk missions or for long solo spaceflights, or those who would work best in a group or crew.

Passing these tests allowed the candidate to be considered for the final selection and to advance to actual spaceflight training. Rejection due to health defects could occur at any stage, but was primarily at the clinical level. The Soviets have stated that in early selections between 25% and 50% of those tested were rejected, primarily due to functional disorders and ailments in internal organs, including the eyes, nose and throat. In addition, X-rays of the spine could reveal injuries from exercise or load stress from jet flying or parachute training. Any blood disorders, adverse results from any of tests of the gastrointestinal pathology, cardiovascular system or respiratory system would also result in rejection from the selection.

The parameters of height (maximum 5 feet 7 inches) and weight (maximum 154 lbs) were also used as a baseline to evaluate a candidate's capacity for workload and reactions under stressful conditions or confined environments.

From the wide range of medical tests, any previous history could be revealed from

Buinovsky undergoes an isolation test. (Courtesy Eduard Buinovsky collection.)

impact, pressure and compression injuries, deafness, or a susceptibility to high g-forces in greying or blacking out. These final conditions are common in pilots of high-performance jet aircraft, and were disqualifying factors, despite a candidate's excellent flying aptitude or accomplishments.

By selecting extremely fit and skilled pilots, the programme would have available the best physical specimens with flight-training experience. Each candidate would also still be young enough to endure a long training programme, and would be able to participate in more than one spaceflight.

The devil's cockpit and the isolation chamber
Part of the selection process and training programme involved a series of one-off tests in a range of facilities. According to the official documents, these tests were designed to 'study the state of the neuro-psychological sphere and physiological reactions, and determine human capacity to accomplish tasks within the framework dictated by the test' – in other words, to put excessive stress and strain on the

A test subject undergoing evaluation on the tilt table.

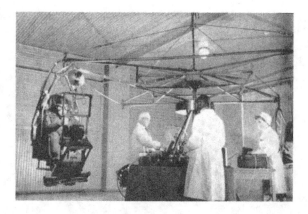

Early testing of the effect of g-forces on a cosmonaut candidate.

cosmonaut's mind and body, to test individual reactions and resilience to meet and overcome these barriers, and at the same time to task them to perform a programme of physical and mental challenges. For the candidates, this was the most vivid analogy to a torture chamber.

Most of these facilities were envisaged as cold, grey steel monsters that seemed to be lying in wait for each unsuspecting candidate. They were designed to increase the candidate's resistance to vibrations and temperature variations (as might be experienced in a rise or fall of cabin temperature in space) and determine individual reactions to a set thermal load. The candidates were subjected to temperatures of 66° C while undressed, then to temperatures of 70° C (158° F) and higher down to well below freezing, while dressed. The vibration devices were used to shake the candidate at 200 vibrations a minute to simulate the stress and vibration likely to be experienced during launch and re-entry.

Centrifuges belonging to the Air Force were used to expose the candidates to acceleration conditions of a variable intensity. They were constantly in use during training, and were called the 'devil's merry-go-round' by the cosmonauts. But they also provided a general strengthening of the body, with candidates performing leg and abdominal exercises during runs to increase muscle strength, and developing breathing techniques to cope with the g-loads. Observations from these runs of 7, 9 and 10 g led to the development of a conditioning system that continues in use to the present day.

In the decompression chamber, a candidate patiently awaited the start of a test, watching the faces of the doctors staring unemotionally through a round porthole. At the command 'lift-off', pumps began to suck out the air, simulating an ascent through the atmosphere. As the 'altitude' dials climbed, so the pressure fell. At first, the candidate did not wear an oxygen mask, and soon experienced oxygen starvation, a rapid heartbeat, dark blind spots before the eyes, and deep, short breathing. As the needle passed several thousand feet (the complete tests reached up to 39,360 ft), many succumbed to the environment and fainted, unable to withstand the stresses. Most pilots had, however, already experienced altitude chambers, and at

A cosmonaut candidate inside an altitude pressure chamber.

prescribed levels the command to put on the oxygen mask was given. Even then the test was not over, because the doctors then observed the reactions of the candidate as they inhaled the life-giving gas after being almost asphyxiated. Once the desired 'altitude' was attained, the command to 'descend' returned the chamber to sea-level pressure, whereupon the door was opened and the candidate's pulse rate and blood pressure were taken and recorded. Another gruelling test was completed.

By far the most demanding device, however, was the isolation chamber. Candidates were locked away for up to ten days in a soundproof chamber, where they lived and worked according to an altered work/rest and day/night cycle determined by the examiners. New methods of recording medical and biological information were also devised during these tests, and these provided valuable baseline medical data for those who would later make spaceflights. Each test was called a 'run', and included memory, concentration, and coordination capacity experiments, and rapid response reactions between states of work and rest and between sleep and activity. During a programme of nine runs, a set of physical training exercises was also tested and developed. This included bicycle ergometers, rubber expanders, and inertial and isometric exercises that were selected for use on later spaceflights. (Photographs of candidates or Air Force testers from many of these simulations later led to Western press reports of 'phantom' or 'missing' cosmonauts). Throughout the tests, functional changes in the emotional/psychological stability and the adaptive capabilities of each candidate were recorded and observed.

A candidate was introduced to the isolation chamber by the technicians, who would relate tales of silence that could drive him out of his mind, and tell him how he would see and do things that he would never dream of doing or want to do again. This was a 'friendly' warning that this test was the toughest of all.

The chamber was a huge steel shell mounted on rubber shock absorbers, with walls 16 inches thick. There was one door and two round double-pane portholes that could be sealed off from the outside. Inside were a table, a steel bed, a chair, and a replica of the Vostok contour couch and control panel. The toilet was also a replica of the one that would be used during Vostok missions.

The first visit was as one of a small group of candidates. With the door shut, the walls seemed to close in, and the silence was overwhelming – but the real test came as each candidate was locked in solitude for ten days. With all sounds and vibrations eliminated, it was a tomb-like existence. They were given small simple tasks at first, then more difficult ones. Hours of inactivity led to boredom and frustration, pushing each of them to the limits of sanity. Loud noises were fed through speakers after hours or days of silence, and lights came on and off at irregular intervals, sometimes staying on for hours and preventing sleep. After a long period of complete darkness, the lights were suddenly turned on, almost blinding the occupant. Gherman Titov once described this part of the selection and training as 'The mirror of truth – when you come suddenly face to face with yourself.'

At the end of the tests, during which the candidates had no notion of time, the opening of the door and the 'avalanche' of sound, light and activity was just as shocking to the system as had been the test. Many of the candidates wanted to rip off the door and run away from the chamber, and many wanted to get back in and close the door against the noise!

The first ones
By 20 February, the first group of Soviet cosmonaut candidates was approved by the Creditional Committee State Commission, represented by members of the Air Force, its military medical service and the political (Communist Party) administration. From a short list of forty, twenty-nine candidates made the final cut, and from these, twenty were named to report for spaceflight training in March.

The formation of the Cosmonaut Training Centre (TsPK) near Moscow was ordered by the Central Committee of the Communist Party on 11 January 1960, and on the same date as the new cosmonauts were selected Karpov was named its first Director with a staff allocation of 250. It consisted of medical specialists, flight and parachute instructors, engineers and technicians. Karpov's deputies were Yevstafiy V. Tselikin (in charge of flight training) and Nikolai F. Nikeryasov (in charge of political activities). The two senior trainers were Colonel Mark L. Gallai and Colonel Leonid I. Goreglad – both Heroes of the Soviet Union – reflecting the importance that the Air Force was placing on this programme. Responsibility for the cosmonaut training programme was assigned to Colonel-General N.P. Kamanin, who remained in this position until 1972. (The members of this historic first cosmonaut selection are listed in the following table. All were serving military officers from various services, but were transferred to the Air Force.)

The group included graduates of Higher Air Force Schools (the equivalent of American Junior Colleges), and two (Belyayev and Komarov) were also Air Force Academy graduates. There were fifteen members of the Communist Party, and five belonging to Komsomol, the Party Youth Organisation. There were no test pilots, although Komarov had performed test-engineer work on new aircraft. Most had flown the older MiG 15 or MiG 17 aircraft, with only Popovich experienced in the high-performance MiG 19. Belyayev had also fought in the Second World War. At

The members of the historic first cosmonaut selection, 20 February 1960

Rank	Name	Date of birth	Age when selected
Senior Lt.	Ivan N. Anikeyev	1933 Feb 12	27
Major	Pavel I. Belyayev	1925 Jun 26	34
Senior Lt.	Valentin V. Bondarenko	1937 Feb 26	23
Senior Lt.	Valery F. Bykovsky	1934 Aug 2	25
Senior Lt.	Valentin I. Filatyev	1930 Jan 20	30
Senior Lt.	Yuri A. Gagarin	1934 Mar 9	25
Senior Lt.	Viktor V. Gorbatko	1934 Dec 3	25
Captain	Anatoly Ya. Kartashov	1932 Aug 25	27
Senior Lt.	Yevgeny V. Khrunov	1933 Sep 10	26
Engineer Capt.	Vladimir M. Komarov	1927 Mar 16	32
Lieutenant	Alexei A. Leonov	1934 May 30	25
Senior Lt.	Grigori G. Nelyubov	1934 Mar 21	25
Senior Lt.	Andrian G. Nikolayev	1929 Sep 5	30
Captain	Pavel R. Popovich	1930 Oct 5	29
Senior Lt.	Mars Z. Rafikov	1933 Sep 30	26
Senior Lt.	Georgy S. Shonin	1935 Aug 3	24
Senior Lt.	Gherman S. Titov	1935 Sep 11	24
Senior Lt.	Valentin S. Varlamov	1934 Aug 15	25
Senior Lt.	Boris V. Volynov	1934 Dec 18	25
Senior Lt.	Dmitri A. Zaikin	1932 Apr 29	27

23, Bondarenko still remains the youngest male ever selected for spaceflight training.

However, they did not all report for assignment together. Popovich was the first to arrive in January, followed by Volynov, then Anikeyev, Bykovsky, Gagarin, Gorbatko, Khrunov, Leonov, Nelyubov, Nikolayev, Shonin and Titov throughout a period of eight weeks. Komarov arrived on 13 March, Belyayev, Bondarenko, Zaikin and Filatyev arrived on 25 March, and the final three – Varlamov, Kartashov and Rafikov – arrived on 28 April.

The group was officially presented to Marshal Konstantin Vershinin, Commander-in-Chief of the Soviet Red Air Force (VVS), on 7 March 1960, a week before the formal training programme began.

PREPARING MEN AND MACHINES

Ground tests

Once the configuration of Vostok had been confirmed, the first elements of hardware in production completed a series of systems tests and simulations. Several full-size mock-up descent capsules (called boilerplates) and scale models were used (both on ground simulations and during altitude drop tests) to test the parachute recovery systems and the landing integrity of the spherical capsule.

Following the November 1958 meeting of the Council of Chief Designers, the basic two-module design had, for the most part, been agreed upon. As there remained a few design issues to be resolved, the construction plant took the unusual step of deciding to set up the manufacturing assembly line. There could not have been many changes in the design that Korolyov had proposed, because the preliminary blueprints of the prototype hull were ready by March 1959, and detailed drawings were completed just two months later.

Using these plans, Soviet engineers established a process of production-line spacecraft construction that has survived for more than forty years. From the design data supplied by Korolyov and his team, two 'electrical analogues' of the capsule were completed by the end of 1959, and were designated 1KP. These 'spacecraft' were outwardly similar to what was planned to be the manned version of Vostok, including instrumentation for the electrical circuitry, attitude control, communications and other systems. Although both of these spacecraft incorporated most of the instruments that were planned for the manned vehicles, neither of them featured the thermal protection (heat shield).

These vehicles were used in ground tests at OKB-1, and were part of a series of engineering mock-ups used as part of the development programme. This was a similar approach to that which the Americans adopted during their development of a new manned spacecraft. These non-flight versions were used for a variety of test and evaluation programmes, including thermal vacuum tests, heat-shield tests at various temperatures, and verification of the subsystems, including parachute deployment, ejection seat operation, electrical and mechanical system tests, nose shroud and spacecraft separation sequence testing, and radio aerial deployment.

As these vehicles were designed to eventually support a human cargo, one of the more extensive test programmes involved the Vostok ejector system and capsule landing integrity. This began in the early months of 1960, and involved models of the spherical capsule being dropped on a variety of ground surfaces from ever-increasing heights. Initial tests would see the instrumented capsule 'drop' from a tower structure. The mock-ups were then taken into the air, slung under helicopters, aircraft and stratospheric balloons, and released to evaluate the parachute recovery sequence and impact velocity. In 1959, several Vostok spherical capsules were constructed for these tests, and five were released from high-altitude aircraft. The final capsule contained a payload of animals.

For the aircraft drop-tests, an AN-12 turbo-prop transport aircraft was used, flying to its maximum altitude of seven miles. It took considerable piloting skill to ensure that the aircraft was still under control when the capsule was dropped, as it suddenly lost more than 2.5 tons of cargo weight. By the time the capsule had dropped to 4.5 miles, it had attained a state of free fall, exactly reproducing the conditions that the capsule would encounter when returning from space. As the capsule dropped, aircraft filmed the jettisoning of the crew-compartment 'Hatch 1' and the seat catapulting clear of the falling capsule. Then 'Hatch 2' would be jettisoned and the recovery parachutes deployed, slowing the capsule towards a 'dust-down' in the test zone.

For human recovery, the Vostok ejection system was tested from static ground

A cosmonaut being tested on an ejector seat tower.

A tester being ejected from a MiG fighter as part of the Vostok ejector seat development programme.

rigs, off the back of high-flying aircraft (including the IL-28 flying laboratory) and from stratospheric balloons. These tests continued to be conducted until just weeks before the first manned flight, and then continued throughout the Vostok programme to test refinements in flight hardware. Subjects used in the tests included mannequins dressed as cosmonauts, more dogs, and human test-subjects. One such test ended tragically in death.

The Volga stratospheric balloon programme conducted by the Soviets in the

Air Force parachute-testers Andreyev (*left*) and Dolgov (*right*) preparing for a high-altitude ascent in the balloon Volga. (Astro Info Service Collection.)

The high-altitude balloon Volga, which ascended to 82,000 feet as part of the Vostok test programme. (Astro Info Service Collection.)

1950s and early 1960s was a continuation of similar programmes carried out during the 1930s. A secondary objective of Volga was to test hardware and systems being developed for the manned space programme. One of the most obvious tests that high-flying balloons could simulate was the cosmonaut ejection seat and parachute recovery sub-system. These tests included the participation, as a back-up crew-member, of Vasili Lazarev, who was later to become a cosmonaut.

On 1 November 1962, Air Force Major Yevgenny Andreyev and Colonel Pyotr Dolgov (both experienced altitude parachutists and record holders) were the latest crew to participate in the Volga/Vostok test programme. Andreyev was to participate in the first part of the simulation. He was wearing a high-altitude pressure suit with helmet, and winter flight-clothing on top, as he lay in the Vostok ejection seat. He was to be the test subject in a non-explosive test of the Vostok ejection system.

At 2 hr 20 min into the ascent, and upon reaching 83,500 feet, he was mechanically ejected from Volga to begin his descent. He separated from the seat 270 seconds later and then began a 79,560-foot free fall, before opening his parachute at 3,940 feet. He finally landed safely after 7 min 30 sec.

Meanwhile Dolgov had remained in the Volga gondola for another 78 seconds, as it ascended to 93,970 feet. Dressed in a full Vostok pressure garment with an independent oxygen supply, he was to test the integrity of the suit during a long parachute descent. This was Dolgov's 1,409th parachute jump, and initially all went well as he left the Volga, even to the opening of his parachute. For 37 minutes, Dolgov descended to Earth, but instead of achieving a safe landing he was found to have died during the descent. His suit had apparently decompressed at some unspecified time, resulting in the loss of one of the USSR's most experienced test parachutists, a Hero of the Soviet Union.

Preparations for a 1962 Vostok ejection seat test from an aircraft. The test subject is shown suited and strapped on an ejection seat prior to being raised out of the top of the aircraft for the test ejection. To the right is Yuri Gagarin. Note that all of them are wearing parachutes. (Courtesy Bert Vis collection.)

Korabl-Sputniks

As well as the ground and airborne tests, the next stage in the Vostok development programme involved a series of unmanned test-flights of the manned spacecraft.

Designated Korabl-Sputniks (KS, spaceship-satellites), these flights would feature the use of a test (boilerplate) Vostok spacecraft (designated 1KP – prototype), and then unmanned versions of the manned craft (designated 3KA).

These flights would begin in Earth orbit, and would be used to evaluate the compatibility of the spacecraft with its intended launch vehicle, and would also provide data on the ability of the design to withstand the rigours of the boosted launch, orbital spaceflight, re-entry and parachute recovery. Evaluating the huge infrastructure would fully test the launch procedures, ground support and tracking and post-mission recovery techniques. There were apparently no sub-orbital flight tests of Vostok, as atmospheric flights provided all necessary data.

Cosmonaut training begins

The first training sessions for the new group of cosmonaut trainees began on 14 March 1960 at Khodyna Airport near downtown Moscow, at a building of the sports club. These initial classroom sessions concentrated on aviation medicine, and became very tedious for the cosmonauts. The early days of training were divided into three days of classroom studies of aviation medicine, followed by three days of intensive parachute training. Unlike their American counterparts, the Soviets trained in secret, and were anonymous even to the Soviet people, let alone the outside world. They also conducted extensive compulsory physical training programmes to build up their physical reserves and to provide constant medical monitoring for the doctors.

This early emphasis on medical issues demoralised the candidates, and when Korolyov, Karpov and Kamanin realised this, they assigned some of the OKB-1 engineers to teach the group astronomy, physics, rocket technology, flight dynamics, spacecraft systems and design. The lecturers included Bushuyev, Tikhonravov, Rushenbakh, and some of the leading members of the bureau, as well as some of the experienced engineers working on Vostok, including Konstantin Feoktistov, Vitaly Sevastyanov, Alexei Yeliseyev and Oleg Makarov, who all later became cosmonauts.

The cosmonaut team (as well as its training and support staff) was given the military number 26566 under the command of Karpov. The senior officer in the cosmonaut 'squad' was Belyayev, who became the team's first commander. Popovich became the unit's Communist Party Secretary, with Nikolayev as his deputy.

A group of experienced test pilots – including Heroes of the Soviet Union Ivan Dzyuba and Mark Gallai – supported the team, although initially there was no available simulator on which the group could train. To compensate for this, the academic load increased to include courses in radio and electrical engineering, spacecraft telemetry, and guidance and navigation, as well as the intensive physical conditioning programme

One aspect of the Vostok mission that could be practised was the parachute recovery at the end of the mission. Several younger members of the team had completed only the five mandatory jumps at the start of their Air Force flying career, but over a six-week period the team completed forty jumps each to rapidly advance them to 'expert' status. High-altitude jumps provided them with experience of stress, weightlessness and accelerated flight. The training was conducted on the Steppes

The first cosmonaut group undergoes parachute training in 1960.

under the control of some of the most experienced Soviet parachute instructors, and was led by Nikolai K. Nikitin, holder of several parachuting records.

The Soviets did not, at first, have their own version of the American KC-135 'vomit Comet', but instead used the back seat of a MiG 15 fighter, flying a similar parabolic curve. This did not provide much experience of true weightlessness, as they were unable to float around in the tight confines of the cockpit. Titov once remarked that on such unpleasant and uncomfortable rides, the most memorable event was having the dirt and dust from the cockpit fly into your face. However, the Soviets eventually adapted transport aircraft to allow the cosmonauts to practise certain techniques in short bursts of weightlessness, both suited and unsuited. This was normally in bursts of up to 20 seconds, but was sometimes as much as 45 seconds. The candidates evaluated the use of food, liquid and waste systems, and the operation of control displays on a simulated Vostok control panel.

The experiences revealed to the Soviets that the reaction to exposure to short-term conditions could be classified as: 1) those who felt good; 2) those who experienced illusionary sensations after 12–15 exposures; and 3) those who developed immediate discomfort and adapted with difficulty (they became sick!). It was found that, in repeated parabolic flights, those candidates with lower tolerance adapted gradually, which meant that a conditioning system could be adopted.

In flight-trained personnel, the disorders were less pronounced, and disappeared more rapidly upon repeated exposure than with non-flight experienced test-subjects. Only twelve or so exposures were required for pilots to adjust, but up to 30 exposures for non-pilots. This experience would be of importance during later training for the longer missions in conditioning the pilot to weightlessness – or so the Soviets thought.

A cosmonaut experiences brief periods of weightlessness in a padded aircraft while completing a series of parabolic curves during training.

Wearing a Vostok pressure suit, a cosmonaut rides the 'devil's merry-go-round' centrifuge, to which is fitted a replica of the ejector seat.

Training facilities gradually improved for developing techniques for future crew assignment in spaceflight conditions and for the development of professional skills in controlling the spacecraft and onboard systems.

The centrifuge, gymnasium facilities, a batut (similar to a trampoline), treadmills, landing simulator test rigs, parabolic flights in aircraft, and continued use of the isolation and pressure chambers, helped to condition the candidates for spaceflight.

For the development of skills to fly a mission, a range of simulators was provided for navigation, communications, orientation, ascent and landing, and for conducting emergency and contingency operations. Proving their skill in manual flight control (but not in orbital manoeuvring), optical orientation, planetarium studies, navigation, computers, trajectory, engine technology, simulated control panels and flight equipment (such as food and waste systems, cameras and recording devices) both in normal and abnormal conditions, led a candidate to the final stages of training, once assigned to a mission.

Group for immediate preparedness

By early May, the Vostok simulator was ready, and was under the direction of Gallai. But it soon became obvious that to train all twenty men to the required high standard would be impractical. So evolved the concept of the training group, which has been used by the cosmonauts ever since. The rationale behind such a decision was outlined by Karpov in an article that appeared in *Izvestia* in 1962, which stated: 'Preparations for the [first] trip [into space] included the selection of five or six cosmonauts from the group to form a sub-group for immediate preparedness.'

On 30 May came the selection of the first group of 'immediate preparedness', consisting of Gagarin, Kartashov, Nikolayev, Popovich, Titov and Varlamov. These six began an accelerated training programme for the first flight. The Soviet Union's first man in space would be selected from these six cosmonauts.

Korabl-Sputnik 1

As the training programme of the cosmonauts intensified, preparations for flight continued with the first 1KP. The plan for Vostok missions envisaged three mission-types of spacecraft to complete the flight programme. These were designated Versions A, B and C, (in Russian, A, B, and V):

- Vostok A – the non-recoverable variant.
- Vostok B – recoverable, carrying biological samples and animal subjects.
- Vostok V – the man-rated version.

By the spring of 1960 the team was ready to put a prototype Vostok into orbit for the first time. In May 1960 the first of these spacecraft – the 'A' variant – was readied for orbital flight. At the time there was still some concern about the integrity of the braking rocket system (TDU-1 – Braking Engine Installation) from the OKB-2 design bureau. For this first flight, it was planned to orbit a 1KP to test the firing of the TDU, and to provide actual flight data of the proposed Vostok mission profile, from launch to just after retrofire. As this capsule did not have a heat shield, however, the spacecraft would disintegrate upon entry, as planned. This ensured that the spacecraft would not accidentally land outside of the territory of the Soviet Union if the TDU malfunctioned or if control was lost, and would also ensure safety from falling debris. However, the real reason – according to designer Konstantin Feoktistov – was so that the spacecraft would not fall into 'the hands of our competitors' – the Americans.

Meanwhile, the Americans were stepping up their coverage of the U2 flights over

Soviet military targets of interest, including Tyuratam. They realised that tests of the rocket into the Pacific heralded a new version of the missile and, from CIA information, that it was probably connected to the suspected manned space programme. By the end of April, CIA intelligence had picked up increased radio traffic at its listening stations located in Turkey, indicating a pending launch from the cosmodrome. A U2 spy-plane was ordered to fly over the launch pad area on 1 May 1960, to take the first photographs of the R-7 on its pad.

The U2 took off from an air base in Peshawar, Pakistan, after which the pilot, Lt Gary Powers, USAF, climbed the plane to its operational soaring height of 82,000 feet, allowing him to fly over the cosmodrome and take the planned photographs, far above the range of any anti-aircraft guns. After completing the photographic run, Powers headed north to Sverdlovsk, before heading back in a wide circle to land in Bødo, Norway. Suddenly, there was a huge explosion in the tail of the aircraft, as a Soviet surface-to-air missile struck it. The badly damaged U2 began to spin uncontrollably towards Earth. Powers bailed out and parachuted to safety, but landed in the forests of the Urals, in Soviet territory.

Powers was captured, and was accused of spying. In Paris, Soviet Premier Khruschev stormed out of a conference, accusing the American government of spying. Powers was put through a well-publicised trial and spent two years in prison behind the Iron Curtain. On 10 February 1962 he was turned over to the embarrassed Americans in an agreed exchange of a 'spy for a spy'. He was just the pilot of a spy-plane, and was exchanged for a 'real' spy, Rudolf Abel, who had been caught in 1958. The event dealt the Americans a serious blow from which it took several years to recover; but U2 operations did not cease, although spy flights over Russia did not occur again. This incident, however, was not the last confrontation with Khruschev's missiles.

On 15 May 1960 the first Vostok – 1KP – was launched from the Baikonur cosmodrome by the 8K72 version of the R-7 launch vehicle. It was placed into an initial orbit of 229 × 194 miles, inclined at 65°, with an orbital period of 91.2 minutes. (Orbital data for all Korabl-Sputnik missions are included in the table on p. 120).

The Korabl-Sputnik (KS-1) was termed Sputnik 4 in the West, and although no illustrations were issued at the time, this was the first occasion that a prototype manned spacecraft had entered orbit – a significant achievement over the American Mercury programme that was still under development. In a statement released on 16 May, *Pravda* reported that after all required data had been received from the spacecraft, 'a pressurised cabin weighing approximately 2.5 tons will be separated from it'. The report also stated that the return of the capsule was not intended. 'Upon a command from the Earth, and after the reliability of its functioning has been tested, [it will] begin to descend and will terminate its existence as it enters the dense layers of the atmosphere.' This statement gave a clear indication that the Soviets were testing a prototype manned space vehicle.

The design of the spacecraft incorporated many of the systems planned for the manned versions, except for the heat shield and life support systems. A dummy 'payload' (weighted to resemble a suited cosmonaut) was placed inside the capsule to

Vostok precursor missions, 1960–1961

Name	Design	Launch	Land/Decay	Mass (lbs)	Period (min)	Inc. (deg)	Apogee (miles)	Perigee (miles)	Orbits	Duration (d:h:m:s)
11F61 (Vostok 1) *Unmanned test spacecraft in preparation for a manned spacecraft*										
Korabl-Sputnik 1	1KP	1960 May 15	1965 Oct 15	10,000	91.2	65	229.2	193.8	?	1979:21:21:??
Simulated load for ejector seat and cosmonaut										
Orientation system failed; firing of TFDU resulted in a higher orbit; instrument module descended after 1979 days										
SA										843:09:36:??
Descent module of KS-1 descended after 843 days										
—	1K	1960 Jul 28	1960 Jul 28	10,140?	—	—	—	—	—	?
Two dogs – Chayka and Lisichka										
First-stage launch failure; dogs killed										
Korabl-Sputnik 2	1K	1960 Aug 19	1960 Aug 20	10,140	90.7	64.95	211.2	190.1	18	01:02:23:36
Two dogs – Strelka and Belka; 12 mice; insects, plants, fungi, cultures; seeds of wheat, corn, peas, onions; microbes; strips of human skin										
First successful recovery of living organisms from orbit										
Korabl-Sputnik 3	1K	1960 Dec 1	1960 Dec 2	10,060	88.47	64.97	154.7	111.8	18?	01:01:45:36
Two dogs – Pchelka and Mushka; probably other biological payload										
Incorrect orientation resulted in steep angle of entry and burn-up; dogs killed										
—	1K	1960 Dec 22	1960 Dec 22	—	—	—	—	—	—	?
Two dogs – Shutka and Kometa; probably other biological payload										
Upper-stage launch failure resulted in aborted mission; dogs recovered 48 hours after landing										

11F63 (Vostok 3) *Unmanned variant of the single-seat manned spacecraft*

Korabl-Sputnik 4	3KA-1	1961 Mar 9	10,360	?		64.93	154.6	114.0	1	00:01:41:06
	Dog Chernuska; cosmonaut mannequin; 40 mice; guinea pigs, reptiles, plant seeds, human blood samples, human cancer cells, micro-organisms, bacteria, fermentation samples									
	Fully successful demonstration of one-orbit mission; dog and mannequin recovered									
Korabl-Sputnik 5	3KA-2	1961 Mar 25	10,350	88.42		64.9	153.4	110.6	1	00:01:40:48
	Dog Zvezdochka; mannequin 'Ivan Ivanovich'; other animal and biological payloads									
	Second successful one-orbit flight; final unmanned precursor mission of initial Vostok programme									
Korabl-Sputnik 6	3KA-3	–	–	–		–	–	–	–	–
	Cancelled, as not required; spacecraft prepared for initial manned flight on 12 April 1961									

simulate the mass of a human passenger. Outwardly, KS-1 resembled the later Vostok manned spacecraft, although the unmanned design featured a small, twin-panel solar array device, mounted on a mast on the top of the Descent Module.

These half discs, measuring 20 inches across, also carried their own orientation devices, enabling them to 'track' the Sun during the orbital daylight pass. They constituted an early investigation into technology that used solar energy, instead of the heavier chemical batteries, to power spacecraft. So, on the very first flight of a Vostok in orbit, the Soviets included a technology experiment that would provide important information for follow-on spacecraft. Indeed, this technology was soon incorporated into the Soyuz spacecraft, with its twin solar arrays.

At 02.52 Moscow Time (MT) on 19 May, as the spacecraft passed over Africa during its 64th revolution, the TDU was initiated for retro-fire as planned. Unfortunately, a failed sensor onboard the spacecraft caused the vehicle to misalign by almost 180°. Much to the disappointment of the design team, instead of slowing the spacecraft it actually increased its orbital speed by 295 ft/sec. Burning for only 26 seconds, it was insufficient for a successful de-orbit and recovery had it been planned. The capsule separated as planned, but instead of burning up in the atmosphere shortly afterwards, the module entered an orbit of 191 × 429 miles. Here it remained, apparently still transmitting data for the next eight days. Natural orbital decay finally destroyed the vehicle on 5 September 1962, 843 days after it had been launched! The Instrument Module followed on 15 October 1965, after a flight of 1,979 days.

Despite the problems experienced with the attitude control system and an excessively noisy communications test frequency, the flight was relatively successful. Misinterpreting the data from the mission, Western journalists reported a failed recovery attempt, whereas the *Pravda* article had clearly indicated that a full recovery was never planned.

Korolyov, although disappointed, remained philosophical, noting that the team now had experience in manoeuvring in space!

Korabl-Sputnik 2

The evaluation of data received from the first KS mission took several weeks, and despite the malfunction of the sensor it was deemed successful enough to plan four test flights of the next ('B') version of the spacecraft, designated Vostok 1K. In addition, confidence in the system was such that on 4 June the launch date for the first manned test was officially set for December 1960 – just six months away!

Each of the four 1K modules would be covered in the thermal heat protection system, and would also have a self-destruct security device installed, in case the capsule should wander off the planned flight path.

The first attempt to orbit the 1K version came on 15 July 1960. A flight of 24 hours was planned, in the first orbital test of the Vostok life support system. The primary payload on this flight was the canine cosmonauts Chaika (Seagull) and Lisichka, the first dogs to be orbited since Laika's flight onboard Sputnik 2 in 1957. Many of these dogs had been selected in the 1950s for the high-altitude rocket research programme, and had 'graduated' to the Vostok programme.

A dog test-subject with canine pressure suit.

Unfortunately, just 17 seconds into the flight, an engine on one of the four strap-on first-stage boosters malfunctioned, causing the booster to break away prematurely, and resulting in the explosion of the launch vehicle, which killed both dogs.

According to one source, after this launch failure there was a move to try to create a cosmonaut ejection capsule. This design would incorporate its own heat shield, in the event of a high-altitude abort or only partial retro-fire. This would allow the crewman to 're-enter' the capsule before using the personal parachute system. Such devices were also evaluated in America in the early 1960s. Apparently, the Soviet system proved too heavy and was soon abandoned.

Little time was lost in reviewing the failure and preparing the next launch, which came on 19 August 1960. Designated Korabl-Sputnik 2 (and Sputnik 5 in the West), this 1K capsule also carried two dogs – Strelka (Little Arrow) and Belka (Squirrel) – inside a small recovery capsule. In addition to the two canine passengers, KS-2 also carried a range of other biological specimens, both on the ejection seat and around the inside of the cabin. These included forty white mice, two rats, numerous flies, and fourteen flasks of a selection of plant seeds, fungi, *chlorella* algae and 'spider' wort. There were also human tissue samples – kindly provided by some of the scientists working on the project!

As with the previous attempt, the flight plan was designed to support a 24-hour mission before retro-fire and an attempt to recover the capsule, hopefully with its passengers still alive.

The capsule of KS-2 had been fitted with two TV cameras, one of them mounted on the ejection seat, providing a frontal view of Belka, and the second on the wall of the capsule, providing a profile view of Strelka. This allowed the scientists to study

Recovery of a biological payload canister from a high-altitude rocket flight in the 1950s.

the reactions of the dogs as they experienced the planned 24-hour mission. In a further experiment, instruments were placed in the cabin to record levels of radiation throughout the flight (to gain information on whether these levels might harm a human crew). Belka also kindly provided the scientists with the first recorded experience of what is now termed space adaptation syndrome – the dog was sick!

On 20 August, during the eighteenth orbit of the Earth, the Vostok descent apparatus was again activated, but this time it worked perfectly. The onboard systems correctly orientated the spacecraft prior to capsule separation and re-entry. When the descending capsule reached an altitude of 4–5 miles above the Earth, the side hatch was blown off and the ejector seat (with the two dogs and other biological payloads onboard) was separated from the sphere, to be recovered by parachute.

After a mission lasting 1 day 2 hrs 23 min, the capsule landed at 33 ft/sec, only 6.2 miles away from the predicted landing site. This was extremely accurate, considering that it was the first time that such a feat had been achieved. A short distance from the capsule, the container and ejector seat carrying the animals landed at 18–25 ft/sec. Showing no sign of harm from their historic journey, the dogs and the rest of the animals and plants were the first living organisms to achieve a complete spaceflight cycle, from launch to recovery.

In post-flight studies of space radiation, the human tissue samples were grafted back on to the original donors so that any adverse effects could be observed.

First casualties

By early July the centrifuge – similar to the Air Force centrifuge used during selection – was ready for the cosmonauts, and Anikeyev was the first to try it. Also in July, the whole group moved to the new purpose-built Cosmonaut Training Centre (TsPK), known eventually as Zvezdni Gorodok (literally 'stellar village', but more commonly known as Star City) and situated near the village of Shchelkovo, 24 miles north-east of Moscow. Although the centre included apartments for the cosmonaut's families, trainers and simulation engineers, and other support staff, the cosmonauts initially told their wives that they were involved in an experimental programme – without ever mentioning space. It was also adjacent to the Chkalov Air Base, the location of the Soviet AF Flight Test Centre.

Any training programme for spaceflight is always risky, especially in the early days, and it was not long before the group suffered its first casualties. Vladimir Komarov, one of the oldest members of the team, known for his piloting skills and his bad luck, was hospitalised on 15 May 1960 for a hernia operation. He was out of training for six months, but tried to keep up with his studies while he recuperated. On 24 July, Varlamov (who, within weeks of commencing training had mastered the intricacies of physics and astronavigation) dived into a lake and fractured his vertebrae. He was in traction in hospital for more than a month, and lost his place in the Advanced Training Group. He would return to the team, but had to leave on 6 March 1961 because of this accident. He continued to work as a cosmonaut instructor, and was described as one of the best candidates of the group.

As the parachute training continued, in August 1960 Belyayev broke an ankle, which delayed his training for more than a year and cost him the chance of an early flight. During the centrifuge training, Kartashov, after pulling 8 g, experienced a bad reaction due to pinpoint haemorrhaging in his spine. He, too, was grounded from further training by what were later termed 'over-cautious medical specialists'. Despite pleas from Gagarin and other colleagues (who described him as the best amongst themselves, and their candidate for the first man in space), he was medically disqualified. He left the team in April 1962 to return to AF duties, and became a test pilot.

The Vostok Advanced Training Group of six was now reduced to four. Replacements were needed, and so Bykovsky and Nelyubov were assigned to join the group to train for Vostok missions.

Alexei Leonov was one of the least senior in rank in the group, and physically one of the largest. He possessed a good nature and quick wit, and his cheerful temperament won him a lot of friends. He was a leading candidate for the first flight until height concerns over the ejection seat and hatch led to the decision to choose shorter candidates. Towards the end of 1960, his career was almost ended when he and his wife were being driven near the entrance of Star City. The car skidded off the icy road and plunged into a lake. Leonov escaped, and not only saved his wife, but also pulled the driver free. At Star City, that stretch of water became known as Leonov's Lake.

At Cape Canaveral, the Americans were also experiencing setbacks. On 29 July 1960, Mercury Atlas 1 was launched – with no test subjects onboard – to test the integrity of the spacecraft's structure during launch and re-entry. Unfortunately, 59 seconds after launch, the Atlas and adapter failed structurally, and the vehicle broke apart. The capsule contained no recovery systems (as this was not part of the mission test objective) and was destroyed upon impact with the ocean.

Disaster at Tyuratam

By October 1960, launch teams were preparing to launch the R-16 on its maiden flight. Developed by the Yangel OKB, it was designed as a replacement for Korolyov's R-7 as a nuclear carrier rocket. The R-7 – soon to be proven as a reliable workhorse for space exploration – was being troublesome in its role as an ICBM. The problem lay in its five-hour preparation for fuelling and launch, because of the cryogenic propellants that were excellent for rocket engines but were difficult to keep cool prior to launch. The bleed-off of fuel was replaced by gas, which threatened to split the vehicle, and so complicated bleed-off valves were required to keep the stages topped up and ready for flight.

With the R-16, Yangel rejected the liquid hydrogen and kerosene that were used in the R-7 in favour of nitric acid and hydrazine. This allowed the rocket to be kept stored fully fuelled, without venting or leakage, for a much longer period, making it much more suitable for silo storage. The only setback was that the so-called 'storable fuels' would not remain stored. Both fuels were excessively corrosive, and the systems were prone to leakage.

The flight of the first R-16 was an important event that needed to be successful. In recognition of this, and of how important it was for Khruschev to demonstrate a growing might in his rocket forces, Marshall M. Nedelin (then Chief of Missile Deployment) went to Tyuratam to witness the final preparations for launch, planned for 23 October. The launch of the first probe to Mars on 10 October (using the upgraded R-7) had failed, as did a second attempt four days later. In New York, Khruschev was at the United Nations, and awaited the news of the successful launch of the first probes to the Red Planet. He received the news of the launch failures with great disappointment, and instead boasted of the missile supremacy of the Soviet Union, which 'was turning out missiles like sausages from a machine!'

To the frustration of Khruschev, OKB-1 could never mass-produce the R-7 fast enough to meet his military requirement. He was also experiencing great difficulty in ordering the Kuibyshev Aviation Factory to be converted to full R-7 production, arguing against A.N. Tupolev, whose bombers were also being built at the plant. This production block, coupled with the design drawbacks of the R-7, conflicted with the operational requirements for constant readiness. Costs escalated tremendously, even to deploy the R-7 on a limited scale.

Khruschev's simple solution was to put more powerful and cheaper hypergolic (nitric acid) engines on the R-7. Korolyov refused, stating that the R-7 was the best they had, and that none of the more powerful engines could handle such propellants. If Khruschev did not believe him, he said, he should ask Glushko. And so Khruschev asked Glushko, who stated that he believed the opposite, and that Yangel had already designed such an engine and rocket (which evolved into the R-16). When challenged, Korolyov stated that his OKB could build such an engine and rocket quicker than could Yangel. This set off a bitter dispute between the designers, and Korolyov called Glushko a traitor, vowing never again to put another of Glushko's engines on one of his rockets.

As early as 1958, Khruschev suggested that his missiles should be placed in silos underground, to prevent damage from attack. Inspired by intelligence reports from the Americans, who were to do the very same thing, the Soviet leader became persistent about the idea, even after Korolyov stated that it was unworkable – certainly for the R-7. Even Yangel and Barmin, the designer of the cosmodrome and all the Russian rocket launch systems, told Khruschev that such a plan was impossible.

However, at OKB-52 the Premier's son worked as an engineer for Chelomei, who had access to the Western aerospace press. He found such designs, and passed them on to his father. Khruschev was enraged, and severely reprimanded his designers for not keeping up with Western technology. Within a year the designers had changed their minds, and one of Yangel's R-12s was launched from a silo. Thereafter, almost all military missiles would be designed to be silo-launched. But Khruschev no longer trusted his 'experts', and as the maiden launch of the R-16 was set up on the pad at Tyuratam, he authorised Chelomei to supply an alternative proposal that put pressure on both Korolyov (for space) and Yangel (for missiles). It was a confrontation that lasted for fifteen years, and probably cost Russia the Moon race, competing amongst themselves instead of against the Americans.

Tragically, the events at the launch pad on 24 October did nothing to help win confidence in Yangel's alternative to the R-7. As the launch time approached, the rocket began to drip nitric acid. The launch was halted, and Nedelin – already under pressure from Khruschev to provide proof of the strength of Soviet missile technology – insisted that dozens of technicians go to the pad to stem the leak and resume the launch. In this situation, the correct sequence of events should have seen a draining of the tanks, followed by the purging of the system with non-flammable nitrogen. Then, about 24 hours later, a few fire-suited technicians could approach the rocket to declare it safe.

In the blockhouse, the command to disarm the firing mechanisms was countermanded by Nedelin, who ordered the launch to be delayed and not cancelled. An incorrect command was sent to the upper stage, which continued to count down. With scores of technicians working immediately around it, the engine suddenly ignited, burning a hole in the stage below it and igniting the fuel. It exploded, killing everyone on the launch gantry.

The consequence of this was dramatic and fatal, as the unsupported stage fell onto the pad in a ball of fire, igniting the rest of the rocket and spreading flame 3,000 feet from the pad. A total of 190 military technicians and officials (including Nedelin) were killed, and many more were seriously burned. The next day, the USAF *Discoverer* spy satellite (one of the replacements for the risky U2 flights, after the Gary Powers incident) overflew the pad and recorded a very large rocket explosion. But as the news was confined to official circles in the Soviet Union, it was not thought that unusual, as in those days rockets had a tendency to blow up – as the Americans knew only too well.

The official statement indicated with great regret that Marshall Nedelin had been killed in an aircraft accident, along with several other senior officers. The cosmonaut team was told that a prototype missile (not the one they were to ride into space) had exploded, and several technicians had been injured. It was many years before the true story of the accident became fully known.

As for the pad itself, only a few of OKB-1's technicians worked on the R-16, while the facilities for the R-7 were undamaged, and although the explosion was a bitter blow to those in and around the cosmodrome, it did not have a serious impact on preparations for Vostok.

NASA, too, was experiencing embarrassment in its Mercury programme, although not with such tragic results. Senator (and presidential candidate) John F. Kennedy had already hit out against the Eisenhower administration by stating that if a man was to be launched into orbit in 1960, 'his name will be Ivan'. On 21 November, an attempt was made to launch Mercury Redstone 1 on a sub-orbital test flight. Premature electrical cut-off of the Redstone occurred, but the safety systems worked as planned and activated the launch escape system when the launch vehicle was just one inch off the pad. The Redstone settled gently back onto the pad with only slight damage. Fearing it might topple, technicians could only watch and wait as the vehicle was safed. Suddenly, the escape tower fired both the escape and tower jettison motors together, shooting the escape tower over half a mile into the sky, to fall just 1,200 feet way. Three seconds later, the forward cylindrical antenna housing 'popped' off the top of the capsule, which dragged the drogue parachute and main parachute out of the container, closely followed by the reserve parachute. With parachutes blowing in the wind, the Redstone could have blown over any moment, but fortunately it remained secured to the pad. It had to be left until the next morning, when it was finally safe to approach the vehicle to prepare it for another attempt. At the Cape, this event became part of spacelore, and is remembered as 'the day they launched the escape tower'.

Korabl-Sputnik 3
As 1960 progressed, it became evident that a manned flight before the end of the year would not be feasible. Two further flights of 1K capsules were planned, before flight-testing the 3KA planned for use with cosmonauts.

On 1 December 1960, Korabl-Sputnik 3 (Sputnik 6 in the West) was successfully launched into an orbit of 117×165 miles inclined at $65°$, close to the orbit to be used by the manned series of missions. The launch vehicle used for this mission was the 8K72, the last to fly before the upgraded 8K72K entered service. The mission was to be a repeat of the previous KS-2 mission, and on board were the dogs Pchelka (Bee) and Mushka (Little Fly), plus mice, insects and plants. The Soviets stated that new systems were being evaluated on this flight, including the transmission of physiological data from the biological payload, and the maiden flight of a computer-controlled spacecraft control system.

Apparently this new system still had some development problems which needed to be resolved, because when the command was given to de-orbit the capsule on 2 December, the engine fired for a far shorter period than was planned. The descent capsule and its payload would re-enter, but over an area outside of the Soviet Union. As a precaution against the unmanned spacecraft falling into foreign hands a self-destruction system was placed onboard each Korabl-Sputnik, being triggered by a sensor that activated the explosive device if re-entry was not detected at the pre-planned time. The device was activated on Korabl-Sputnik 3 as the vehicle began its re-entry, and destroyed the vehicle and its passengers. The flight had lasted 1 day 1 hr 45 min. In post-flight news releases, the Soviets announced that the capsule had unfortunately burned up due to an incorrect re-entry angle that was too steep.

Undaunted, plans went ahead to try to fly the last 1K mission before the end of the year. The No.4 capsule (onboard the new 8K72K launcher) carried the dogs Shutka and Kometa. Unfortunately, this time the RO-7 engine on the upgraded upper stage sustained a failure upon ignition, and inflicted damage to the spacecraft, which immediately headed for an emergency recovery. The capsule containing the two dogs separated successfully, but the self-destruct system apparently failed and the capsule landed by parachute, intact, in the remote Tunguska region of Siberia, 2,175 miles downrange.

On board, a 60-hour timer had been installed in the capsule's emergency destruction system, and by the time rescue teams reached the remote landing site 60 hours had elapsed, but the vehicle had not exploded. In sub-zero temperature the team disconnected the arming device before finding that the detonation cables had burned through. Although the descent hatch had been ejected, both dogs were still alive – just – and, to the recovery team's surprise, were still inside the ejection seat capsule inside the sphere, which had not been jettisoned as expected. The dogs had managed to survive two days at –45° C. Due to the conditions it took several days to recover the capsule, but post-flight examination revealed that the descent and instrument units separated only because of the re-entry heating, and the ejection seat fired as the hatch was released instead of 2.5 seconds afterwards. This caused it to smash against the hatch porthole and buckle, thereby preventing ejection. With these malfunctions and the failure of the destruction system added to the difficulties on the previous flight, any chance of a manned flight in February was lost.

By the end of 1960, the Soviets had attempted to launch five 1K versions of the Vostok, with only one full success – not an inspiring start to the flight programme. At that time, according to cosmonaut Gherman Titov, the psychologists were worried that news of events such as exploding rockets and faulty recovery systems would have an adverse effect on the confidence of the cosmonauts in training for the first flights on the Vostok, and it was therefore decided to not tell the group about the setbacks until several weeks later. By then, it was hoped, unmanned flight tests of the manned variant would have ironed out the problems, so that such news would not be so unnerving for the men risking their lives with this technology.

On 19 December, NASA successfully launched the Mercury MR-1A spacecraft on an unmanned ballistic trajectory into the Atlantic. The spacecraft had been reworked after an aborted attempt to launch MR-1, when the launch vehicle had malfunctioned.

Korabl-Sputnik 4

Despite the setbacks of 1960, during the first weeks of 1961 Korolyov and his team pressed ahead with preparations for the first flight of a cosmonaut and with plans to fly up to three unmanned versions of the manned capsule. These were designated 3KA-1, 3KA-2 and 3KA-3. It was decided that if the first two flights were successful, the third could be prepared to support the first manned flight.

Despite evidence that the overall design of the Vostok was good, there were still major concerns over the ability of the TDU to correctly orientate the capsule for entry. A cosmonaut could ensure that correct attitude had been achieved before retrofire, but there would be insufficient time to confirm the position. The original plans for the first manned flight envisaged 6–18 orbits (9–27 hours), but this was changed to limit the first flight to just one orbit (90–120 minutes) and the return of the pilot.

On 17 January 1961 the six men of the training group passed their first examinations, and on 25 January all six were confirmed as potential candidates for the first flight. During the two days of written, oral and practical tests, the Commission also evaluated the earlier performance tests and medical fitness reports, as well as each candidate's overall and communication skills. The final ranking was: 1, Gagarin; 2, Titov; 3, Nelyubov; 4, Nikolayev; 5, Bykovsky; and 6, Popovich. All were then designated 'Military Cosmonauts', and were no longer 'candidates'. The other members of the enrolment gained their 'promotion' to Cosmonaut on 8 April 1961, with the exception of Anikeyev, Filatyev and Zaikin, whose promotion was delayed until 16 December 1961. In late February the cosmonauts of the 'Top Six' were invited to Korolyov's design bureau to see the Vostok capsule for the first time.

On 31 January the Americans scored a success by launching Mercury Redstone 2 (carrying the chimpanzee Ham) on a sub-orbital flight down the Atlantic missile range. Mercury Atlas 2 followed on 21 February, on a sub-orbital test of the spacecraft's heat-shield. That same day, astronauts John Glenn, Virgil Grissom and Al Shepard were selected to begin special training for the first manned sub-orbital Mercury spaceflight. The race was hotting up.

The first of the Soviet 3KA spacecraft was launched on 9 March 1961, and was designated Korabl-Sputnik 4 (Sputnik 9). Weighing 10,360 lbs, it was placed into an orbit of 155 × 114 miles inclined at 65°. The capsule carried a payload of rats, guinea pigs, biological specimens, and one dog – Chernuska (Blackie).

This time the payload was located in the capsule, as occupying the seat was a dummy mannequin wearing a prototype Vostok spacesuit, with the label 'Makat' ('Dummy') placed behind the visor. The space suit, designated SK-1, was being flown to test its integrity. At that time there was a fear for the health of a returning cosmonaut, so the suits were developed with fully automated systems for use in an emergency. These included the closing visor, seat ejection, the personal parachute, automatic opening of a breathing valve, and inflation of the rubber neck-collar, designed to keep the cosmonaut's head above the water in the event of a splash-down.

For the Vostok programme, six manned launches were initially envisaged, and so seven suits would be manufactured, including those for the back-up pilots. The first fittings for these suits by the 'group of immediate preparedness' probably occurred in the late summer of 1960. However, as the test programme developed, a number of additional modifications were needed, and after every modification a new batch of pressure garments was produced in small quantities. The exact number of suits manufactured was probably in the region of two dozen, as between ten and fifteen were manufactured for ground tests before the first manned flight.

KS-4 was a full dress rehearsal of the proposed first manned spaceflight, with the

Recovery of the cosmonaut mannequin from the snow after ejection from the Korabl-Sputnik 4 Descent Module, 9 March 1961.

capsule completing just one orbit of the Earth. It also demonstrated a successful orientation, entry and recovery phase, and at the prescribed altitude the seat containing the mannequin was ejected to complete an independent parachute recovery. Meanwhile, the capsule containing Chernuska and the biological payload also made a safe landing a short distance away. The Soviets claimed that should a cosmonaut have ejected in the seat or elected to remain in the capsule, he would have landed without harm. The flight time was reported as 1 hr 41 min.

The loss of Bondarenko

On 23 March 1961, cosmonaut Valentin Bondarenko, the youngest member of the team, was completing a ten-day run in the isolation chamber. It had been reduced in pressure, requiring a higher oxygen content. These tests were extensive and gruelling, and placed a strain on both the cosmonauts and the test conductors. All the cosmonauts found this to be one of the most demanding elements of their training, pushing their stamina and character to the limit. During a similar test, Titov was nearly dropped from the programme after rebelling at what he called 'silly questions' during psychological tests. At the end of each run in the chamber, the cosmonauts were tired and were prone to a lack of concentration.

Wired up to medical monitoring devices, Bondarenko was removing the stick-on sensors and cleaning his skin with cotton wool swabs soaked in alcohol. He casually tossed one aside, and it fell onto a heating ring – and in the oxygen-saturated environment of the chamber, it instantly ignited, creating an inferno. Bondarenko, sealed in the chamber and wearing his woollen training suit, suffered 90% burns from the fire. It took several minutes for the chamber pressure to be equalised and the door to be opened. The severely injured cosmonaut, accompanied by Gagarin, was taken to the nearby Botkin Hospital. Suffering from shock and severe burns,

Senior Lieutenant V.V. Bondarenko, who, on 23 March 1961, died from injuries sustained in a training accident three weeks before Gagarin flew in space.

Bondarenko died eight hours later, just three weeks before Gagarin's historic flight. It was to be 25 years before the full story of Bondarenko was released, but Gagarin and his colleagues carried his memory into space. The pressure to succeed was very telling on the cosmonauts, as well as on Korolyov's team.

Korabl-Sputnik 5

Despite the success of the KS-4 flight, Korolyov ordered one more unmanned flight before the decision would be taken to commit to a manned launch. Consequently, on 25 March 1961, just 18 days after KS-4 had flown, Korabl-Sputnik 5 (Sputnik 10) took off from Baikonur in the final unmanned test flight. The launch was witnessed by members of the cosmonaut team, just two days after the loss of Bondarenko and during their first visit to Tyuratam. Vostok capsule 3KA-2 carried the dog Zvezdochko (Little Star) and the second dummy cosmonaut, this time called 'Ivan Ivanovich'. The mannequin was again dressed in a full Vostok pressure garment and helmet.

As with the previous flight, the mission lasted for just one orbit, and was a complete success. Mission duration was reported as 1 hr 40 min. The landing site was near the town of Izhevsk, and the capsule landed in snow five feet deep. To recover the capsule and its payload, a team of thirty paratroopers was sent to guard the site, while the rescue engineers used a ski-plane and a horse-drawn sleigh to reach the

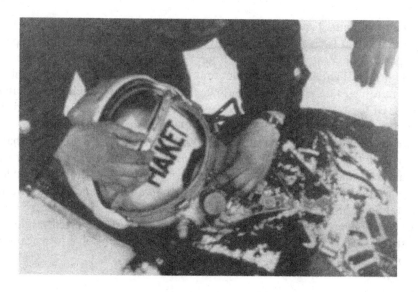

Korabl-Sputnik 5 occupant, the mannequin 'Ivan Ivanovich', is recovered after ejecting from the descending capsule, 25 March 1961.

landed capsule. They found the half-scorched capsule lying in a gully, still steaming hot from the re-entry, and with snow melting around it.

Lying nearby was the mannequin Ivan Ivanovich. Local peasants who had witnessed the landing were dismayed that the paratroops had not gone to the aid of an obviously injured parachutist lying motionless in the snow. It was only when one of them was allowed to touch the face of 'Ivan' that they were convinced that he was not human.

Reportedly on hand at the landing site were some of the cosmonauts, including Gagarin and Titov, who were to be the next Vostok passengers. The Americans were indicating that they would soon be launching a Mercury capsule on a sub-orbital trajectory. The Soviets, however, intended to launch a lone cosmonaut in April, not on a sub-orbital trajectory but on a one-orbit flight around the Earth. With the success of the two 3KA flights, the order was given to prepare 3KA-3 for the first manned flight.

The work of 3KA-2, however was not complete at the landing. One report indicated that the capsule was used for cosmonaut training at TsPK. The development of the manned Vostok was conducted at the same time as the basic Vostok design was being developed for unmanned military applications. Once the 3KA-2 capsule had completed its post-flight analysis, it was shipped to Kuybyshev and the Central Specialised Design Bureau, where KS-5 became the design model for the spy satellites, beginning under the Zenit code name. In 1967, after the removal and destruction of instrumentation and equipment in the interest of Soviet State secrecy, 3KA-2 moved to the Kuybyshev Training Institute as a space demonstration model. The secret classification was removed from the capsule in 1986, and ten years later, Sotheby's offered it for sale.

At the dawn of manned spaceflight

In little more than twelve months, Korolyov's team had launched a prototype and several production Vostoks into orbit. Four 1K variants had also been launched to complete orbital flight and recovery, but only two had achieved orbit, with only one of those returning safely. But the two flown 3KA spacecraft were both outstanding successes. Of the seven orbital missions, two were only partial successes, and there were two launch failures. Only three could be termed successful.

However, the flights had proven that the Vostok system was sound, and that the problems with the TDU had been resolved, allowing the manned launch to proceed. Important data had also been recovered – on flying a variety of biological payloads in space; on the environment through which they were flying; and on the effects of the orbital flight on the structure of the capsule, as well as mission support progress in other areas. Even with the impending flight of an American, mixed results from the Vostok precursor missions could not upset Korolyov's resolve to place a man into orbit for the first time.

First man and first day

VOSTOK 1

The formation of a training group of six cosmonauts was a significant step towards the selection of the one man who would make the first flight on Vostok. As the training progressed, an extensive programme of ground and airborne tests was supplemented by the Korabl-Sputnik missions. Progress in both the training and hardware areas led to formal approval to take the next logical step of putting a cosmonaut in space.

The decision to proceed

The authority to proceed with the first manned spaceflight of Vostok evolved from a top secret/special importance memo (dated 19 September 1960) sent to the General Department of the CPSU Central Committee. The memo was signed by Dmitri Ustinov (Central Committee Member and Chairman of the USSR Council of Ministers), R. Malinovskiy (Central Committee Minister and Minister of Defence), K. Rudnev, V. Kalmykov, P. Dementyev, B. Butoma, V. Ryabikov (USSR ministers), M. Nedelin (Commander-in-Chief of the Strategic Rocket Forces), S. Rudenko (Deputy Commander-in-Chief of the Air Force) and M. Keldysh (Vice-President of the USSR Academy of Sciences).

The memo also bore the signatures of the 'Top Six' group of chief designers involved in the manned space programme: S. Korolyov (OKB-1), V. Glushko (OKB-456), M. Ryazanskiy (NII-845), N. Pilyugin (NII-855), V. Barmin (State Special Design Bureau for Special Machine Building) and V. Kuznetsov (NII-855).

The memo stated that, following the successful launch, flight and landing of the spacecraft (Korabl-Sputnik 2 – article 'Vostok-1') on 19 August 1960, preliminary results had 'shed new light on the dates for performing a manned flight into space'. It was therefore proposed that a spacecraft (article 'Vostok-3A') could be prepared for launching a man into space by December 1960. The memo added that the R-7 launch vehicle (8K78) had already been modified to carry a payload of 7–9 tons into low Earth orbit.

As a result of this memo, a decree dated 11 October 1960 stated: 'The suggestion

The Council of Chief designers at Tyuratam in 1959. This heavily retouched photograph is typical of photographs of the era, during which it was often Soviet practise to add to the confusion and hide certain aspects of the true nature of the photograph. Here there are ghostly shadows in the background. The personalities featured are (*left to right*) Bogomolov, Ryazanskiy, Pilyugin, Korolyov, Glushko, Barmin and Kuznetsov.

is adopted concerning the preparation and launch of a spacecraft (article 'Vostok-3A') with a man aboard in December 1960, his mission being of special importance.' According to this plan, the first manned flight of Vostok should have occurred in December 1960 and not April 1961. So why was there a delay?

In January 1960, Khruschev had reportedly instructed his Chief Designers to propose more urgent measures that would meet the American challenge to Soviet space supremacy. With the possibility of a new US President in office from 1961, Khruschev was concerned about the capabilities of the American space programme, despite there being no official manned project after the one-man Mercury series. However, there were reports from the US of a trend towards an expanded manned space effort, which included proposals for a manned flight to the Moon and a strengthened military presence.

By June 1960 a new decree reflected the Soviet leadership's change of attitude to space, with a new emphasis on military rather than scientific projects. The December 1959 decree that had initially adopted the manned space programme was revoked by the June 1960 decree. After the success of Korabl-Sputnik 2 (carrying the dogs Belka and Strelka), it took the 19 September memo from Ustinov, advocating approval for manned spaceflight, to push the Central Committee towards a firm decision. That decision came with the 11 October 1960 memo.

Supporting the decision was the growing body of evidence from the Korabl-Sputniks that the Vostok design would operate in space, although a lot more work was required on the de-orbit and entry system. However, Belka's sickness in orbit raised some concerns about the possible implications for flying the planned 24-hour duration on the first manned flight.

This optimistic expectation was not realised, for several reasons. First, the rush to launch two unsuccessful probes to Mars in early October 1960 was followed by the huge explosion on Pad 41 at the cosmodrome on 24 October. The government commission enquiry into the explosion of the R-16, headed by Leonid Brezhnev, recommended that a manned launch by December 1960 would not be possible due to the number of changes necessary after the 24 October explosion. Further unmanned launches of the Vostok spacecraft were authorised, to provide much-needed flight experience of the hardware and entry profile before committing to a manned launch.

In January 1961 the strain and pressure finally took its toll on Korolyov, who suffered a heart attack. Coupled with preparations to launch the first (initially successful) probe to Venus (Venera 1) in February, this diverted his full attention from the manned programme until later that month, adding to the delays in launching the manned mission.

Manned flight profile
Two spacecraft of the Vostok-3A (manned) variant were to be prepared for the first manned spaceflight. The first would be used for ground testing and cosmonaut training, and the second would be the flight vehicle. The mission profile was changed to just one orbit, reflecting the concerns raised by the illness of Belka, and would end with a landing within the territory of the Soviet Union, along a ground track running between Rostov, Kuibyshev and Perm. Although always promising a successful flight, the mission was nevertheless planned to include several contingency procedures to preserve the safety of the cosmonaut. These included the choice of an orbit which, in the event of a failure of the retro-rocket system, would decay naturally, resulting in re-entry in between two and seven days, and landing between the latitudes of 65° N and 65° S. In the event of a forced landing on foreign soil, or rescue by foreign forces, the cosmonaut would be provided with 'appropriate instructions'. The craft was also to be stocked with a ten-day supply of food and water, and an emergency portable supply of three days.

Final authority
In March 1961 a full programme review was conducted, including the latest findings from ground testing, the status of cosmonaut training, results from the Korabl-Sputniks, and analysis of ground and support equipment. The review also examined all safety issues and concerns, and cited an expected high reliability from the R-7.

This attention to the safety of the cosmonauts was contrary to growing Western rumours of a Soviet programme that wanted to maintain space supremacy even at the risk of losing cosmonauts on 'doomed' missions. In truth, there was no-one more aware of crew safety in technologically high-risk programmes than Khruschev. Twenty-five years earlier he had been one of those accused by Stalin of contributing to the cause of the crash of the *Maxim Gorki* on 18 May 1935. This was the largest aircraft in the world at that time, and was often promoted by Khruschev and others as a demonstration of superior Soviet aircraft technology. During a flight display, it crashed, killing all of the crew and 36 passengers in what was labelled a 'technological stunt'. In light of this, and despite wanting to upstage the American's in space once again, Khruschev would not have wanted to answer for another fatal accident.

The previous decree (covering a now abandoned launch in December 1960) had expired, and its replacement incorporated a number of amendments. Most notable was the lack of the signatures of the Chief Designers, for whom Korolyov merely signed in their absence. Also significant by their absence were the signatures of Minister of Defence R.Ya. Malinovskiy, and the former head of the State Commission V.M. Ryabikov. These two men had probably refused to endorse the project, which could have been reflected later in the post-flight awards following Vostok. They were the only known members of the State Commission for Spaceflight not to be honoured as Hero of Socialist Labour. As always, when presented with the chance of beating the Americans in space, Korolyov could not resist, even to the point of having to override or ignore any opposition.

The resulting memo, dated 30 March, concluded that everything was finally ready to perform the first flight of a manned spacecraft. The memo bore the signatures of Ustinov, Rudnev, Kalmykov, Dementyev, Butoma, Keldysh, K. Moskalenko (who had replaced the deceased Nedelin), K. Vershinin (Deputy Commander-in-Chief of the Air Force), N. Kamanin (Deputy Chief of combat training of the Air Force, and also in charge of cosmonaut training), P. Ivashutin (First Deputy Chairman of the KGB) and Korolyov.

The final approval for the flight was dated 3 April, and the following day Korolyov reported to the government commission, convened at Tyuratam, on the state of readiness for the mission. On 6 April, Korolyov assembled the Council of Chief Designers to review an agenda of final technical matters, as well as the progress of pre-launch operations for both the launch vehicle and the spacecraft. The selection of the prime crew-member for the first flight would be revealed at the State Commission meeting on 8 April.

'My little swallows'

From the time that the members of the accelerated training group were chosen in May 1960, they followed a concentrated preparation schedule leading towards the first spaceflight. On 18 June the cosmonauts were taken to see Korolyov at OKB-1 to examine the hardware for the first time. Though he was concentrating on the development of the spacecraft, Korolyov continued to have influence on who would fly the missions – especially the first mission. He had reviewed the files on each man, and had met them several times.

The cosmonauts were naturally apprehensive about meeting the Chief Designer at his design bureau, but Korolyov wanted to put the cosmonauts at ease. He told them that what his OKB was trying to do was, relatively, the easiest thing in the world. After they invented something, they sourced the right people to build it correctly and then had the best factories all over the country build the components for them and deliver them back to OKB-1. According to Korolyov, all they had to do then was to put the pieces together. Realising his gross understatement, the cosmonauts immediately took to Korolyov's friendliness and warmth towards them, and he began referring to them as 'my little swallows'.

The Top Six training group inspect flight equipment prior to the first manned Vostok mission.

Korolyov took to several of the group immediately, but one in particular seemed to be taken under his wing. Yuri Gagarin emerged from the crowd by making a good impression on Korolyov with his attention and questioning on that day.

During the tour of the facility where the Vostoks were being constructed, as jet pilots the cosmonauts wondered for the first time how a ball without wings could ever fly, let alone orbit the Earth. The team was told that, in time, they were to be instructed and tested on all aspects of the spacecraft and systems.

In the assembly shop, the cosmonauts were invited to examine first-hand one of the ground test versions of the capsule, and Gagarin was the first to volunteer to ascend the ladder and climb inside, first taking care to remove his shoes. So engrossed was he inside the confines of the capsule that, after several minutes, he had to be called out to give the others a chance. Many of the other cosmonauts realised that Gagarin had made a lasting impression on the Chief Designer.

By August, Kartashov and Varlamov had been medically stood down from the 'six', to be replaced by Nelyubov and Bykovsky. As the workload increased, so did the strain on both the cosmonauts and their families. Years later, Gagarin's wife Valya told a Soviet journalist that Yuri would often return late from work, and often took long trips away from home, although he did not tell her many details. When she asked, he would always dismiss the work with a smile and a joke. As time went on it seemed to Valya that Star City was increasingly taking him away from his family. 'I tried to make it seem that I hadn't noticed, but from time to time I would be overcome by a strange anxiety'.

The cosmonauts were ordered to keep the true nature of their role secret from their families. For some of those who would never fly in space it was a secret that was kept for years, with no hint of their work in pioneering their nation's space programme. This was in sharp contrast to the American astronauts, who were national heroes on the day they joined the programme. This placed enormous strain on the astronauts' immediate families. They came largely from the secluded military family life to the forefront of the world's press almost overnight, and for some, marriage did not survive the pressure.

The Top Six becomes the 'prime three'
The two-day state examinations were completed by 18 January 1961, and by 25 January the results had established a flight order of Gagarin, Nelyubov, Titov, Nikolayev, Popovich and Bykovsky. Though it is not clear when the cosmonauts were made aware of this ranking, Gagarin, Nelyubov and Titov must have been aware, from the intensity of their training, that they were the prime trio of candidates, although at the time none of them were aware who would be assigned to the flight. Of the three, it seemed that Gagarin remained the favourite, although Titov also scored well in the tests.

For Nelyubov, this was a bitter disappointment. He was one of the most talented pilots in the first selection, he had an unusually dynamic and witty character, and he was also athletic and very competitive, often setting endurance records in the isolation chamber. Unfortunately, he was also outspoken, and often openly claimed that he should be the first to fly, gaining some support for his desire to become the first man in space from some members of the team. However, he had also made a number of enemies – most notably, Kamanin – during previous months. He had also embarrassed Korolyov at their first meeting by stating that he thought the physical training programme was being conducted at the expense of flying. By February, Kamanin had had enough of him, and exchanged him in the ranking with Titov. This move virtually took him out of the race for the first flight, which left only Gagarin and Titov.

On 24 March – the day after the loss of Bondarenko in the isolation chamber fire – the six, led by Kamanin, left for their first visit to the Tyuratam cosmodrome. During this visit, Gagarin, Titov and Nelyubov completed a joint training session in the MIK assembly facility, where they practised donning the Vostok pressure suit and entering the spacecraft, as they would on the launch pad. The visit also coincided with final preparations for the launch of the last Korabl-Sputnik and, to take advantage of the opportunity to practise activities on the pad, the three cosmonauts suited up and were taken to the launch complex by bus. There, they rode the lift to the hatch area of the unmanned spacecraft, to practise their own launch-day activities up to the moment of entering the capsule.

That first visit to the place where they would leave Earth impressed the cosmonauts – none more so than Gagarin, who looked at the giant launch tower and deep flame trench with a strange mixture of awe and excitement.

The selection

On 3 April, Kamanin announced at Star City that it was the decision of the Soviet Government to put a man into space. The following day the authorisation papers were signed for the flight, with Gagarin the preferred choice as expected, although the prime crew-member had yet to be confirmed. The Top Six again flew to Tyuratam with some of the other members of the team, many of whom would be performing support and CapCom assignments for the flight.

On 5 April, Kamanin wrote in his diary of some doubts about the choice of Gagarin over Titov. Titov had performed better in several aspects of training, and was more self-assured. Although both were excellent candidates, there were frequent comments about Titov being the favourite of the trainers, and how he never seemed to question any of the training requirements. Gagarin frequently questioned some of the training activities or flight procedures, although this could also be interpreted as Gagarin the pilot inquiring all about the 'craft' he was to test-fly. The one reason that Kamanin withheld nominating Titov as the first to fly was to suggest him for the next flight of sixteen orbits, or one day in space. Here, a strong character would be needed to endure such a hazardous mission, and Titov would be a fine candidate for that flight. It was also obvious to Kamanin and others that the first flight would result in world attention. The fame that would go with it required a person of outstanding ability and character, as the name of the first cosmonaut would never be forgotten by history. As for the second or the third, they could easily be forgotten as each succeeding flight set new records to surpass them.

During 7 April, reports were received from America that an astronaut could be launched in a Mercury capsule on a sub-orbital mission, but not until the end of the month. With almost all the preparations completed at Tyuratam, it appeared that the Soviet Union would still become the nation to put the first man in space. But which man would it be?

On 8 April, members of the State Commission gathered at the cosmodrome to authorise the selection of the crew for Vostok. The Commission included Rudnev as Chairman, Korolyov, Glushko, Pilyugin, Barmin, Kuznetsov, and others. During the subsequent meeting, Kamanin proposed to the Commission that Gagarin should make the historic flight, with Titov acting as his back-up and Nelyubov as the second back-up. In the event of Gagarin being unable to make the flight, the prime position would fall to Titov, and Nelyubov would support him as his back-up. The State Commission unanimously approved the decision.

Shortly afterwards the other cosmonauts were told of the assignments of their three colleagues. The Chief Designer added that although Gagarin would be the first, others would follow, explaining: 'We will have new flights that will be in the interest of science and to the benefit of mankind. Soon we will have a ship for a crew of two or three.' The team had been made aware of the flight ranking since January, and so the news was not a surprise. Gagarin was visibly happy, but Titov, although smiling, showed some natural disappointment. Gagarin offered encouragement to his friend, that his chance would come soon enough; and Nelyubov must also have thought his would too – but when?

At the same meeting there was discussion about exactly when to announce the flight to the world. There were some who wanted to delay the announcement until

The State Commission selects Yuri Gagarin as the prime candidate for the first mission: (*standing, from left*) Kamanin and Gagarin; (*seated*) Titov, selected as back-up pilot, and Nelyubov.

Gagarin had been recovered, but it was decided to make the announcement shortly after Vostok had been placed in orbit.

In the event of an emergency, official requests would have been placed with foreign countries to seek help in recovering Gagarin and the spacecraft. But with just two weeks to go to the launch, the decision was made not to equip Vostok with a self-destruct system, which was a standard feature on Soviet recoverable spacecraft, to prevent 'state secrets' from being retrieved by foreigners.

The inclusion in the State Commission of Ivashutin, from the KGB, was in part due to the suggestion to broadcast the first Tass report immediately after the spacecraft entered orbit. This was for two reasons: to help speed up the organisation of an international rescue effort; and to preclude any foreign state declaring that the cosmonaut was a military reconnaissance scout.

Confidence in flying a single orbit manned mission was heightened by the successful flights of the final Korabl-Sputnik missions in March, but even so there were still concerns about the ability of the Vostok life support system to maintain a cosmonaut if the descent was delayed. Kamanin stated in his diary that serious problems in removing humidity from the cabin atmosphere of the spacecraft remained, although it was 'certified' safe for 6–7 days.

There was also apparent concern should Gagarin have to land in water. The life support pack was not quite as waterproof as at first thought, although Gagarin would be provided with a small dinghy that was designed to keep afloat for over 24 hours. However, trying to spot a lone cosmonaut on a large expanse of water, even in the bright orange suit, would be difficult, as there were no plans to carry signalling devices. The survival pack therefore included smoke canisters which would enable the cosmonaut to attract the attention of search and rescue teams.

Water survival training in the Black Sea.

Mercury delayed

Rumours of a pending manned launch by the Soviets had been analysed by the CIA for some days prior to informing the White House. At the Cape, preparations continued to launch the first American astronaut at the earliest opportunity.

Alan Shepard had won the race to be selected for the first Mercury sub-orbital flight, with Virgil 'Gus' Grissom selected for the second. John Glenn would provide back-up duties for both missions, before probably flying the third mission. On 20 January 1961, having won his own race for the White House, Senator John F. Kennedy was sworn in as the nation's 35th President.

Just over a week later, on 31 January, Mercury Redstone 2 had been launched carrying astro-chimp Ham on a sub-orbital trajectory. The plan was to have seen the Redstone take the capsule to 115 miles high and a speed of 4,400 mph before the escape rocket fired, separating the capsule for splashdown. Ham would be employed in watching lights on the console. These would indicate when he was to pull a toggle to receive a banana pellet, as an evaluation of the chimp's actions during powered flight, zero g, and re-entry. If he did not respond, he would be 'zapped' with a small electric charge across his feet to encourage him to contribute to science. The maximum g force was expected to range from 9 g for the boosted phase, to 12 g briefly during entry, with approximately five minutes in zero g.

In reality, the launch experienced a high level of vibrations and an excess of thrust from the propulsion system, making it a rougher ride than predicted. This additional thrust saw the vehicle follow a steeper climb to 157 miles altitude. Ham had to endure 17 g on the ascent, seven minutes of weightlessness, and 14.7 g during entry, splashing down 132 miles past the target point. Sealed in a pressured unit, Ham was fine, despite the capsule taking on water through the open-air valve during recovery. After the chimp was recovered, he kept smiling (as did all good test pilots), and was eager to eat a piece of fresh fruit. It was reasoned that if Ham could survive the flight with the increased g-loads, then Shepard certainly could – and without the added incentives of banana pellets and 'zapped' feet!

Despite the astronaut's desire to proceed with a manned flight in late March, von Braun and his team were not so confident. They wanted one more unmanned flight to test the changes incorporated for controlling the vibrations in the rocket, and to resolve a host of other small niggling problems. In Washington, the President was ordering his own review of NASA and its current and long-term programmes, at a time when rumours were circulating that the agency was soon to be dominated by the military. The result of the over-cautious approach by both von Braun's team and the politicians in Washington was that Shepard's launch slot would be taken by an unmanned Mercury Atlas 3 mission. If this 'booster development' launch was a success, Shepard would receive approval to launch on 25 April. The MA-3 mission was flown on 25 March, and was indeed a much needed success, clearing the way for the launch of Shepard on a sub-orbital trajectory. Shepherd then knew that no-one could prevent him from being the first American in space and possibly the first man in space, as long as the Soviets stayed on the ground.

Phantom cosmonauts

With coverage of the American attempts being featured regularly in the news, rumours about Soviet activities gathered pace. This also generated a host of totally fictitious stories about earlier launch attempts and doomed cosmonauts involved in several space accidents. These rumours – both before and after the first Vostok mission – were mainly the creation of the journalists who wrote them. But they also played upon the secret nature of the Soviet programme, and it was assumed that any unidentified men observed in cosmonaut attire or in training were deceased members of the team.

In fact, many of these men were either Air Force Testers evaluating space systems, cosmonaut candidates who were not selected, or cosmonauts who never flew a mission. Over the years they were gradually identified, from film and print, by a dedicated group of 'space sleuths', and were eventually officially identified in news releases by the Soviets themselves. To fuel the rumours: of the twenty men selected in 1960, only twelve actually made a spaceflight, and it took more than 25 years to reveal the other eight, leaving them open to journalistic flights of fancy.

Some of the most famous stories of 'lost' cosmonauts included reports by the Italian brothers Judica-Cordilla from their listening post in 1958. Their sensational and highly imaginative accounts reported that in late 1957 Alexei Ledovsky had been lost after reaching 200 miles into space, and losing communications; Serenty Shiborin had supposedly been launched from Kapustin Yar in 1958, but never returned; and Andrei Mitkov was reported killed when his rocket blew up 500 feet above the launch pad.

Other rumours centred on a group of cosmonauts named Alexei Grachev, Gennedy Zavodovsky, Gennedi Mikhailov, Ivan Kachur, Alexei Belokonev and A.N. Ischak. Despite photographs from the Soviet Press, which reported that they were trainees and technicians engaged in the testing and development of space hardware, Western journalists continued to spread rumours that they had all died in space accidents. In one account, Belokonev was said to have spun out of control in orbit for 'several days'. Although he and many of the others were identified on Moscow Radio or at Soviet Press offices after they had reportedly been killed, the rumours continued.

Perhaps the most 'famous' account of a 'lost' cosmonaut was that of Vladimir Ilyushin, the son of the famous aircraft designer. Just two days before Gagarin launched, the Moscow Correspondent of the *Daily Worker*, Dennis Ogden, interpreted a report that Ilyushin had survived a bad car-crash as a cover story for an ill-fated spaceflight. Ilyushin was supposed to have been launched onboard spaceship *Rossiya* on 7 April and completed three orbits (one of the plans for the first Vostok). According to Ogden, he had returned to Earth 'in a bad way', suffering from physical and mental problems, and 'an announcement was expected from the Kremlin'. According to NORAD space tracking stations, no such launch took place (and they were certainly looking for one). Ogden was adamant that he had had contact with a reliable and informed person in a government department. Another story from French broadcaster Eduard Bobrovsky had 'discovered' that Ilyushin, who was a Lt Colonel and test pilot in the Air Force, had used his influence to fly in space in March 1961, but had returned to Earth badly shaken and had slipped into a coma.

Certainly these rumours were fuelled by the mood at the time, and the announcements that two Korabl-Sputniks carried mannequins aboard. The decision to supply a taped message of a human voice to test ground receiving stations initially caused some concern amongst KGB officials, who thought that Western listeners might assume that the Soviets were flying an intelligence officer on a secret spying mission. It was then decided that the test message should be in the form of a song, but this too was soon dismissed, as it was thought that observers would think that the cosmonaut had gone mad and had decided to sing songs instead of completing his mission! Finally, in order to prove that this was nothing more than a taped message, they decided to use a recording of a whole choir and a voice reciting Russian soup recipes. After all, no-one would believe the Soviets had put a cook into space, let alone a choir!

Pre-launch

Once the Government Board had met and had approved Gagarin as the prime candidate on 8 April, Korolyov instructed Gagarin one final time on Vostok's various systems, stressing their importance and reliability and his confidence in both his spacecraft and his cosmonaut. The cosmonauts were to remain overnight near the launch pad, and a short distance away a pair of small wooden cottages was being prepared. One of them would be accommodation for the two cosmonauts, and the second would house Korolyov.

At 05.00 am MT on 11 April, the R-7 carrying the Vostok capsule trundled down the tracks on the back of a locomotive carriage, towards the 'stadium' and 'pit'. All day and all night, service personnel fussed around the vehicle as it was lifted and prepared for its short flight into the history books.

That morning, both of the cosmonauts met the launch team at the launch pad.

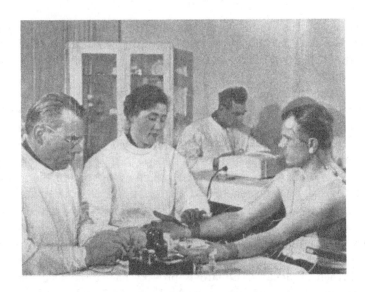

Gagarin undergoes medical tests before his mission.

They had also taken the opportunity to complete another training session inside the Vostok. Lunch consisted of the kind of food that Gagarin would be sampling in orbit the next day. The cosmonauts had been eating this type of food for a couple of days to familiarise themselves with it and to provide pre-flight control data for monitoring. For the rest of the day the two men relaxed, reviewed the flight plan, and underwent medical examinations to record pre-flight parameters. Later that evening, Korolyov visited them at the cottages, and all three chatted for some time about the flight, their childhood, their schooling and their military service. The two cosmonauts also watched the film *Beloye Solntse Pustinny (White Sons of the Desert)*, and played a few games of pool before retiring for the night.

In the second cottage, Korolyov found it difficult to sleep. He tried to read, but was disturbed by a telephone call from his wife wishing him good luck. He had become so fatigued over the previous few days that his heart was again disturbing him. He took one of his pills and lay on the couch, but still could not sleep. His head ached from checking over in his mind that everything possible had been done, then checking just once more to determine whether there was anything that he had missed. He eventually told himself that everything humanly possible had indeed been done. As he rested, his chest pain subsided, but sleep still eluded him as the Sun rose the next morning. It was time to write a new page in history.

Launch day
At 05.30 MT, Yevgeni Karpov, Director of the cosmonaut training centre and an experienced aviation doctor, woke Gagarin with a plain 'Time to get up, Yuri.' Asking how he had slept, Gagarin replied, 'As you taught us.' Karpov then woke Titov. A brief medical examination of both men declared them fit for the mission, and after an exercise session, they ate breakfast.

Riding on the cosmonaut bus, Gagarin smiles on the way to the pad for his historic launch. Seated behind him, also suited, is Titov, while Nelyubov stands in the background.

Unlike the American astronaut pre-spaceflight breakfast of steak, eggs and orange juice that would soon become a tradition, Gagarin and Titov ate breakfast from tubes consisting of meat purée, blackcurrant jam and cold black coffee. Gagarin commented that although the food would be acceptable for orbit, it would not provide sustenance on Earth.

Both men were then suited up – Titov before Gagarin, to reduce the time that the prime cosmonaut would be inside the suit without air conditioning, which was located in the transfer bus and spacecraft. After a pressure check, the two men – accompanied by Nelyubov and Nikolayev, both dressed in their military uniforms – moved to the transfer bus. It was during this bus-trip to the pad that a tradition began. Gagarin wanted to answer a call of nature before entering his spacecraft, so the bus stopped along the route and, using a toilet tube connected to the pressure suit, he wet the tyre of the bus. Since that date, almost every cosmonaut who has followed him into space (apart from the females and some of the international cosmonauts) has followed the same tradition.

Another short ride then took the party to the launch pad by 6.50 am. At the launch pad they received best wishes from fellow cosmonauts, officials and pad workers; and fellow cosmonaut Nikolayev, apparently forgetting that Gagarin was wearing his helmet, embraced him and bruised his head on the visor.

The first cosmonaut reported to the Chairman of the Government Board that he was ready to perform his assigned mission, and then headed for a short flight of stairs on the launch gantry. At the top, near the door of the elevator that would take him up to the capsule hatch, Gagarin turned and waved to onlookers. 'See you soon,' he called.

Exiting the lift, Gagarin was met by one of the Vostok designers, Alexei Ivanov, and was placed in the seat by the pad team, headed by Oleg G. Ivanovskiy (who

Korolyov speaks to Gagarin just prior to the cosmonaut boarding Vostok, 12 April 1961.

Gagarin waves from the gantry to pad workers – 12 April 1961.

ensured that all connections and attachments were secured). By 7.10 am, Gagarin was inside Vostok and was attached to the communication circuit to talk to launch control. Gagarin used the call-sign Kedr (Cedar), and the ground used Zarya (Dawn).

For Vostok – as with the Korabl-Sputniks that preceeded it – a network of ground communication sites had been set up for the whole programme, as there were no Soviet dedicated ocean-going communication and tracking vessels available until the late 1960s. These ground stations were in Central Control (TsUP – pronounced

'Soup') at Moscow (call-sign Vjezna-1 on short wave and VHF), the Tyuratam cosmodrome (Zarya-1, using VHF), Novosibirsk (Vjezna-3, on short wave), Kolpashevo (Zarya-2, on VHF), Khabarovsk (Vjezna-2, using short wave) and Yelizovo in Kamchatka (Zarya-3, on VHF). The short-wave radius of these stations was just over 3,000 miles, while those with VHF covered about 1,000 miles.

Each ground station was called a Command Point (KP), with a communications team consisting of a KP chief, a communications officer, a cosmonaut, a doctor and a representative of the Ministry of Communications.

Gagarin's first task was to check the pressure of the suit, followed by a check on the communications line – which improved when the piped music was turned off! During preparations for launch, Gagarin communicated with several cosmonauts serving as communication controllers (CapCom – CAPsule COMunicators, using the American term), as well as Korolyov and Kamanin. Checks of the spacecraft systems, the engine propellant tank pressures, and of Gagarin's actions during the count-down and flight, were also completed. Korolyov also reminded Gagarin that at T–1 minute, there would actually be a planned five-minute hold before the count resumed.

As Popovich, the primary CapCom, came on to the communications link, he asked how Gagarin was doing, to which Gagarin replied, 'Like they taught me.' Laughing, Popovich replied that the whole cosmonaut team sent their collective regards and wishes for a successful flight, and added that all preparations were proceeding normally.

Gagarin had been inside Vostok for about an hour when the hatch was finally closed. Suddenly it was opened again, and he knew something must be wrong. Korolyov came on the radio to inform him that one of the contacts in the hatch had not secured, and that the ground crews needed to recycle it again. Korolyov also mentioned that he would count down the time left to launch in figures only, and would not mention seconds, thus: '150, 100, 50' and so on.

As the time to lift-off ticked away, Korolyov informed Gagarin that officials in Moscow had asked how everything was proceeding, to which the Chief Designer had replied, 'Very well.' When the hatch had been replaced, Popovich asked Gagarin if he was bored, as he did not have that much to do except sit and wait. Gagarin requested that the music be turned on. With the portholes covered by the launch shroud, Gagarin was unaware – until he was told – that the service platforms were being lowered.

In the blockhouse, Korolyov's chest was hurting again. He had asked to be informed of everything that was happening on the pad, so that he could follow the events as they occurred. Korolyov spoke to Leonid Voskresensky while they waited, said that he too looked strained, and offered his deputy one of his cardiac pills. Korolyov began by monitoring the countdown from his chair, then tried pacing up and down; but he could not stop thinking about what he might have overlooked. He was reminded by others in the room that Gagarin, sealed in the Vostok, was perhaps much calmer than they were, shut in the blockhouse.

At T–15 minutes, Gagarin put on his flight gloves and sealed his helmet. At T–1 minute, he was aware of slight movement, as the R-7 seemed to sway a little on the pad.

S.P. Korolyov in the control bunker talks to Gagarin onboard Vostok prior to launch.

At 09.05 Gagarin announced: 'Roger. I'm in the mood, I feel fine, and I'm ready for the launch. I felt the working of the valves.' The rocket suddenly came alive, as propellant passed through valves and the engines began their firing sequence. Gagarin heard a faint rumble far below him, followed by the build-up of noise. But it was not so loud that it deafened him or interfered with his actions.

At 09.07, Korolyov reported: 'Preliminary stage ... intermediate ... main ... LIFT-OFF! We wish you a good flight. Everything's all right.'

Gagarin: '*Poyekhali!* [Off we go!] Goodbye, until [we meet] soon, dear friends.'

The first flight of a manned spacecraft had begun.

Ascent

Gagarin did not notice the initial movement, and felt only a slight shiver through the structure of the R-7. He heard Korolyov report on the progress of the ascent, and in response to being asked how he felt, Gagarin indicated that all was well – and countered by inquiring how the Chief Designer felt!

As the vehicle rose, so the g-load increased, and Gagarin was aware of being pinned further back into the couch and of how difficult it was to talk. At 09.09 the four first-stage strap-on boosters fell away, followed one minute later by the separation of the launch shroud.

At 09.12, five minutes into the flight, the second (core) booster shut off and separated, followed by the ignition of the third, upper stage, to push Vostok into orbit. Inside the spacecraft, Gagarin described the experience of staging for the first time. American astronauts would later liken it to a 'great train wreck'.

Ignition of the main engines of the Vostok launch vehicle, 12 April 1961. A hold-down arm restrains the vehicle while thrust builds to the point of release to begin the mission.

Gagarin suddenly felt a sharp drop in the g-loads on his body as the core booster shut down, pushing him forward against his seat harness. 'It felt as if something had suddenly separated from the rocket. I felt something like a knock, and then the noise dropped sharply. The state of weightlessness seemed to emerge, although the g-load was [still] about 1 g at that time. Then the g-load came back and began to increase. I began to be pressed back to the seat, but the noise level was substantially lower.'

At 09.21 the third stage shut down abruptly, separating ten seconds later and placing Vostok in orbit around the Earth. Gagarin felt a decrease in g-load and a sharp thump and jolts as the spacecraft began to slowly rotate. At this time, communications between the ground and Gagarin had been lost for several seconds, leading to fears that the pioneering cosmonaut had been lost or had fallen into unconsciousness due to the stress of launch. Suddenly, Gagarin's voice was heard again and, now in orbit, and using the Vzor optical device, he gave his first impression of the view outside the spacecraft. 'I see Earth! I see the clouds It's beautiful, what beauty!' Gagarin later wrote that he could clearly see mountain ranges, large forests, islands and coastlines 'I saw the Sun, clouds and light shadows on my dear, far Earth.'

At ground control, the news that Gagarin was in space was greeted with great enthusiasm in the control room – so much so that the normally reserved Korolyov flung his arms around a surprised Feoktistov and exclaimed: 'What a day!', as though a great weight had been lifted from his shoulders. Then, as communications were lost, Korolyov suddenly fell silent again in a solemn mood, his burden returning. When Gagarin resumed contact with the flight control team, Korolyov sighed deeply as he dropped heavily into a chair. 'It's moments like those that shorten a designer's life,' he commented.

The first orbit

With communications restored, Korolyov informed his first cosmonaut that all was well and that the flight should continue. Onboard Vostok, Gagarin was looking through the Vzor porthole, and could clearly see the curvature of the Earth's horizon as 'a pretty, light blue.' He continued to describe the dramatic change from the blue of Earth to where stars shone in the blackest black he had ever seen. 'At the very surface of the Earth, a delicate, light blue gradually darkens and changes into a violet hue that steadily changes to black.'

Returning to his duties, the cosmonaut commented on his adaptation to his new environment as 'a normal reaction to free fall. I feel fine. All equipment and systems functioning normally. Everything floats. It's great!'

Throughout the flight, Gagarin continued to report on his own physical condition and the parameters of his spacecraft, either to the ground stations when they were in range, or into the onboard tape recorder. 'The lights are on the Descent Module monitor, I'm feeling fine and in good spirits, Cockpit parameters: pressure 1; humidity 65; temperature 20; pressure in the compartment 1; first automatic 155; second automatic 155; pressure in the retro-rocket system 320 atmospheres' Over the sea, the surface appeared grey and not light blue as he had expected. 'The surface was uneven, like sand dunes in photographs,' he reported.

Gagarin's scientific experiment work-load was almost zero. On this first trip beyond Earth, he was merely a passenger. Such was the fear that adverse effects from weightlessness might cause him to try to take over control of the spacecraft, the manual controls had been locked before launch. However, a 'secret' three-digit code (1-2-5) was set on the manual control logic clock. Gagarin was not told of the code, which would be issued only if needed; but it never was needed, and Gagarin provided no actual input into the 'flight' of Vostok. Although it was meant to be a restricted code, if a communications link had dropped out losing the link to command retro-fire, Gagarin had a sealed envelope with the secret code inside it! He had only to rip open the envelope to reveal the code he was not supposed to know of or use. To some of the training staff – especially the test-pilot instructors – this situation was ludicrous.

Eavesdropping on the TV coverage

Live TV transmissions of the cosmonaut were relayed to the control centre, and were used to monitor the state of the cosmonaut during the flight – similar to the way in which the canine cosmonauts were observed during the Korabl-Sputnik missions. It was these TV signals from Vostok that were picked up by the US National Security Agency electric intelligence (Elint) station operators as the spacecraft passed near Alaska twenty minutes after launch. Thirty-eight minutes later they had interpreted these signals, which clearly showed a man moving in front of the camera. Before Gagarin had landed, the US intelligence network had technical confirmation that the Soviets had indeed been able to put a man in space, and that he was still alive.

Actual confirmation of this has been the subject of several research papers over the years. A review of the ground track of Vostok indicated that it passed close to the NSA's Shemya station in Alaska. It was known that there was a radar installation here, for tracking the Soviet missile test launches out of Russia and into the Pacific. It appears that analysis of the data received by that station took some time, and that it was not read successfully. Swedish space-tracker Sven Grahn concluded that a second NSA station must have picked up the signals, but was unsure of the location of this second station as the spacecraft passed overhead at 58 minutes into the flight, allowing a read-out of the TV pictures to confirm that Vostok was indeed carrying a man. He reasoned that it could be in South America at Cape Horn on the tip of Tierra del Fuego, but also offered a more geographically ideal site directly under the spacecraft ground track at a US Antarctic Research Station called Palmer, which could have picked up and analysed the signals from Vostok. Whether there was, or still is, a NSA station at this location has never been confirmed or denied.

In any event, US intelligence confirmed that a man was in space, and that he was a Soviet citizen. The news was wired to the CIA in Washington, who prepared a report for President Kennedy.

As Gagarin continued his flight, he requested information on the status of his mission and orbit from ground data: 'What can you tell me about the flight? Give me some results on the flight.' Ground control responded: 'There are no instructions from No:20 [Korolyov's personal call-sign], and the flight is proceeding normally.'

As he passed over the Pacific, Gagarin entered Earth's shadow at 9.49, and later recalled: 'Entry into Earth's shadow was very abrupt. Up to now, I had at times observed intense illumination through the windows. I had to turn away from it or cover my face so the light would not reach my eyes.'

Over the Pacific and out of direct communications, Gagarin read his report into the onboard tape recorder: 'I am transmitting the regular report message: 9 hours 48 minutes [Moscow Time]. The flight is proceeding successfully. Spusk 1 is operating normally. The mobile index of the descent mode module is moving. Pressure in the cockpit is 1; humidity is 65, temperature 20.'

At 09.51 MT, Gagarin reported that the Sun-seeking attitude control system had been switched on. Then, at 09.57, he received the awaited descent system command transmitted from Khabarovsk: 'By order of No:33 [Kamanin's callsign], the transmitters have been switched on, and we are transmitting this [the descent commands]. The flight is proceeding as planned, and the orbit is as calculated.'

Gagarin reported that the commands had been received, and that he was passing over the continental United States. In fact, he was over the South Pacific approaching Cape Horn. Five minutes later, Soviet Radio announced to the world the Tass bulletin on the first manned spaceflight, indicating that Senior Lieutenant Gagarin had been promoted to Major while in flight. As the spacecraft continued over the Atlantic, news of the historic event spread just as quickly around the world. The broadcast was written by the KGB, and was one of three that had been prepared – one for a success, and two for failures. It was 25 minutes after the launch and well into the orbit before tracking confirmed a stable orbit, allowing Tass to select the appropriate announcement.

Onboard Vostok, Gagarin had the opportunity to evaluate the habitability of space. 'I ate and drank normally. I noticed no physiological differences. The feeling of weightlessness was somewhat unfamiliar compared with Earth conditions. Here you feel as though you are hanging in a horizontal position in straps,' he stated.

As he ate from the tubes, he released the writing pad (with its attached pencil) with which he recorded his report, marvelling as it floated before his eyes. When he proceeded to write the next report, he grabbed the pad out of mid-air – but the pencil was nowhere to be found, as it had detached and floated off beyond his reach. 'I closed the journal and put it in my pocket. It wouldn't be any good anyway, because I had nothing to write with.' Nearly all space explorers who followed him would discover for themselves that if items are not strapped down, attached, or watched, they can float off to hidden corners of the cabin.

Almost an hour into the flight, and all too soon for Gagarin, it was time to prepare for entry and landing. The first command was given at Ground Elapsed Time (GET) 56 minutes into the mission. As the world learned of the flight of Gagarin and Vostok, he was preparing to come home and tell of his unique experiences.

Re-entry and landing

As Vostok emerged from the Earth's shadow at 10.09, the spacecraft was rolling slowly. Inside Vostok, Gagarin closed the shutters on two of the three viewports, closed his helmet visor and tightened his harness. The retro-rockets ignited at 10.25, and inside his spacecraft, Gagarin felt a sharp bang and a bump as the rotation continued while Vostok descended into the atmosphere.

Gagarin felt the braking rockets ignite, and heard a buzzing in his helmet. He also noted a drop in pressure readings on the braking engine. After 40 seconds, as the braking rocket fired, he noted the gradual increase of g forces at cut-off, and a return to weightlessness for a few seconds.

Anticipating the separation of the Instrument Module, Gagarin realised after the expected 10–12 seconds that there was a problem. At shut-down, all monitoring lights from the Instrument Module went out – as they would after separation – but the indicator lights preparing for entry were not lit when they should have been. Suddenly, the indicators from the instrument stage lit up again, which could only mean that it had not separated correctly, and that he was probably dragging the Instrument Module behind him, although he could not see it out of the portholes. There was nothing that he could do except ride out the situation. He knew the braking rocket had worked normally and that he was on his way down, but was dragging the extra mass behind him, and he quickly estimated that his landing could overshoot the planned recovery area. He was 3,700 miles from the Soviet Union, over Africa, and with the borders of the Soviet Union being about 5,000 miles apart he reasoned that he would land somewhere in the Far East!

Believing that it was not an emergency and that the connections would soon burn through, Gagarin transmitted the 'all normal' signal (the number 5) with the keypad. (Numbers 1–4 would have indicated different states of his condition.)

As the spacecraft entered the atmosphere, Gagarin noted that the spacecraft's orientation '8-ball' continued to revolve around its axis at a rate of 30° per second. Vostok was unable to find its natural centre of gravity while dragging the Instrument Module behind it. Gagarin began to feel his spacecraft oscillating 90° left and right.

As the glow due to re-entry appeared around the side of the porthole shutters, Gagarin became aware of a crackling sound outside the capsule, but was unsure if it was the protective thermocoating or the spacecraft beginning to break up. The crackling lasted about a minute, and he became visually aware of the extremely high temperatures surrounding him as the glow around the porthole blinds intensified. The spacecraft was rotating around its axis at a very high rate, spinning the cosmonaut inside, and as he looked out of the unshaded Vzor porthole, he could see a land mass, then the horizon, then the black of space, and then Earth again. Suddenly, the Sun came into view, and a blinding light entered through the Vzor porthole. Gagarin barely managed to move his feet to block out the intense light as the g-force increased. After ten minutes, the Instrument Module cables burned through, and as the stage finally separated, the rotation began to decrease.

The g-force had increased to about 10 g, and for a few seconds the instrument readings appeared blurred to the cosmonaut as he began to black out. As Vostok dropped through the sound barrier, the g-force eased, and all became clear again. Gagarin felt sure that even from inside the sealed capsule he could hear the rushing sound of the air as the hot sphere dropped towards Earth.

According to IAF rules, to qualify for and establish a new aeronautical record a pilot had to take off and land inside his vehicle. Initial Soviet reports of the landing indicated that Gagarin had stayed inside Vostok all the way to landing, and it was not until 1978 that reports emerged that he, like all the other Vostok cosmonauts, had used the ejection system and descended by separate parachute, as was part of the flight plan.

At 23,000 feet, the hatch behind Gagarin's head was detached with a bang, as explosive charges initiated separation. Realising that he had not ejected, he looked up at the same moment as the ejection motor fired and ejected him overboard. It happened very quickly, but smoothly, and Gagarin was relieved that he did not hit anything on the way out. Still strapped in the ejection seat, he recognised, from previous training jumps, the landing area below him. The large river he could see was the Volga, and the town was Saratov, near to the pre-planned recovery area.

Separating from the seat, Gagarin deployed his parachute. As he descended he thought that he would be blown into the river and would actually make a splashdown, but he managed to steer himself towards a dry landing in a nearby field. Suddenly, the reserve parachute opened, but thankfully it did not deploy or become entangled with the main parachute, and instead just hung below him. Although he could not look straight down due to the restrictions of the suit, Gagarin could see where the Vostok had landed, about 2.5 miles from where he was heading – which was straight into a small gulley next to a house.

Workers on the farm looked up, and as they ran towards the descending figure dressed in orange clothes and a white helmet, a breeze blew him away from the gulley

Gagarin's Vostok craft after landing, 12 April 1961, showing the effects of re-entry heating.

Gagarin, at the beginning of worldwide fame, (*left*) shortly after landing, and (*right*) being welcomed back by local people.

and straight towards a ploughed field – near to a woman, a little girl and a small calf. Gagarin returned to Earth by planting his feet firmly in the field at 10.55 MT. Man's first flight into space had ended in one of the fields of a collective farm, nineteen miles south-west of the City of Engels, near the village of Smelovka. Gagarin was surprised at the softness of the soil, which allowed him to stay on his feet upon landing. After removing the parachute harness and opening his helmet, he took a breath of fresh air. He had left Earth 108 minutes earlier, witnessed by some of the top officials of his nation's space programme, and had flown once around a world which was beginning to learn of his feat. Now his historic landing was witnessed by peasant woman Anna Takhtarova, a small girl and a cow!

The frightened girl backed away from the stranger in orange with a large white head, and other onlookers were at first unsure who this stranger was. Having learned of the shooting down of Gary Powers' U2 the year before, they thought that this was another invading American. Realising this, Gagarin flashed his broad smile, explained that he was a Soviet just like them, and told them not to be afraid. He then asked to use a telephone to report his safe landing! Once they realised who had landed in their field, they enquired if he really had come from space as the radio reports had indicated. Gagarin smiled again and told them that it really was true.

The first Vostok mission logged 108 minutes flight duration, reached a maximum altitude of 187.6 miles, and travelled a total of 24,000 miles around the Earth. Gagarin's Vostok mission already had a place in the history books, and it would forever remain one of most memorable milestones in human history. A few days after the flight, a monument was placed on the spot where the empty Vostok sharik landed.

The aftermath

Al Shepard was awakened by a ringing telephone in his motel room at the Cape, early in the morning of 12 April. The call did not bring him good news. Shepard was both angry and disappointed with those who were too conservative to allow him to

A tired-looking Gagarin shortly after becoming the first man in space.

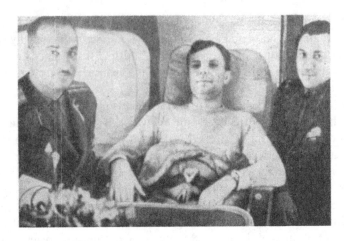

Shortly after landing, Gagarin rests in the aircraft on the way to the dacha near the Volga.

fly his Mercury Redstone 3 mission in March. In public, he was gracious in his comments on the outstanding achievement of the Soviets, but privately he was far from happy. 'What's done is done,' was his normal reply (through gritted teeth) to those who asked for his comments on the event.

President Kennedy expressed the feeling of the nation in his comments on the achievement of Gagarin: 'No one is more tired than I am [in seeing the US second to the Soviet triumphs]. We are, I hope, going to be able to carry out our efforts this year, but we are behind. The news will get worse before it gets better, and it will be some time before we catch up.'

The first cosmonaut to greet Gagarin after he landed was his friend and back-up, Gherman Titov. To reach the landing site, Titov had travelled from the cosmodrome to a nearby airfield, and he passed on the congratulations of the whole cosmonaut group. That evening, they celebrated with other members of the recovery team in a dacha (country cottage) located in woods on a hill overlooking the Volga River. Throughout the evening, each of them offered many speeches and discussed the prospects of further space flights by other cosmonauts

Post-flight debriefing began almost as soon as Gagarin had landed. Professor Boris Viktorov recalled that the cosmonaut reported all events in great detail, very calmly and vividly. The scientist was pleased to see that the flight had not in any way changed Gagarin's character one bit. Everyone concerned with the flight felt relieved that all the tension, nervousness and fear were over, and it was also noted how Gagarin's eye's, although tired, shone brightly with happiness and exhilaration.

On 13 April, the day after the mission, Gagarin rested and, accompanied by Titov, took a boat trip on the river and long walks as an informal debriefing session. Gagarin was offering Titov his advice and recollections while it was still fresh in his mind, and Titov was grateful to receive all the information. Titov was now in training to be the next Russian in space, and for much longer than the brief time that Gagarin had flown.

The space pilot speaks

Later that same day, Izvestia Special Correspondent G. Ostroumov interviewed Gagarin. In this and other post-flight press conferences, Gagarin's obviously evasive answers did not include exact details of the flight, the hardware, or future missions. The political propaganda machine still needed to retain as much secrecy as possible from the Western world.

In response to the question of how long he could have stayed in space, Gagarin commented: 'In the spaceship I could have stayed much longer, I could have kept flying in space as long as was necessary.' To the question of the size of the cosmonaut team, he replied: 'I believe there are more than enough to undertake important flights.' And when asked when the next 'important flight' might occur, he smiled and said: 'Our scientists and cosmonauts will undertake the next flight when it is necessary.'

For the *Izvestia* article, Gagarin commented: 'I was happy and proud that the flight in space was to be made by me. At the same time, I had a sense of responsibility for the flight, where there was so much that was unknown. I was proud for our people, who were able to build ships powerful enough to lift a man into outer space. I certainly did not feel lonely at all [in space] for I knew that my friends, in fact all the Soviet people, were watching my spaceflight. I can hardly describe my feelings when I stepped again on our Soviet soil. I was happy to have carried out my assignment. As I was descending I kept singing the song 'My country hears, my country knows'. When I returned to the Earth, I rejoiced at the warm reception I was accorded by the Soviet people. I was moved to tears by the telegram I received from Nikita Sergeyevich Khruschev.'

Gagarin concluded the interview with a comment on foreign press reports that the US also intended to send a man into space: 'We shall, of course, rejoice at the achievements of the American spacemen *when* [authors' emphasis] they fly into space. There is enough room for everybody there, but space must be used only for peaceful and not military purposes. The American space flyers will have to catch up with us. We shall hail their accomplishments but we shall always try to keep ahead of them.'

Gagarin: a new Hero of the Soviet Union

On 14 April, Gagarin flew with a small group of correspondents and officials to Moscow to receive an official welcome by Party leaders and by the largest crowd since the Victory Day celebrations in 1945. During the 2 hr 40 min flight (almost an hour longer than it took to fly around the world), Gagarin answered a few questions and took the controls of the aircraft for a few minutes. He seemed relaxed as Kamanin informed him that the award of Hero of the Soviet Union had been conferred upon him, and for most of the flight he read newspaper accounts of his feat. He had already been promoted while in orbit, and he also received a new award – Pilot Cosmonaut of the Soviet Union. As the aircraft circled low over central Moscow, he could see the crowds gathering in the streets leading to Red Square, and to some observers he seemed worried. It appeared that, for the first time in three days, Gagarin felt uneasy about his new fame. By the time the aircraft had landed at

Gagarin takes his 'long walk' in front of the crowds at Vnukovo airport during his triumphant return to Moscow.

Vnukovo Airport he had composed himself, but the scale of what he had achieved was beginning to dawn on him.

'There is hardly anyone in the world who had to go through as much as I had to on that day of celebration,' he would later recall. 'The radio which endlessly repeated my name, and the papers with all the photos of me and articles about my spaceflight were only the beginning, [although] they both pleased and embarrassed me. There were still more experiences in store for me, which not even the most fanciful imagination would have thought up, and which I never would have guessed would happen. [Flying above Moscow] I looked below and gasped. The streets of Moscow were flooded by people: human rivers seemed to be flowing in from every part of the city.'

From inside the aircraft at the airport, Gagarin glimpsed the huge enthusiastic crowds, surrounded by bright mountains of flowers and a long bright red carpet leading to the grandstand. 'I had to walk to the platform, and walk alone. Not even on the Vostok just before take-off was I as nervous as I was then. The carpet was

Gagarin is welcomed by his family and by Premier N. Khruschev at Vnukovo airport.

With Gagarin and Party leaders on the Lenin Mausoleum, Red Square, fellow cosmonauts Leonov, Khrunov and others enjoy the celebrations and the anonymity in the crowd, 14 April 1961.

endlessly long. I managed to get a grip on myself as I walked down it under the glass eyes of the cameras. I knew everyone was looking at me.'

On the grandstand he recognised the faces of the Presidium members of the Communist Party's Central Committee, and caught sight of his parents and his wife. 'Then suddenly I felt something that no-one else could notice – one of my shoelaces had come undone. At any moment [I felt that] I was going to step on it and fall flat on my face on the carpet in front of everyone. That would be a funny mess; he managed all right in space, but fell over once on solid ground.'

Gagarin did not trip over, and reached to the podium. It was the beginning of global admiration for the young pilot. He embarked upon a highly publicised world tour, starting with visits to places of his childhood and then military air bases across the country, followed by hundreds of invitations from Kings and Queens, Presidents, heads of state, governments, and people around the world.

Life for Gagarin, and those close to him, would never be the same again – a situation he often reflected on as the price of instant fame: 'So now what happens? A lot of articles and reports are being written about the flight. Everyone is writing about me [and] it makes me uncomfortable to read such things. I was far from alone in this achievement; there were tens of thousands of scientists, specialists and workers who participated in preparing for this flight. I feel awkward because I am being made out to be some sort of super-ideal person. In fact, like everyone else I've made lots of mistakes and have my weaknesses too. It's embarrassing to be made to seem like such a good, sweet little boy. It is enough to make one sick.'

VOSTOK 2

For the next flight it was agreed that a longer mission should be the aim, although the duration of such an attempt was not easily decided upon. Korolyov favoured a sixteen-orbit mission that would last 24 hours – the original plan for the first manned flight before it was cut to just one orbit after the dog Belka had become sick during its mission.

Kamanin thought that to leap from one orbit to sixteen was still too risky and, supported by Dr V.I. Yazdovsky, suggested that they should complete a three-orbit mission first. This would enable them to cautiously gather further information on the effects of weightlessness before committing to a complete day. The cosmonauts and engineers supported Korolyov in favour of the day-long mission. They wanted the opportunity to fully test the spacecraft systems (including an improved air-conditioning system) on a longer mission.

After three orbits, the ground track would pass beyond landfall in Soviet territory until orbit sixteen the next day. In order to increase the duration over Gagarin's single orbit, the mission would either have to consist of two or three orbits, or at least sixteen. Korolyov was adamant that they should not mark time and should push ahead in a bolder step. He insisted they should keep the next flight in orbit for 24 hours. It has long been thought that Korolyov was a very cautious and methodical man, and that it was the influence of the Kremlin that dictated the drive to achieve one space spectacular after another. While political influences were there, Korolyov himself had a strong desire to attain notable achievements in space, and Kamanin often referred to him as 'adventurist' in driving his space planning forward with almost cavalier vigour.

Korolyov, with the support of his Council of Chief Designers, won the argument, authorising the next Vostok mission to last sixteen orbits. This would allow them to evaluate the systems over one day in order to plan for longer flights, and again demonstrate Soviet space capabilities to the West. According to Feoktistov, it took great courage to commit to a 24-hour flight after just one mission of 108 minutes. Most of the main concerns were medical in nature – eating, sleeping, and dealing with body waste.

Gagarin had managed to take some liquid refreshment during his one orbit, but he had had little time to eat a 'meal' in zero g. On the day-long mission, the cosmonaut would take with him some test samples of solid food, to determine whether he could swallow them in the absence of gravity. As muscles, rather than gravity, take the food down to the stomach, swallowing was known not to be a problem. But there were still questions about how gravity influenced the passage of food through the body, and how long-term weightlessness affected the digestive system. The results from these tests would have an influence on food selection for even longer missions being planned.

Another major concern was sleep. Doctors were not totally sure that a cosmonaut could be woken up again once asleep under weightlessness. As strange as these fears seem today, at the very start of the programme these concerns were just part of a range of unknown factors that needed to be resolved before extended-duration flights over several days were attempted. Some of the wildest fears were of the blood boiling, ear drums shattering, space travellers choking as their tongue floated to the back of their mouth, and eyeballs being ripped from their sockets and turning inwards! If that was not enough to deter any spaceflight volunteer, recovery from a space mission was thought to induce madness from high radiation exposure, and broken bones and ruptured organs from excessive vibrations and high g-loads.

As the preparations for the first flights in the USSR and America continued, hundreds of ground-based and airborne tests on mechanical dummies and mannequins, animals and human test subjects, dispelled these rumours one by one. Stratospheric balloon ascents, time-delay altitude parachute jumps, rocket sledge tests, centrifuge programmes, and the animal-crewed spaceflights, had provided a baseline of medical data allowing the space planners to confidently predict that man could survive in space. The unanswered question was, for how long? Gagarin had proved that a man could complete one orbit of the Earth without too much difficulty, but what happened after prolonged exposure and a night's sleep, or a lack of sleep? Until someone tried to go further, not even the doctors could really say what the outcome would be. The flight of Gagarin had finally quashed many of the wildest ideas of the hazards of spaceflight, but many unknowns still remained.

Cosmonaut number two

Selecting the right candidate for this mission would be as important as choosing the pilot for the very first flight. Korolyov had identified Gherman Titov as the probable candidate for an extended mission, during training for the first flight. Titov was an excellent gymnast (although he hated running), and was ideally suited for the experience of being alone for 24 hours in space, having admirably passed the hardships of the isolation chambers during training. For a while this also made him a strong contender for the first mission.

Titov's role as first back-up to Gagarin also meant that he could extend his training to the next prime position. His own back-up should then have been Nelyubov, the third cosmonaut on the Vostok 1 'crew', who could have then expected a rotation to fly the even longer Vostok 3 endurance mission. However, Nelyubov's overall attitude certainly did not help him secure an early flight assignment, and neither did his comment that he wished to return to operational Air Force flying after only one flight. He was to be disappointed a second time, when the back-up position on Vostok 2 went to the quieter Nikolayev, while Nelyubov was relegated to a CapCom role.

Specific training for the mission began almost as soon as Titov had completed his role as Gagarin's back-up, which he accompanied his friend during post-flight discussions the day after landing. Too soon, it seemed, Gagarin had been whisked away to satisfy the political leaders, when Titov felt he needed Gagarin's support to prepare for his own mission.

Timing of the second mission

As Gagarin was celebrated as the hero of the space age and Khruschev basked in

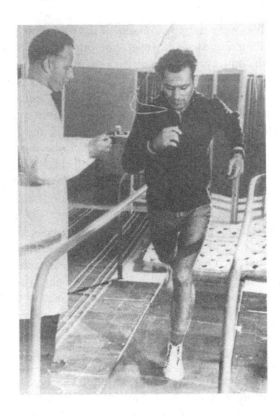

Titov on a treadmill as part of the fitness regime.

the glory, Korolyov remained in the background, anonymous to most. He published some articles on the Vostok mission under the pseudonym Professor G. V. Petrovich, and many reports of the flight mentioned only the title Chief Designer, but not his name. Korolyov and many of his colleagues were awarded the title of Hero of Socialist Labour, reflecting their part in the success of the mission, but he had resumed work to immerse himself in the next mission almost immediately. He started planning Vostok 2 in detail while on holiday in June.

Korolyov believed that to capitalise on the new experience in space technology, the training of the cosmonauts should be extended to accumulate new skills in actually living in space, rather than merely surviving. In order to do this, Korolyov knew that he had to wait for the opinions of the medical specialists who were analysing the results of the first flight.

The second flight would last sixteen times longer than had Gagarin's, and would study the impact of weightlessness on the work capacity, and whether such an environment is harmful to the human organism. Titov would take over the controls of the Vostok for the first time in space to move the spacecraft about the axis, demonstrating the capacity for a cosmonaut to have input into the flight profile of his mission.

During May, the team took a holiday with Korolyov and his family at Sochi on

Sixteen of the first twenty cosmonauts with training officials at Sochi, on the Black Sea coast, in May 1961. Front row (*left to right*): Popovich, Gorbatko, Khrunov, Gagarin, Chief Designer Korolyov, Korolyov's daughter Nina holding Popovich's daughter Natasha, training officer Karpov, parachute instructor Nikitin, and medical chief Fedorov; second row, Leonov, Nikolayev, Rafikov, Zaikin, Volynov, Titov, Nelyubov, Bykovsky and Shonin; third row Filatyev, Anikeyev and Belyayev. (Missing: Varlamov and Kartashov, who had left the group by May 1961; Bondarenko, who had died in March 1961; and Komarov.)

the Black Sea coast. It was a time to rest and to reflect on the success of the previous twelve months. The next mission was planned for later in the summer, but by late July solar storms were recorded, raising additional concern over increased radiation levels. Global observations provided Soviet scientists with important data about the levels, to predict the time of the Sun's minimum activity for the duration of the flight. The earliest launch date would now be August, with Khruschev dictating the exact timing of the flight. This time he was planning to use it as a diversion to a larger political project in the Soviet sector of Berlin, in Germany.

American success

Lying in his spacecraft Freedom 7 on the morning of 5 May, Al Shepard (like Gagarin a few weeks before him) wanted to empty his bladder before he flew into space. Shepard had been lying on his back, waiting for technicians to clear some minor problems and launch the Redstone. But the holds dragged on, and now, unable to hold his bladder much longer, he asked if he could get out and visit the bathroom. The reply was a simple 'No'. Well, if he could not get out, he would have to let it go in the suit. He informed the control centre of his decision, but did not ask permission. He reasoned that the thermal underwear would soak up the moisture and that the steady flow of oxygen through his suit would dry everything out before he launched. So he relaxed and let go, experiencing a dampness that moved to his lower back for a while! More relaxed now, Shepard suggested that he was a lot cooler than the launch team, who, he suggested, should, 'fix your little problem and light this candle.' They did so, and shortly afterwards Al Shepard took America's first step into space on a 15-minute ride.

Covered on live TV, Shepard became a new hero and America's first man in space – a worthy opponent to the Soviet cosmonauts. Kennedy saw the mood of the country was right, after Shepard's flight, to put the disappointments of Gagarin and the Bay of Pigs behind them, and to move forward. The President knew he could rely on the support of his Vice President, Lyndon B. Johnson, and asked the head of the Space Task Force, Bob Gilruth, what they could do in space that the Russians would be unable to do. Gilruth suggested a task so difficult and new that they would have to make a fresh start. The feeling that the Russian flights were nothing but gimmicks on big rockets led to the suggestion that perhaps a big rocket could be developed to take Americans to the Moon. Such a programme would require new technology, but it was felt that if the President wanted to do it, America would win, because both countries would have to begin again. In harnessing such technology, America already had an advantage. Such a bold venture would restore faith in Kennedy's Presidency, and the time was right to take the space challenge laid down by the Soviets to a new higher level – 240,000 miles away.

On 25 May 1961, Kennedy began one of the most historic speeches in space exploration, believing that his nation should commit itself to landing a man on the Moon before the end of the decade. Two weeks later, Kennedy and Khruschev met in Vienna for summit talks. Over lunch, Kennedy suggested that it might be possible for the two space powers to cooperate in the lunar programme. Khruschev at first refused, then considered that it might be possible, and finally changed his mind again with an emphatic 'No'. In the press, the Soviet leader was criticised for having a negative attitude, but when asked by his son Sergei why he turned down such an offer, Khruschev replied that if they accepted, America would gain knowledge of their missile programme and would realise that they had far fewer weapons than was widely thought. At the time distrust and misunderstanding were such powerful factors in the relationship between America and the USSR that Khruschev believed that once his secret was out, America would attack Russia within days. It was not that they had anything to hide; it was more a case of 'We have nothing, and must hide it!'

For Mercury, the next event was the 21 July sub-orbital flight of Virgil Grissom onboard Liberty Bell 7. Virtually a repeat of Shepard's mission, the flight is best remembered for the loss of the capsule due to a prematurely blown hatch, and Grissom almost drowning in the ocean. Despite this, both flights demonstrated the success of the Mercury design. A third flight was planned for John Glenn, but it was considered that they could cancel further manned sub-orbital flights and progress to manned orbital flights to catch up with the Russians. After all, they had put only one man in orbit for 108 minutes, while the first manned Mercury was scheduled for at least 3–7 orbits.

Gherman Titov boards the cosmonaut transfer bus to take him to the launch pad on 6 August 1961.

Titov and Nikolayev inside the cosmonaut bus on their way to the pad.

Return to Baikonur

Gherman Titov slept in the same bunk that he had occupied on the night before Gagarin's mission. He offered Gagarin's bunk to Nikolayev, telling him it would bring him luck, but not until after Titov's own launch. Both cosmonauts had visited the pad with Korolyov earlier in the day and now, in the quiet of the cosmonauts' cottage, Titov recalled how in April it had been difficult to come to terms with leaving Gagarin at the pad and watching the launch from the control room. Yuri was sitting in his spacecraft on the top of the rocket on the pad, and then just a few short minutes later he had gone and was in space! What was more astonishing to Titov was that tomorrow it would be his turn to leave the Earth.

Titov was not as relaxed as he had been in April. It was hot and stuffy in the cottage, so he got up and opened the window, and heard the sounds from the assembly shop and pad area. As the night drew on, the desert temperatures fell, and he got up again to shut the windows and turn off the fan. All too soon, Karpov woke him and, looking across at Nikolayev on the other bunk, he noted that he was lying in the same position in which he had fallen asleep. At least Titov could look forward to some sleep in space.

On the morning of the launch, the two cosmonauts followed an almost identical routine to the previous April – of medical checks, exercise, breakfast, and documentation by V.A. Plaksin (Sports Commissioner) to sign the certificates for flight. The next task was to suit up and take the ride on the bus out to the pad, following the established 'traditions' before climbing into the Vostok capsule.

The launch of Vostok 2

Strapped inside Vostok 2, Titov could hear and feel all sorts of noises coming from deep within the booster below him as the final seconds counted down. At 09.00 (MT) on 6 August 1961, the R-7 left the pad and headed for orbit. Although he had heard the words 'Lift-off', Titov initially hardly noticed the upward movement. As the speed increased, he became aware of the vibrations and build-up of g-forces against him. Each of the launcher stages separated as programmed, and the separation of the stages was felt by the cosmonaut, as his heart rate reached 132 beats per minute. He reported that his vision and breathing were unaffected by the ride.

As soon as the shroud separated, he stole a glance out of the porthole, but with the launcher still heading for orbit he did not have time to admire the spectacular

Inside the cramped confines of the Vostok sharik.

view of the Earth's curvature. After the upper-stage engine cut out and separated, Vostok 2 was in orbit. The change from the rough, noisy, boosted flight to the serenity of orbital flight impressed the cosmonaut: 'Suddenly the noise stopped, silence fell, and I had the sensation that I was assuming a head-down position. The objects that surrounded me seemed to float up. The first orbit around the Earth had begun.'

'I am Eagle'

Shortly after the spacecraft had entered orbit, Tass announced: 'A new launching into orbit of an Earth satellite – the spaceship Vostok 2 – piloted by citizen of the Soviet Union pilot-cosmonaut Major Gherman Stephanovich Titov.' The cosmonaut's call-sign was 'Oryel' ('Eagle'). At the time of Titov's launch, Gagarin was in Canada as part of his world tour, and was woken by his hosts. Upon being told the news that his friend was in space, he smiled and said 'I know.' During his tour, he had avoided giving any hint of his knowledge of the next flight and its pilot during his tour. Although fully aware that his own back-up would also soon experience the thrill of spaceflight, he subsequently sent his congratulations to Titov by way of a telegraph to Mission Control.

The announcement of a second cosmonaut in orbit was accompanied by the objectives of the mission, reported by Tass as an investigation into the effects of prolonged orbital flight on the human organism, and studies of a man during a long period of weightlessness. The cosmonaut's personal tasks were to observe the performance of on-board equipment, complete two tests of the manual controls, conduct visual observations through the portholes, maintain direct radio communications with the Earth twice an hour, and conduct physical exercise.

Upon entering orbit, Titov established radio communications with Earth, pulled off his gloves, and opened his helmet. As he looked out of the porthole he became aware of the movement of the planet below him, as he entered Earth's shadow for the first time. 'I noticed that in moonlight, our planet seemed to be dark grey. The horizon was visible all the time I was in shadow; it looked like a hardly noticeable bright edging. While emerging from the shadow, I watched the dark sky, then the blue border and purple band surrounding the very black Earth. From a lighted cabin while travelling in the Earth's shadow, our planet does not give the impression of complete blackness. It appears to be covered by a grey sheet and, if you look at it more intently, you can even notice its curving surface.'

The world would have to wait for the first cosmonaut artists and poets, but Titov provided an impressive account of the experience:. 'Marching across the planet in a circle of deep red–orange, the vivid sunset yielded unwillingly to the dark surrounds that advance to envelope the light of day, flashing colour to the horizon where the blue halo increased its richness of hues until, by some magic of transformation, night reigned supreme.'

Emerging from the shadow, and enjoying an equally impressive sunrise, he commented that dawn was, 'an explosive arrival of dazzling brilliance.' Titov then took manual control of Vostok 2 for the first time. He grasped the attitude control handle with his right hand and, firing the gas thrusters, was able to turn the spacecraft about its axis, but not change its orbital parameters. He was impressed with how easy it was to control the spacecraft travelling at 17,500 mph – a feat he would repeat on the seventh orbit. Titov may not have been the first in space, but he was the first to 'fly' in orbit around the Earth. 'I am Eagle! I am Eagle,' he called over the radio.

Above the Earth

As Vostok 2 approached the coast of Africa, Titov noted that it was clear of clouds, with the desert sands standing out quite clearly against the rich blue of the Mediterranean Sea. Then the weather closed in, and he recorded the cloud cover on camera. He also reported that he could see the shadows of the clouds on the ground as they passed over snow, and that without the cloud cover he could easily distinguish rivers and vast areas of cultivated land. Later, while flying over the Gulf of Mexico, he became fascinated by the, 'beautiful pale green colour of the coastal waters, which further on were changing into the greenish glassy surface of the sea.' Over night-time Brazil, he could clearly see the lights of the great cities. He saw the Moon a couple of times and noted that from outside the atmosphere it looked further away, and that he had the feeling that it was moving across the porthole of his stationary spacecraft, although he knew his Vostok was moving over the Earth's surface. He also noted that the stars were brighter, and did not twinkle.

Titov used hand-held Zritel cameras to record motion picture photographs of the Earth below him – a task that would see him honoured as the first cinematographer in space. He was impressed by the stark contrast of the green of the land and the blues of the oceans. The repeated sunrises and sunsets he witnessed seemed unusually beautiful, with sharp contrasts between the brightness of the sunlight and the darkness on the night pass, broken by twilights of pink clouds.

Space menu

During his third orbit Titov ate dinner, on the seventh he consumed his supper, and on the fourteenth he ate breakfast. Gagarin had managed to take refreshment during the thirtieth minute of his mission, but Titov expanded the tests over a full day. He did not have any difficulties in eating, nor did he notice any significant change in his taste sensitivity, but as the first two flights were of such short duration it was not possible to evaluate how such food affected the physiological requirements. That could take much longer flights of several days.

The day before the mission, all food items had been stowed in a metal container along with a key for unscrewing the caps and a package for collecting the left-overs and empty containers. The containers were closed with a removable textile cover for ease of access when in flight. Inside was a selection of food in tubes that could be squeezed directly into the mouth without the need for warming. Like Gagarin before him, Titov had been eating similar preserved foods from tubes up to two days before launch, to help his system adapt to the unique style of cuisine. Various purées were included, consisting of meat on its own, or mixed with sorrel, vegetables or oats. There were prunes, liver paté, a selection of fruit juices, processed cheese, and chocolate sauce for dessert, as well as cold coffee with milk. Small quantities of solid food were also provided, including bread and smoked sausage, and confectionery, as well as multivitamin pills, all vacuum packed in synthetic film.

Titov found that to eat in space required a particular technique. 'While opening the tube containing blackcurrant juice, I spilled a drop of the liquid, which was suspended in front of my nose and I had to hunt it'. To save weight and volume, children's utensils (tied together with twine to stop them floating away as had Gagarin's pencil) were provided for the solid food.

On the ground, the dieticians listened to the cosmonaut's reports to evaluate whether the system worked, and whether his natural body-clock, accustomed to accepting three meals a day governed by time on Earth, would work in space. Titov reported that, with practise, eating would not be a problem on longer flights, although he thought that the food was 'joyless'. What was only beginning to be understood was a different problem that could have an influence on the amount of time a human could spend in space.

Space sickness

Upon entering orbit, Titov had found that he assumed a head-down position, and that during supper he did not have a good appetite. The feeling began on the fourth orbit, and between the sixth and seventh, he experienced a change in his mood when he moved his head abruptly. He had loosened the straps to float just above his couch, but then experienced unpleasant sensations that reminded him of seasickness. He was becoming nauseous, although he did not vomit. Titov found that by remaining still in his couch, the sensations subsided a little, but did not disappear completely.

This was the first indication of a condition that would affect about 50% of all space travellers during the ensuing twenty years, as their inner ear otoliths registered what seemed to be a loss of balance. Inside the otoliths, minute stones float in liquid and are orientated by gravity. Any change of position triggers signals to the brain to compensate for the new attitude. In weightlessness, the stones lose all sense of reference, and cause the confusion in balance that leads to the feeling of nausea. Today this is known as 'space adaptation syndrome', and it disappears after about 48 hours as the body adjusts to its new environment. However, in 1961 after a flight of only 24 hours, this seemed to be an insurmountable hurdle to planners of extended space missions.

Medical investigations

The onset of space sickness (SAS) was not totally unexpected, but doctors had also discovered that flying high-g manoeuvres on aircraft or in the centrifuge could lead to greying out and then blacking out. Several candidates had been disorientated in the centrifuge tests. To examine this problem further, the 24-hour flight was to collect the first rudimentary medical data from extended spaceflight.

Attached to his skin, Titov had a range of medical monitors which recorded parameters for the doctors. He had an electrocardiograph, a pnuemograph and a kinetocardiogaph. A new TV system was also installed in the cabin, to provide visual monitoring of the cosmonaut, in addition to the telemetry and voice data. The older system of 100 lines used in the Korabl-Sputniks and on Vostok 1 was replaced by a 400-line system, although both could only transmit pictures at only ten frames a second. In completing his physiological and psychological tests, Titov initially noted that no disorders were detected, and apparently did not report the nausea!

Also on board was a biological payload – similar to that carried by Gagarin – consisting of *drosphilia* fruit flies, dry seeds, and lysogenic bacteria. According to the Tass report, these were carried 'for obtaining supplementary data on the influence of cosmic radiation on living organisms and biological subjects.'

It was not immediately clear how the symptoms affected the work programme, as reports indicated Titov's 'excellent spirits'. However, the cosmonaut later agreed that, 'maybe I should not have called my mood actually excellent.' Despite feeling nauseous, he was still able to maintain his flight log, record all his responses on the onboard tape recorder, complete the manual operation of the spacecraft, perform observations, take photographs, and consume food and water. He also exercised by using rubber pull cords, working his abdominal muscles and limbering up his joints and muscles

While Titov was flying over the USA during the sixth orbit, Ed Correl, – an electronics technician at the Wright-Patterson AFB, Ohio, – called Dr Duane Graveline (one of the first scientist-astronauts selected by NASA in 1965) to his laboratory to assist in the identification of a strange signal that he was receiving. They determined that it was a high-frequency device that also transmitted the cosmonaut's heartbeat. This simple device was probably a prototype, for use on later flights, of a transmission device for heart and respiratory rates. This was the first time that the Americans had monitored Soviet biomedical data in real time.

Breakfast on orbit
During the seventh orbit, the cosmonaut reported using the 'excretion-removal arrangement', (space toilet), and reported his 'excellent spirits'. It obviously worked. Next came time to prepare for sleep. It had after all been a long 'day', with several sunrises and sunsets. In radio communications with Earth, Titov informed Mission Control: 'You may do what you like, but I'm going to sleep.'

First fixing his hands under the harness straps to prevent them floating in mid-air, he fell in to a light slumber, trying to remain still to remove the feeling of nausea. Initially he did not sleep well, but he then managed to drop into a sound sleep, without dreaming. He slept 35 minutes longer than called for by the flight plan, and at no time felt the need to turn from side to side as he did when sleeping on Earth.

On the ground, a shift system was operating to ensure that all posts were maintained during the course of the mission, monitoring the functioning of the spacecraft and the state of the cosmonaut (by his pulse and respiration rates) while he slept. It was reported that, during the mission, there were 20,000 arithmetic (computer) operations per second in space tracking and data handling.

During the thirteenth orbit, ground control tried to wake Titov, but could not raise him, leading to concern about his condition. When he finally awoke, it took him a few moments to realise where he was, as he opened his eyes to objects floating in front of him. He found he had rested well and that the sensation of sickness had passed. The relative ease with which the flight was undertaken made an impression on Western observers.

As a visual reminder of the apparent superiority of Soviet rockets and spacecraft, Americans were able to watch the cosmonaut's spacecraft passing overhead as a pinpoint of light. The fact that they would also wake up under the gaze of a Soviet pilot was also a little disconcerting, as Titov pleasantly reminded them after the flight. 'While flying over Washington, the capital of the USA, and sending my best regards to the American people, I thought that after several minutes I would be flying over the capital of the USSR. I covered the distance between Washington and Moscow in only eighteen minutes. Cosmic space brought our cities closer together.'

Throughout his flight, Titov also sent greetings to other countries. This was seen as both a humanitarian message and a communist statement of the vulnerability of the countries under his flight path. TV pictures showed the calm and smiling face of the young cosmonaut, who, some forty years later, still remains the youngest person to have flown in space. His flight took place just five weeks before his 26th birthday.

Return to Earth

Titov later wrote that the yearning for Earth struck home hard, and without warning. He realised that this was only a day-long flight, but even so he felt that every positive view and experience was accompanied by a negative emotion of loneliness and monotony. But even these could not possibly compare, 'with the exultation of anticipating the feel of solid earth beneath your feet and the desire to see the heavens where they have always been – up above me!' Apparently, for Titov the experience of spaceflight was impressive, but the experience of life on Earth was even more so.

During the seventeenth orbit he prepared for entry, and stowed all the objects that he had used in flight. He checked the survival equipment, and prepared his suit by closing the visor and putting on his gloves. The automatic orientation and braking system worked perfectly and Titov was on his way home.

The retro-rocket fired at 09.52 on 7 August, and Titov felt this in the increase of g-loads. Outside the porthole, he saw what appeared to be snowflakes floating by. He had heard the crack that signalled that the two modules had separated, but then heard a light rapping sound that seemed to indicate that he was dragging the Instrument Module into the atmosphere with him, as had Gagarin. Like Vostok, the vibrations and heat build-up eventually severed the straps, and the module separated, to be burnt up in the atmosphere as Titov continued to descend inside his sharik.

The cosmonaut began to enjoy the increased g-loads, 'either because the unpleasant sensation of nausea had disappeared or because of the fact I was returning to Earth.'

Violating regulations, Titov did not put up the covers on the portholes, as he was curious about what he would see. His reward was a, 'raging purple flame, as the portholes turned yellow and the glass acquired a thin coating.' As he watched the sky turn from the blackest black, through a blaze of delicate pink to scarlet, purple and crimson, inside the capsule he was a comfortable 72° F. Outside, it was 10,000° F.

Loads of 8-9 g pushed him into the seat, and hampered his breathing. All the way through the fireball, he confidently shouted to himself, 'I'll be home, I'll be home, I'll be home'

Emerging from the black-out and dropping like a stone, Titov grabbed the ejection handle, his eyes glued to the chronometer, awaiting the red light to flash on the indicator panel to begin the ejection sequence. At 21,325 feet, 'thunder crashed in my ears, and at the same time I felt a tremendous force beneath me. The ejection shell exploded on schedule, and a blur of daylight flashed in my eyes as I burst away from the spacecraft.'

Titov felt like a prisoner released from his cell after 24-hour solitary confinement. At 8,000 feet he separated from the seat and looked up at the welcome sight of his canopy deploying. He then began to descend under his parachute, hoping that a wrist injury – sustained many years earlier, and which he had hidden from the selection doctors and trainers, – would not succumb to the roughness of his landing. Had the injury been discovered during selection or training, or during pre-flight medical examinations, he would not have made the flight.

As he descended, Titov could see the Vostok 2, under its own parachute, gently descending to Earth. As he looked down, he saw small figures running to where he would land, near a haystack. He also saw a rail line with a locomotive on it. In the distance was a river and two large cities. After he landed, just 600 feet from the rail lines and on his side, Titov sat up and opened his face-plate. He had landed near the village of Krasny Kut, 450 miles south-east of Moscow, near to where Gagarin had landed four months earlier. After travelling 703,438,761 miles in 25 hrs 18 min, the second cosmonaut was sitting, laughing in a ploughed field. He had grabbed a

Titov in his Vostok spacesuit after a parachute jump during training.

handful of Earth, and took in its aroma as he answered the questions of the local peasants around him, who were asking if he was the Titov that they had heard of on the radio.

Post-flight

Nikolayev was the first to greet Titov at the landing site, eagerly asking about his experiences in orbit, as he was now the next to fly. 'Very interesting,' Titov replied. As in April, Titov stayed at a cabin in the Zhiguli Hills overlooking the Volga, where Gagarin joined them for the post-flight debriefing. The official celebration would begin later, in Moscow and although not as grand as Gagarin's, it would still be impressive. Like his predecessor, Titov received the highest state honours, including the Hero star and the Pilot Cosmonaut medal.

As a result of Titov's flight, the phenomenon of space adaptation syndrome (more commonly known as space sickness) became apparent and as a result, vestibule training was altered and increased for all future flights. Titov seemed to recover quickly after the flight, although he was grounded for a time, as his symptoms were not fully understood. Titov reported to doctors Yevgeni A. Pobydonostsyev and Vladimir I. Yazdovsky, who were at the landing site, that he felt unwell for a time. This was a further indication that adaptation to flying in space and re-adapting to life on Earth would require extensive studies.

Despite concerns about the inner ear problem, the Soviets planned for even longer cosmonaut missions. The second orbital mission by the Soviets in less than four months had increased their spaceflight log to 1,626 minutes, compared with the two American sub-orbital missions, which totalled just over 30 minutes. By spending a whole day in space, Vostok 2 had demonstrated the ultimate goal of the Mercury programme, which was still two years away. If the Soviets could achieve that on their second mission, what lay ahead in those two years did not bear thinking about for the Americans.

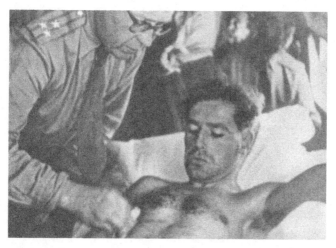

Titov undergoes medical tests.

On 18 August, NASA publicly announced that the analysis of data from the sub-orbital flights had met all requirements and that no further Redstone flights were planned. On 13 September, the unmanned Mercury Atlas 4 was launched and became the first Mercury spacecraft to enter orbit. Carrying a mechanical crew-man simulator, it was recovered after just one orbit, and the system was declared ready to support a manned spaceflight in orbit.

As a direct result of the Vostok 2 mission, the Soviets demonstrated a capacity to sustain human life in orbit and in relatively good condition for 24 hours. This provided them with the confidence to plan even longer flights of the Vostok system, to further study long-term spaceflight. However, during a speech presented at a Committee on Space Research symposium (Cospar) in May 1962, Titov warned: 'At the present time, it would be ill-advised to assume the problems of weightlessness have been understood.'

Obtaining that understanding would fall to new cosmonauts and longer missions in the future. A few nights after his return, Titov took a walk with Gagarin to share some experiences that were unique to the two of them. They talked of the future, of new flights both would like to make – perhaps to the Moon, and one day, maybe, to Mars. Just six months before, such talk would have been the realm of science fiction; but now these two cosmonauts had taken those dreams one stage further, into reality.

First group flights and first woman

Germany and Gemini

On the day that Titov was launched into space, Khruschev authorised the German Democratic Republic to begin the construction of the Berlin Wall. The launch of the second Vostok was perfectly timed by the Premier – to dilute the international reaction to the Wall with the success of the latest cosmonaut in space. To help support the East German regime, Titov's first foreign trip after the mission was to the German Democratic Republic in September.

In America, after two successful Mercury sub-orbital manned missions, all efforts were focused on preparations to launch John Glenn into orbit. This was the first step in fulfilling the Presidential commitment to go to the Moon, but exactly how to undertake such a journey remained uncertain. Three methods were short-listed for the flight profile to take Americans to the Moon. The all-out direct ascent mode would require a huge multi-stage rocket, capable of generating a monstrous 11 million pounds of thrust. Once the rocket left the pad, each stage would fire in sequence until sufficient velocity had been reached to fly directly to the lunar surface. After exploring, the astronauts would launch from the Moon and fly straight back to Earth for recovery. The design of such a rocket, called Nova, existed only on the drawing board, and was so expensive and so far into the future that it would not be completed by the end of the decade.

The second method was to launch two or three smaller rockets, carrying parts of the lunar spacecraft into Earth orbit. The Moon-bound spacecraft would be assembled in orbit, and would fly a direct path to the surface and then home again. The major problems with the Earth Orbital Rendezvous (EOR) method were to achieve coordinated launches on time, and to bring items of space hardware together while travelling at 17,500 mph. Nothing like this had ever been attempted before, and NASA would have to determine how to practise such manoeuvres in Earth orbit before committing to lunar missions.

A third option being discussed at the time was the Lunar Orbital Rendezvous (LOR). This method would involve a launch into Earth orbit by a three-stage rocket, followed by a flight to lunar orbit. Once in lunar orbit, a smaller landing craft would separate from the main spacecraft and make the landing, leaving the main spacecraft in lunar orbit. Once the surface exploration was completed, the lander would rendezvous with the main spacecraft in lunar orbit, and only the main spacecraft would make the return trip. This would save on launch weight, and would also remove the need for heat shields on the landing craft. The problem would be the mastering of rendezvous and docking in lunar orbit.

On 11 July 1962, the decision to adopt the LOR concept for the lunar mission was approved. Concurrent to the studies into just how America would reach the Moon were discussions on what would follow Project Mercury. Project Apollo had been on the NASA drawing board since 1960, as a three-man spacecraft that had the capacity to fly the lunar distance, and this was therefore the programme that NASA assigned to complete the lunar landing goal by 1970. But before Apollo flew, NASA realised that several techniques – including the all-important rendezvous and docking operation – had to be evaluated and mastered.

Chief Designer of the Space Task Group, Max Faget, had considered improvements to the design of Mercury to include a second astronaut. The new vehicle would be launched by the larger Titan II booster, and the Atlas Agena B combination would be used to place a target vehicle in orbit for practising rendezvous and docking, – the technique required by the EOR and LOR proposals. The larger capacity of the improved Mercury spacecraft would also allow for missions of up to fourteen days – well in excess of a projected manned lunar flight.

A Project Development Plan, completed by 27 October 1962, included twelve missions (eleven manned) to be flown between June 1963 and March 1965. This plan became known as Mercury Mark II, but by 3 January 1962 it had been officially renamed Project Gemini – the successor to Mercury, and the pathfinder for Apollo. NASA documents and aerospace journals were full of articles and illustrations of this new programme that would take America a step closer to the Moon. The openness of the American programme presented the Soviets with the chance to compare the competition with their own plans to send cosmonauts to the Moon in the spacecraft that was to follow the Vostok series.

Korolyov had for some years been evaluating ideas to modify and adapt the Vostok capsule, and although he had no formal flight plan after the Gagarin mission, there were about eighteen spacecraft of the Vostok type on order – half for the Zenit reconnaissance satellite programme, and half for manned spaceflight. What he did not wish to do was merely use these spacecraft to repeat previous missions. There was an unwritten rule that required each new mission to be a significant advance over the previous flight, and so plans were being formed for a possible multi-vehicle launch. As the news of the Gemini missions, – including a crew of two astronauts completing rendezvous and docking, EVA, and 14-day missions – began to emerge, Korolyov sought further support in expanding his plans for Vostok.

On 20 February 1962, America achieved a new success with the three-orbit flight of astronaut John Glenn onboard *Friendship 7*. At last, America could celebrate a man in orbit, and although the mission was longer than Gagarin's mission, it was much shorter than Titov's mission. But this did not prevent the Americans from referring to the mission as a major advance – which for them, it was. In May 1962, as Scott Carpenter flew a second three-orbit mission onboard *Aurora 7*, America cheered its latest heroes. It was also noted that nothing had been heard from the cosmonauts for almost a year. Rumours circulated concerning what space spectacular the Soviets were planning as a 'come-back' to surprise the West.

The emergence of Vostok spy satellite hardware

Preparations for the next Soviet manned spaceflight were delayed by the decision to begin the programme of unmanned spy satellites of the Zenit class, based on the Vostok design. These were planned for June–July 1962, and with only one pad capable of launching R-7/Vostok derivatives from Tyuratam, the military satellites took precedence following a high-level decision. Due to the enforced delay between Vostok launches, the Khruschev administration began to play down the significance of manned spaceflights.

The initial Zenit missions were planned before the next cosmonaut flew in space, using the Vostok 'bus' to test improvements to the manned spacecraft design after the flights of Gagarin and Titov. These improvements included refinements to the cosmonaut's temperature control sub-system – after the prime and back-up ventilators on Vostok 2 had been turned on at the same time, dropping the temperature to a chilly 10° C – and measures to ensure correct separation of the Instrument Module for re-entry.

The first Zenit was lost in a launch failure in December 1961, but it was followed by a first successful flight (designated Cosmos 4) on 26 April 1962. A further spy-satellite was launched on 1 June, but the premature separation of one of the strap-on boosters brought the R-7 down just 1,000 feet from the pad area. The damage resulting from this accident led to further delays while the pad was repaired. The qualification for a manned launch occurred with the success of Cosmos 7 (Zenit) on 28 July, the stated aim of which was 'to investigate the radiation caused by the explosion of an American nuclear device in the atmosphere on 28 June 1962'.

After a further high-altitude nuclear test by the Americans on 9 July, the Soviets tried to score additional propaganda points by requesting that the US refrain from experiments that could hinder the exploration of space for peaceful purposes or could endanger a cosmonaut's life. As a result, the Soviets announced that, until residual radiation returned to permissible levels, no cosmonaut would fly in space – which disguised the real facts concerning the long delay in Vostok launches. Although no further tests were planned, the Americans also reassured the Soviets that the cosmonauts would be in no danger from the radiation levels created by American nuclear tests.

Bold plans for a new mission

The flight plan for the next mission had evolved over the weeks following Titov's mission. Korolyov wanted to fly no less than three separate manned Vostok missions in orbit at the same time, in what was termed a 'group flight'. Each could be launched on successive days, orbit for two or three days, and then land. Korolyov wanted all three in space at the same time at some point during the mission, to demonstrate a 'significant step forward' in Soviet cosmonautics'.

The launch was targeted for November 1961, but this proved to be too optimistic, and the priority of the Zenit flights pushed Korolyov's plans well into the summer of 1962. Over the next few months, the cosmonauts continued training for a mission that had yet to be scheduled with a firm launch date.

On 21 February 1962, Ustinov unofficially notified Kamanin to prepare the cosmonauts for a spaceflight planned for around 10–12 March. Pre-flight preparations required that the cosmonaut who would make the flight should be at Tyuratam between 2 and 3 March. It was the day after the high-profile Glenn mission, and Khruschev demanded that the next Soviet space mission be immediately brought forward. Kamanin recorded the situation in his diary. 'This is the style of our leadership. They've been doing nothing for almost half a year and now they ask us to prepare an extremely complex mission in just ten days time, the programme of which has not even been agreed upon.' This target was not met due to the short time-scale, but it does reflect the Soviets' use of this programme to support political aims.

Following the successful Zenit launch of 28 July, the group flight was planned for mid-August. It was decided to launch two, instead of the more complicated three Vostoks. What was more difficult to decide was the duration of each flight. Kamanin was concerned about Titov's motion sickness, and suggested flights of one day each, with a maximum extension of two days. Korolyov did not wish to mark time while the Americans gained ground on their early achievements, and pressed for three days for each flight. The disagreements continued until only a few weeks before the launches. Gradually, Korolyov persuaded other designers to support his wishes, and on 18 July 1962, two flights lasting up to three days were confirmed.

Kamanin accused several people of bowing to Korolyov's pressure and of putting him in a 'no-win' situation. If they were successful, he would be likened to a 'coward', and if they failed, he would be accused of poorly treating the cosmonauts.

The Vostok was unable to manoeuvre in orbit, so orbital rendezvous and station keeping was not possible. Instead, the second vehicle would be launched at a precise time to match the orbital trajectory of the first Vostok. For a short time during the early orbits of the second Vostok, the two spacecraft could claim to be relatively close to each other.

The selection of cosmonauts three and four

In September 1961 it was decided that members of the first cosmonaut selection should commence studies at the Zhukovsky Air Force Engineering Academy in Moscow, to broaden their engineering and academic skills. All were to attend – with the exception of Belyayev and Komarov, who had already graduated from engineering academies. Of the group of eighteen, the eleven still assigned to the team graduated in February 1968. Gagarin had been named Commander of the cosmonaut team (replacing Belyayev), and both he and Titov had been elevated to State Commission members involved in all pre-mission planning.

By October 1961 a new top training group was confirmed, consisting of Nikolayev, Popovich, Nelyubov and Bykovsky, from the original training group for Vostok 1. Komarov joined them (back from medical disqualification), as did Volynov.

In March 1962 the first member of the 1960 group – Rafikov – was forced to leave, for disciplinary reasons. He was experiencing marital problems, and was found to be absent without leave from the training centre. Kamanin, and the group's political officer Nikeryasov, filed a report stating that he was politically unreliable and should be returned to the Air Force to resume his flying career. Prior to the selection of the top six to train for the first mission, each of the twenty cosmonauts was asked to submit a peer rating written evaluation of their colleagues, answering the question, 'Who should be first in space?' Rafikov stated: 'I should be sent, although they will not send me, but my first name [Mars] is 'cosmic' and this could sound good.'

In February the training group was expanded to include Shonin, and so seven cosmonauts were prepared for the group flight. However, it was not long before Shonin was removed from the training group, due to his low tolerance for high g-loads on the centrifuge, although he remained in the cosmonaut team. Following orders from the Kremlin to fly a mission in March at short notice, a restricted training programme required a smaller training team. This consisted of Nikolayev, Popovich, Nelyubov, Bykovsky and Komarov. Training was aimed to support a 'volley' Vostok mission of two launches, planned for the 10 and 12 March, each lasting two days. It was during this period that Nelyubov failed his centrifuge tests and was dropped from the training group, to be replaced by Komarov in May 1962. It was a bitter blow to the cosmonaut

Cosmonaut Nelyubov in a Vostok pressure suit.

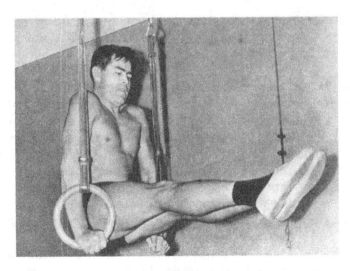

Nikolayev uses the rings in the gymnasium to maintain his fitness.

after the disappointment of the previous year. He was now to lose his back-up role, and with it, perhaps, his last chance to progress to a flight of his own.

Andrian Nikolayev was selected for the flight of Vostok 3. He had previously served as Titov's back-up, with Valeri Bykovsky as his alternate. Pavel Popovich would fly the second Vostok (Vostok 4), with Vladimir Komarov as his back-up. Boris Volynov would handle the support role on both missions. These men were chosen for their reliability in endurance tests.

The flight of the Vostok pair was to continue the studies of extended spaceflight by doubling the duration of Titov's mission to gain further information on the effects of space adaptation.

Nikolayev, a quiet and reserved character, had shown incredible stamina during the early medical tests, and constantly surprised doctors with his ability to endure silence, total isolation, and extremes of temperature, setting records unmatched in the environmental chambers. If anyone could push beyond the limit of Titov's flight, Nikolayev would lead the way. Popovich was a more outgoing personality, and was the first to arrive for cosmonaut training. He became the unofficial greeter and quartermaster of the new arrivals, and possessed a fine tenor voice that he often put to good use. He had been the CapCom during Gagarin's launch.

Bykovsky was also a highly experienced veteran of the isolation chambers, and had been one of the early test subjects for that demanding piece of hardware. In recognition of such pioneering work, he had been awarded the Order of the Red Star. As Nikolayev's back-up, he was already a step closer to flying on the next mission after Popovich.

Komarov was one of the older and more experienced members of the team and his selection was only to last a short while. Training revealed a heart irregularity that grounded him. It was a problem similar to that which had grounded American astronaut Deke Slayton from the Mercury 7 mission the same year. Despite dozens of tests by a number of cardiologists, it would be some time before Komarov was declared fit for spaceflight, although not the ten years that Slayton had to wait on the ground. He was later removed from the training group, and Volynov stepped in to replace him in back-up duties for the Popovich mission. The mission planners saw no reason to assign another cosmonaut to these missions so late in the training cycle.

New cosmonauts

Plans for a successor to Vostok had originated as early as 1958. The plans would encompass manned circumlunar flights to the Moon over the next decade, mastering the techniques of rendezvous and docking that could lead to the creation of large space platforms that, in turn, could lead to the long-term goal of Korolyov and his team – a Soviet manned flight to Mars. These plans were developed, over the next five years, but there were significant technical problems that delayed the first flights until the mid-1960s. With these plans on the drawing board it became evident that the original cosmonaut team would not be enough to support further flights, and so a new recruitment drive was initiated by the Air Force.

In March 1962 the selection of the first females to train for spaceflight was completed, with the selection of five women, who reported to Star Town during the preparations for the first Vostok group flight. They were selected due to political pressure, and were placed under the control of the Air Force.

In December 1961, Kamanin was authorised to begin the selection process, for an intake of up to sixty new cosmonauts beginning in 1962, although the selection of a second intake of male pilot cosmonauts was delayed until 1963, as the selection of female candidates began.

On 12 April 1962, the first anniversary of Gagarin's flight was declared a national holiday in the Soviet Union, and was renamed Cosmonautics Day. This is an

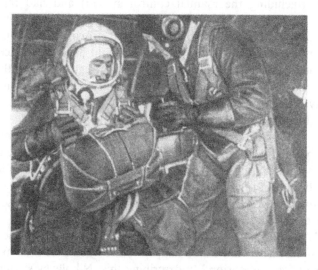

Nikolayev is assisted with his parachute prior to a training jump.

important day for all workers in the space industry, and it continues to be marked with a number of ceremonies in central Moscow.

The launch of Vostok 3

The night before launch, Nikolayev and his back-up Bykovsky occupied the cosmonaut cottage. Nikolayev chose the bed that Titov had slept in twice before. He was unaware that Korolyov visited the cottage during the night to ask how both men were sleeping; even though sleep did not come easily to the Chief Designer himself. The signs of stress were showing on Korolyov's face as he tried to take on more projects and personally drive forward every aspect of the space programme. On launch-day morning, Popovich accompanied Nikolayev and Bykovsky on the bus to the pad. That night he would be staying at the cosmonaut cottage with his back-up, Volynov, before his own launch the next day.

At 11.30 am Moscow Time, Vostok 3, carrying Nikolayev, rocketed from the pad at Tyuratam and into an orbit of 113.7 × 155.9 miles, inclined at 64° 59'. Nikolayev, using the call sign 'Sokol' ('Falcon'), orbited the Earth every 88.5 minutes. The flight was designed to obtain further information on the human organism, and to investigate man's ability to work under weightless conditions. The cosmonaut would also conduct specific scientific observations, as well as performing tests on the communications, guidance and landing sub-systems. There was no mention of the pending launch of a second vehicle.

Shortly after Vostok 3 entered orbit, Nikolayev established communications with the control centre. Some of the transmissions were relayed via Soviet fishing trawlers located off the US eastern seaboard, and were intercepted by Western observers in the US.

'I feel fine! All is well aboard. [I] can see the Earth clearly through the porthole,' called Nikolayev as he began his first orbits. By the third orbit he had checked out his spacecraft (including the manual control system) and had reported on his physical condition. Apparently, the new vestibular training programme which every cosmonaut had to complete before their flight, was proving beneficial, as Nikolayev explained after the mission: 'I expected that I should have experienced discomfort due to the reaction of the vestibular apparatus through the weightless condition. However, neither during the first day, nor at any time throughout my flight did I experience any discomfort.'

Nikolayev then prepared his first meal, which he reported to be 'earthly, nutritious and tasty'. After he had eaten, at the beginning of the fourth orbit, TV pictures were beamed to Earth, and were also shown for the first time on Soviet TV. During the programme, Nikolayev initially appeared in a trance, with his eyes closed and hands motionless. But gradually the viewers saw him working the controls of the spacecraft, moving his head, and replying to well-wishes from Khruschev, who was talking to the cosmonaut by radio-telephone.

Nikolayev's next task was to release his straps and float about the cabin for the first time. 'No-one knew what would happen to me when I released myself from the belt in a state of free suspension.' In describing this, Nikolayev stretched the story, noting his fear that he may not get back to the seat without help, or that bumping

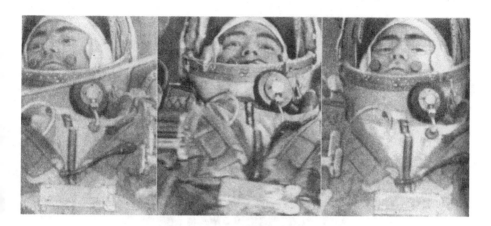

Nikolayev in orbit onboard Vostok 3.

into things could result in him being unable to return to his couch. 'The result would be an emergency with unpredictable and perhaps very grave consequences. The whole flight programme might be disrupted. This called for caution above all.' What was not made clear at the time was that the crew compartment had a limited internal volume and that he was only able to float a few inches above the seat. The account was merely a ruse to suggest that the actual spacecraft was large and comfortable, indicating a 'ceiling' and an 'armchair', and containing significant 'home comforts' compared with the cramped American Mercury spacecraft.

Nikolayev did not experience the nausea that had plagued Titov, and after about an hour of 'free-float' he returned to his seat and recorded his experiences in the log-book: 'I touched the cabin wall with my finger, [and] I drifted off the opposite direction. From the ceiling I could shove myself gently back into my armchair. The devil, it seemed, was not as black as [Titov] had painted him.'

At the end of his first day in space, Nikolayev settled down to eight hours sleep; but he needed only six hours, and woke up fully refreshed, despite having woken up three times during his rest period. As he woke, and ate breakfast, at the cosmodrome preparations were being completed to launch the next Vostok to join him in orbit.

The launch of Vostok 4
The orbit of Vostok 3 was tracked precisely and the lift-off of Vostok 4 was timed for when Nikolayev passed over Tyuratam on his sixteenth orbit. At 11.02 am MT, on 12 August, just 23 hours and 32 minutes after Nikolayev left Earth (and as he was about to surpass Titov's endurance record), Popovich began his flight from the same pad as Vostok 3. The Tass press announcement indicated that the purpose of the mission was to place both spacecraft in orbit close to each other, in order to obtain experimental data 'on the possibility of establishing direct contact between the two ships.' Two-way communications had been established within the hour, with Popovich using the Callsign 'Berkut' ('Golden Eagle') from an orbit of 111.8 × 152.9 miles, 65°, 88.5 minutes.

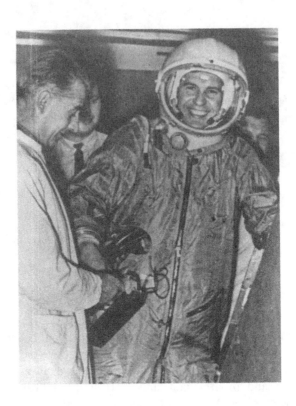

Cosmonaut Popovich during preparations for Vostok 4.

Other tasks for the joint flight included obtaining data on direct communications between two orbiting spacecraft, co-ordinating the actions of the cosmonauts, and checking their actions and reactions to spaceflight (thus providing two sets of biomedical data simultaneously).

'I watch the Earth in the clouds. I see the black, black sky. My spirits are wonderful. Everything goes excellent,' called Popovich from orbit, the TV views showing him making entries in his log-book. His descriptions of the view out of the porthole were relayed to the State Central Television network, and became part of regular broadcasts.

'Attention! Attention! I am speaking to you from the Soviet spacecraft Vostok 4. The spacecraft is passing over the Pacific Ocean. Outside it is night – I can now see the Moon. What a beauty! She looks more bulky than from down on Earth – a ball surrounded by emptiness. The spacecraft is coming out of the shadow. What a view! A person on Earth will never see anything like it. This is the cosmic dawn! Just look! The Earth's horizon is a vivid wine colour, and then a dark blue band appears; next comes a bright blue band, shading off into the dark sky. Now this band keeps widening, growing, spreading out, and the Sun appears. The horizon turns orange and a more delicate, lighter blue. Beautiful!'

Both spacecraft were functioning normally, with Nikolayev seeing the distant Vostok 4 as a point of light through his porthole, as Popovich entered orbit to begin the joint flight. Interpretation of the Soviet announcements led some Western analysts to predict that the two spacecraft would actually link up in space, but they were unaware that Vostok was unable to do so, and could not complete the intricate rendezvous manoeuvres necessary for docking. It was also reported that both cosmonauts had two-way TV to view each other in the capsules, enabling them to co-ordinate their activities. Post-flight analysis of the orbital data revealed that the two spacecraft came no closer to each other than four miles during the early orbits of Vostok 4, and with no capability for orbital manoeuvring, they soon drifted further apart. By the beginning of the thirty-third orbit of Vostok 3 they were 528 miles apart, and by the sixty-fourth orbit the distance was 1,770 miles.

Space brothers

As Popovich became accustomed to weightlessness, he heard Nikolayev calling him from Vostok 3. Although he was anticipating the call, it still amazed him that he could talk with his colleague directly rather than via ground control.

Nikolayev: 'Berkut! Berkut! This is Sokol! [Do you] hear me?'
Popovich: 'Andryusha! I'm here along side you! You're coming in fine'

Popovich was actually several miles from Nikolayev, but this did not stop the two men sharing their experiences and impressions as though they were side by side, as Gagarin interrupted them with congratulations: 'Everything is going very well, friends. I congratulate you. We'll soon meet again on Earth.' On Vostok 4's fourth orbit, Khruschev also added his congratulations on their achievement.

The two cosmonauts now coordinated their activities to provide two sets of data throughout the joint flight programme. Even meal times were taken at the same time, and in their first meal together the two cosmonauts toasted each other's success – with cold fresh water!

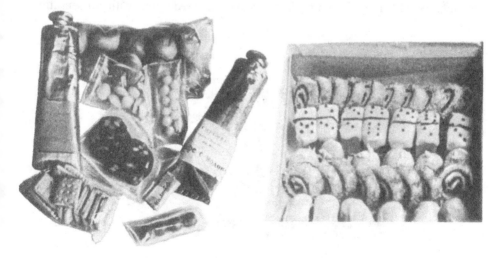

Examples of the food supplied to Vostok cosmonauts (Vostok 3 and Vostok 4) in tubes, packages (*left*) and bite-sized chunks (*right*).

Rations for both men included the puréed and liquid foodstuffs in aluminium tubes, but was supplemented by a variety of meat dishes. These included hamburger, roast beef, roast veal, chicken fillet, and beef tongue, as well as red caviar sandwiches, pieces of fish (Caspian roach), confectionery, bread and fresh fruit (oranges, lemons and apples). In order to ease intake and to prevent contamination of cabin air, all food was provided in bite-sized chunks.

Neither cosmonaut reported problems in chewing or swallowing the pre-cooked food, which had been packaged in cellophane/polythene film and hermetically sealed under vacuum. The packaging was prepared under bacteriological conditions, and was arranged in a complete menu of breakfast, lunch, dinner and supper.

Having practised eating the solid foods during parabolic training flights, they both found them completely acceptable for short flights once in space. Nikolayev, however, was disappointed that the menu did not include the national Chuvash dish called khuram-kukly (onion patties), and Popovich missed the flat Ukrainian pumpkin pies.

Popovich was soon ready to repeat the 'experiment' of floating out of his seat. After a few words of encouragement from his colleague – 'don't be afraid Pavel, I've done it already It's not bad really' – Popovich was a little over-eager to try, and, releasing his straps quickly, floated straight up and banged his head on the cabin ceiling. He quickly learned that in space, caution was imperative. Like Nikolayev, Popovich did not report any unpleasant feelings of nausea.

For the benefit of the TV, both cosmonauts demonstrated the new phenomenon of weightlessness by releasing pencils, log-books and food packages in mid-air. They showed that larger objects (such as cameras) weighing several pounds on Earth apparently weighed nothing in orbit.

Sleep came easily once they turned off the lights and closed the porthole shades. The training they had completed in the isolation chamber on Earth was found to be of huge benefit as they floated in the silent enclosed cabin with no sensation of

Popovich, onboard Vostok 4 in orbit, demonstrates the effect of weightlessness on a pen.

movement, even though they were orbiting at 17,500 mph! Nikolayev did tend to awaken during the sleep period each night, and on the fourth day, thinking he was over the Soviet Union, found that he was actually over North America.

Each morning they exercised by jamming their bodies against immovable objects, then stretched and bent to work their muscles – in particular, the stomach muscles. Throughout the working hours they exchanged greetings and reported to Earth on the status of the spacecraft.

Observations and experiments

Due to the restrictions of flight duration, scientific experiments were limited on the first two Vostok missions, although observations through the portholes were completed on both missions. For Vostok 3 and Vostok 4, with the extended flight time, additional activities were included in the cosmonauts' flight plans. They observed the stellar background and the Moon, commenting on its appearance as a sphere rather than a disc. The lunar light was so bright that the cosmonauts turned off their bright cabin lights and worked by the soft half-light of the Moon. It reminded the cosmonauts of working by the light of a 'Leningrad white night' (reflected light from the snow).

They also observed lightning storms on Earth, which seemed like a strange, silent light show. From their peaceful orbit a storm was logged, raging 125 miles below them. The two cosmonauts performed navigational measurements using the Vzor device, determining that their orbit was close to the prescribed orbit.

They also completed a daily series of psychological tests, including simple calculations. Using a special chart they identified geometric forms in 'chaotic disorder'. They drew spirals and stars for hand/eye coordination, and practised writing with the left and then the right hand (with their eyes open and then closed), testing their control, concentration, and levels of fatigue.

Each day, five or six minutes were spent in a medical self-check-up, taking their own pulses and listening to their heartbeats and breathing as they took their body temperature. Throughout the mission, the cosmonaut's flew wearing their spacesuits, although they did remove their gloves and open their helmet face-plate. Even so, it was restricting and cumbersome, even while unbuckled from the seats.

Both cosmonauts seemed relaxed, and exchanged pleasantries. On one occasion, Nikolayev asked Popovich what he was having for supper: 'The same as you, and a piece of dried fish besides,' Popovich replied. This was greeted with surprise from Vostok 3: 'Dried fish? Couldn't you spare one little slice for me?' asked Nikolayev. 'Well, come a bit closer, and we'll share what I've got,' Popovich teased.

One experiment conducted by Popovich concerned the observation of air bubbles. A hermetically sealed flask was two-thirds filled with water. When it was undisturbed, he reported that the water gathered in the middle of the flask around the walls. When shaken, a large bubble split into hundreds of smaller bubbles, gradually merging back into one large bubble again. Popovich also sprayed a small amount of water into the cabin, which formed small globules that gravitated to and settled on the wall. This was an early experiment to understand the dynamics of fluids in space, which had applications in the design of spacecraft fuel systems and also highlighted the potential danger of globules of water floating behind the electronic circuits.

Popovich also commented on the glowing particles outside the spacecraft, as observed by Gagarin and Titov (as well as the American astronauts Glenn and Carpenter). The Soviets indicated that on their missions these were exhaust particles from the rocket motors; but the Americans claimed that they were formed from ice particles on the outside of their spacecraft.

Biomedical research

Both flights were to gather further information on the response of the human organism to extended-duration spaceflights. An electroencephalogram recorded if the subject was awake or asleep, fresh or tired, and how the central nervous system reacted to various influences. An electro-oculogram recorded eye movement, using small silver electrodes fitted into the inside of the helmet. An electrocardiophone recorded the state of the cardiac muscle, the rhythm of the heartbeat and the passage of the rhythm through the various sections of the heart. For this, electrodes were placed on the right and left sides of the heart on the cosmonaut's chest. Two silver electrodes were attached to the foot and the lower part of the right shin to record dermogalvanic reactions, the ohmic resistance of the skin to various exertions that affect the body.

As the signals from the sensors were quite weak, they were amplified before being transmitted by the radiotelemetric system. When direct transmission was not possible, onboard recorders were used to store biomedical data.

The cosmonauts' pulse and respiration rates were monitored from four hours before lift-off, to entry, and then after landing. The weight of each cosmonaut was also monitored, from two days before the flight to fourteen days afterwards, with all food intake monitored for calories, weight and wastage. Flying for more than a day required the repeated use of onboard sanitary equipment, as well as the assessment of hygiene conditions and habitation of the capsule.

Dr Vassily Parin, Director of the Institute of Normal and Pathological Physiology, USSR Academy of Medical Sciences, revealed that after the flights, it took 7–12 days for the cosmonauts' cardiovascular response to return to pre-flight levels.

Arguments and misunderstandings

From the censored and carefully worded reports from Soviet news agencies, the progress of the twin Vostok flights appears to have been carried out without incident. However, two interesting accounts of communications from the Vostok indicated that was not necessarily the case. During 14 August, the Swedish Enkoping Telecommunications Administration monitoring station picked up a heated exchange from Nikolayev to the Vensa 1 ground station:

Nikolayev: 'Vensa 1, ya Sokol. You have made an error of five minutes at your latest report time. I have checked and the time was not 15.12 Moscow Time as you said, but rather 15.07. Now give me a new time. Can't you hear me? Now get started with your time report. No, you are getting it wrong again! You do not do this the right way! Now listen to me! LISTEN! I said! I will teach you how to do this perfectly. Correct time is 15.09...Attention...NOW! That is the way to run the show and not the sloppy way that you did... Please convey my thanks to 'Vensa 6', their time report was really good. Yes OK, your time report is correct now thanks.'

Popovich also seemed to have convinced those on the ground that he was having problems in orbit and should return early to Earth. During 13 August, the decision was made to extend the Vostok 3 mission by 24 hours, despite opposition from Kamanin. The same decision was made for Vostok 4 during 14 August, after seeking personal approval from Khruschev.

On Vostok 4, Popovich was experiencing uncomfortable conditions, reporting that the temperature had dropped to 10° C and the humidity to 35% – similar to the conditions experienced by Titov. Despite efforts to solve the problems, the levels did not increase. This raised concern about continuing the mission and the likelihood of landing on the original day planned. The Soviets were aware that Western eavesdroppers were monitoring their voice transmissions with the cosmonauts, and therefore issued a series of code phases that would indicate the condition of the cosmonaut in orbit. If the cosmonaut said 'I'm feeling excellent', this meant that the mission should continue, whereas 'I'm feeling well' meant that the mission should be terminated. One other phrase was 'I'm observing thunderstorms', which was the coded message that the cosmonaut was experiencing space sickness and needed to come down immediately.

This use of such a phrase was misinterpreted during the forty-ninth orbit of Vostok 4 – the original landing orbit before the mission was extended. With only forty minutes left to make a decision about entry on that orbit, Popovich reported (without thinking) that he was currently 'observing thunderstorms'. Suddenly realising his mistake, he corrected the statement adding, 'I'm feeling excellent. I observed *meteorological* thunder and lightning.' Mission Control misunderstood this as a plea to come down and ordered him to descend on that orbit.

Only after he had landed did Popovich explain that he was actually looking at thunderstorms on the Earth. Kamanin and Gagarin were sceptical of his claim, however. In his diary, Kamanin wrote: 'Perhaps (Popovich) experienced a brief bout of motion sickness, which passed quickly and then, having regretted his initial report about thunder, he decided to beat a retreat. It will be difficult to find out what kind of thunder he actually observed, but find out we will.' Whatever the outcome, Popovich was back on Earth.

Landing

In America there was surprise at the Soviet achievement of orbiting two manned spacecraft within 24 hours – a feat that the Americans could not accomplish with Mercury. From his weekend retreat in Boothbay Harbor, Maine, President Kennedy was once again forced to pass on the best wishes of the American people to the cosmonauts for their 'exceptional feat and courage', and wishes for a safe return. The success of the mission was interpreted in several reports as a clear intention that Russia intended to put a man on the Moon within a few years. According to Ken Gatland of the British Interplanetary Society, 'Once [the Soviets] had achieved orbital rendezvous, they had taken a vital step towards lunar flight.'

In the final orbits prior to landing, both cosmonauts increased their exercise in preparation for the increased g-loads that they would encounter during re-entry. During post-flight reports the cosmonauts indicated that Korolyov had instructed them to choose their own method of landing – either inside the capsule, in a repeat of the 'official' landing method used by Gagarin (to establish the IAF record), or ejection by personal parachute like Titov. Since both were fond of sport parachuting, they said that they elected to eject. Prior to initiating the de-orbit burn, both cosmonauts wished each other good luck, saying that they would soon meet on Earth.

In reality, landing inside the Vostok capsule was not possible, and this statement was added to confuse the West about the design characteristics of the spacecraft as well as reconfirming the original statement that Gagarin reportedly landed in his capsule.

Vostok 3 deorbited first, at 09.24 MT on 15 August, followed six minutes later by Vostok 4. As the atmospheric density increased, communication with Mission Control was lost as the antennae burned off. The cosmonauts required some effort to raise a hand or turn the head, and a dazzling glare was observed outside the portholes, as they both decided not to cover them up. The cracking outside alarmed Nikolayev for a few seconds, as he thought that the heat shield was breaking up. Each man watched as flames of many colours raged outside the fireproof glass. Although expected, both men realised they were in a fireball of several thousand degrees, and felt that this was perhaps the most trying period of their flight. Re-entry reminded them of riding springless carts down a bumpy farm track, and they realised why they had had to endure such vibrations during training on the vibrostands.

Both men ejected from their capsules, and to their relief their parachutes opened

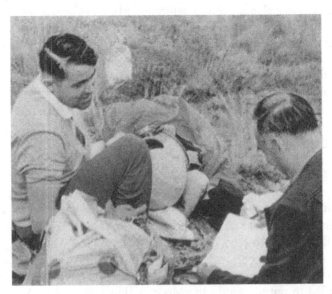

Nikolayev at the Vostok 3 recovery site after ejecting from the craft, 15 August 1962.

and they floated gently to Earth. The delay of six minutes code separated them by approximately 295 miles, so they could not see each other as they landed.

At the two landing sites, they were checked by medics from the recovery teams. The spacecraft were also inspected, and were pronounced usable for another spaceflight. Vostok 3 had logged 94 hrs 10 mins travelling 1,639,190 miles over 64 orbits. Vostok 4 had flown for 70 hrs 44 mins and had travelled 1,230,230 miles over 48 orbits. This was an impressive record.

Post-flight

Western analysts later noted a flight duration similar to Zenit reconnaissance satellites and Korabl-Sputnik 1. The landing zone was similar, at the 48th parallel, compared to that of the later Vostok 5 and Vostok 6 at the 53rd parallel, and those of Zenit spacecraft at the 52nd parallel. This indicated that the latter Vostoks were also testing recovery techniques (but not the ejection systems) for the Zenit spy satellite programme.

Korolyov and his team were pleased that the flight went better than they thought, and it helped to alleviate some of their fears about conducting a proper rendezvous operation. But this would not be available until the deployment of the new generation of spacecraft that were under development, which was still some years away. The Americans were working on rendezvous and docking with Gemini, within two or three years. Clearly, other Vostok missions might help steal some of the thunder from Gemini, before the Soviet docking hardware was ready.

In the planning for the mission it was reasoned that if the second launch was precisely one day after the first, then as the spacecraft flew over pad they could achieve a close proximity of just few miles from each other.

During a 1990 interview, Vasily Mishin admitted that 'with all the secrecy, we didn't tell the whole truth. The Western experts who hadn't figured it out thought that our Vostok was already equipped with orbital approach equipment. As they say, a sleight of hand isn't any kind of fraud. It was more like our competitors deceived themselves on their own. Of course, we didn't shatter their illusion.'

The enthusiasm generated by the success of Vostok 3 and Vostok 4 initiated several articles in Moscow newspapers. They forecast that 'the 1960s would witness a flight to the Moon, [and] there can be no doubt that in the 1970s man will visit Venus and Mars.' During a visit to London at the time, former US President Eisenhower attempted to diminish such claims, and added that there was no longer any gap between the two nations in space.

During the post-flight press conferences, in a response to a question on the nature of the next spaceflight by cosmonauts (and if it would be like theirs), Popovich stated: 'Analysis of all the material of our group flight will show whether it is necessary to repeat this kind of flight, or whether we can go further, onto something else.'

That 'something else' was already in training.

VOSTOK 5 AND VOSTOK 6

Upgrading Vostok

Korolyov's plans for a new spacecraft to replace Vostok had met with repeated disapproval since 1960. These plans included designs for Earth-orbital and lunar spacecraft, and in order to gain approval Korolyov sought support from the Air Force in March 1961, in a Tactical–Technical Requirement (TTT) proposal for a two-person version of Vostok. If Korolyov could gain the support of the Air Force, he also hoped to be able to defend his position, as the leading manned spacecraft designer, from other design bureaux in the aviation industry.

A leading competitor in this struggle for contracts was V.N. Chelomei, of OKB-52. By 1963 his design bureau had both the experience and support infrastructure to provide a real threat to Korolyov's plans. Chelomei had produced several missile and space projects that had met with some success, and his plans for anti-satellite systems and shuttle-type aerospace aircraft were gaining support in the Air Force.

Korolyov's efforts to improve Vostok and to limit the work of Chelomei were hindered by the bureaucracy of the military leadership. Authority to order new spacecraft lay with the Main Administration for Missile Armament (GURVO) of the Strategic Rocket Forces (RVSN). Korolyov's attempts to win support from the Air Force were frustrated in that only the GURVO held authority to propose TTTs for military spacecraft. In addition, Minister of Defence Malinovskiy initially refused Kamanin's 1962 request for more Vostok flights, and insisted that an 'appropriate military vehicle' should be developed before any series production of spacecraft could begin.

Despite authority to develop a number of experimental Vostok spacecraft, each manned mission required approval by the Presidium of the Central Committee, and confirmation by a decree of the Council of Ministers. Unlike the weapons programme, the manned Vostok was never intended to be put into mass production, although that is exactly what occurred with the unmanned military Zenit derivatives.

Following the Gagarin mission, the Vostoks performed two 'space spectaculars', with a twelve-month interval between them. These were used to great political effect by the Soviet regime in a clear demonstration (in their view) of the superiority of the Communist system over the decadent West. The first day-long mission, followed by the first group flight, appeared to be the beginning of Soviet capability for orbital rendezvous and docking. This created some concern in the American programme, and added a driving force to inspire them to move ahead. This then gave Korolyov cause to argue for a strengthened Soviet programme to counter these new Western advances.

American developments

By the beginning of 1963 the Mercury programme was nearing its conclusion, with a day-long flight. The next major effort would see the start of the Gemini programme in 1964. After the flight of Nikolayev and Popovich in August 1962, in October the Americans had orbited Wally Schirra onboard Sigma 7 on an 'extended' eight-orbit mission. In May 1963, Gordon Cooper (onboard Faith 7) would complete the primary objective of the Mercury programme – a 34-hour, 22-orbit mission. Follow-on plans for a three-day Mercury flight were not approved, and the Mercury programme subsequently ended after two sub-orbital and four orbital manned missions.

With the Apollo lunar programme under development, NASA wanted to divert as many resources as necessary to complete the series of Gemini flights by 1966. This would allow the Apollo Earth-orbit mission to possibly begin that same year. To support these programmes, a new group of pilot astronauts had been recruited in September 1962, and a third recruitment was planned for October 1963. Although these groups would include the first civilian pilots, selection still consisted of male jet-pilots with military backgrounds.

In a 5 December 1962 update on the Soviet space programme, a CIA intelligence report indicated that 'The underlying motives of the Soviet leaders in planning their space program are to enhance the security of the USSR. Military and political gains from space accomplishments will see the Soviets broaden the scope of their program, but attempts to accomplish spectacular 'firsts' will continue.'

The report noted a significant increase in the Soviets' unmanned space activities. They noted that some of the Cosmos satellites used an ICBM booster similar to the manned missions, and payloads of 'Vostok-type' vehicles weighing approximately 10,000 lbs were recovered from a 65°-inclination orbit. These were the first of the Zenit reconnaissance satellites, which were indeed based on the Vostok design.

The report also stated that there was no confirmation of a firm commitment to a manned lunar landing programme, but did indicate expectations that such a programme could support a manned landing two years after the initial 'recognisable' test-flights.

The CIA also believed that there would be a considerable increase in manned space activity within the next year. Such an increase could see a manned satellite with space manoeuvring capability, to perform rendezvous and docking and crew transfer operations. An increase of flight duration (with a demonstration of a ten-day life support system) was also expected, as was the possibility that the Soviets could orbit a two-man Vostok capsule at any time. 'By 1963/64, a small space station or a manned circumlunar flight could be attempted,' the report predicted.

As for military goals, the document highlighted the probability of developing space systems for military support–both offensive and defensive weapons – and that the recovered Cosmos satellite had probably accomplished photographic, electronic and nuclear reconnaissance. Finally, the report added that, 'statements by the aircraft designer Mikoyan indicate Soviet interest in a suborbital 'cosmoplane'.' This information clearly indicated a military as well as a scientific build-up of the Soviet programme that required a similar response from the United States. Under the control of the USAF, the Americans were developing their own aerospace plane, called X-20 ('Dynasoar'), and were evaluating the concept of military space platforms (which evolved into the Manned Orbiting Laboratory (MOL) Pro-gramme), as well as unmanned reconnaissance satellite programmes to replace the vulnerable U2 spy-plane missions.

A woman in space

All members of the first cosmonaut selection were male military pilots, but for some time Korolyov had thought of sending civilian cosmonauts into space, which might include some members of his own design bureau (and perhaps include himself).

On 24 October 1961, Kamanin noted in his diary the hundreds of letters from Soviet citizens (many of them women) who wrote to TsPK after Titov's flight, asking to join the cosmonaut team. Kamanin used these letters to put forward proposals for selecting female cosmonaut candidates, and suggested that training should start immediately. 'We cannot allow that the first woman in space will be American. This would be an insult to the patriotic feelings of Soviet women. The first Soviet women cosmonauts will be as big an agitator for Communism as Gagarin and Titov have turned out to be.' In fact, despite attempts by a group of thirteen American female pilots to be accepted for astronaut training in Project Mercury, NASA had no plans to select or fly anyone who was not a test pilot or jet pilot – and at that time there were no American female jet test pilots.

The Central Committee of the Communist Party gave its approval for the selection of up to sixty new cosmonauts – of which five would be women – on 30 December 1961. Although the selection of the male candidates was delayed until 1963, the selection of female candidates proceeded on time.

The idea of flying a woman in space appealed to Khruschev, who thought that such a flight would demonstrate an 'above average' level of Soviet workers, as well as the reliability and simplicity of Soviet spacecraft. The preparations for the women's Vostok flight would include an abbreviated training programme, and candidate selection would be restricted to experienced pilots or parachutists, who would be prepared for a flight in the late summer of 1962.

Kamanin had hoped to find a hundred candidates suitable for medical testing from a master list of four hundred names supplied by aviation clubs across the country. He had to settle for only fifty-eight who met the minimum requirement. In January 1962, the women completed medical evaluations, and on 28 February the credentials committee recommended five candidates (listed in the following table). Three candidates were approved on 12 March, with the other two receiving approval on 3 April.

Women cosmonaut candidates, 28 February 1962

Name	Date of birth	Age when selected
Tatyana D. Kuznetsova	1941 Jul 14	20
Valentia L. Ponomaryova	1933 Sep 10	28
Irina B. Solovyova	1937 Sep 6	24
Valentina V. Tereshkova	1937 Mar 6	24
Zhanna D. Yorkina	1939 May 6	22

Three of the five women cosmonauts during training: (*left to right*) Tereshkova, Yorkina and Solovyova.

All were experienced parachutists, and Solovyova was a world-champion member of the national team. To disguise their new role, the five were told to inform their families that they were joining a specialist parachuting team. All were civilians, but upon selection were enlisted as privates in the Soviet Air Force and began a qualification programme to fly as passengers on MiG 15 trainers. Only Ponomaryova was a qualified pilot, having learned to fly through a DOSAAF air club. Kuznetsova remains the youngest female to be selected for spaceflight training.

In October, Ponomaryova, Solovyova, Tereshkova and Yorkina were officially designated as cosmonauts, following the completion of state examinations. Due to health problems and a later marriage, Kuznetsova fell behind in her training and did not officially receive her cosmonaut designation until January 1965.

The second Air Force enrolment

The enlarged cosmonaut team was needed to provide sufficient flight crew candidates for missions in the planned (but not yet approved) follow-on missions of the Vostok-3K spacecraft (and its successor – designated Vostok 7K, then under development). With the expanded goals of these programmes, although only military officers would be considered, the age limit was raised to forty and the selection was open to engineers and navigators as well as pilots. A further change to the selection criteria was the requirement for being a graduate of a civilian university or a military academy.

Screening produced sixty-five applicants, who were invited for medical examinations. Kamanin and the credential committee examined twenty-five candidates on 8 January 1963, and nominated twenty-one for enrolment as cosmonaut candidates. However, VVS Chief of Staff, Sergei Rudenko, had only authorised TsPK to train twenty cosmonauts in 1962 and 1963, including the five women. He therefore authorised only fifteen candidates. Those who reported to TsPK on 11 January 1963 are listed in the following table:

Cosmonaut candidates, 11 January 1963

Rank	Name	Date of birth	Age when selected
Engineer Capt.	Yuri P Artyukhin	1930 Jul 22	32
Sr. Engineer Lt.	Eduard I. Buinovsky	1936 Feb 26	26
Engineer Lt.-Col.	Lev S. Demin	1926 Jan 11	37
Major	Georgi T. Dobrovolsky	1928 Jun 1	34
Major	Anatoli V. Filipchenko	1928 Feb 26	34
Major	Alexei A. Gubarev	1931 Mar 29	30
Sr. Engineer Lt.	Vladislav I. Gulyayev	1938 May 31	24
Engineer Capt.	Pyotr I. Kolodin	1930 Sep 23	32
Sr. Engineer Lt.	Eduard P. Kugno	1935 Jun 27	37
Major	Anatoli P. Kuklin	1932 Jan 3	30
Engineer Captain	Alexandr N. Matinchenko	1927 Sep 4	35
Major	Vladimir A. Shatalov	1927 Dec 8	35
Major	Lev V. Vorobyov	1931 Feb 24	31
Capt.	Anatoli F. Voronov	1930 Jun 11	32
Sr. Engineer Lt.	Vitaly M. Zholobov	1937 Jun 18	25

There were seven pilots (Dobrovolsky, Filipchenko, Gubarev, Kuklin, Matinchenko, Shatalov and Vorobyov,) with Shatalov – an Air Force inspector – being the senior. The seven engineers included a researcher at the Ministry of Defence's NII-30 (Demins selected on his thirty-seventh birthday); a member of the Zhukovsky AF Engineering Academy staff (Artyukhin); four candidates (Buinovsky, Gulyayev, Kolodin and Zholobov) who had served at RVSN bases at Tyuratam, Plesetsk and

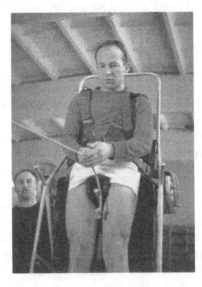

Cosmonaut Dmitri A. Zaikin undergoes medical tests.

Bearded cosmonaut V.A. Shatalov is several days into an extensive isolation test at the cosmonaut training centre.

Buinovsky – a 1963 candidate – in the centrifuge. (Courtesy Eduard Buinovsky collection.)

Candidates for the 1963 cosmonaut selection, during the summer of 1962, undergo medical examinations at the Central Air Force Hospital, Sokolniki, Moscow. Of the six, three were selected in 1963: V. Zholobov (*standing, left*), E. Buinovsky (*standing, right*) and B. Belousov (*sitting, second from right*) who was held back until October 1965. (Courtesy Eduard Buinovsky collection.)

A group of 1963 cosmonauts relaxing with some of the trainers. In the back row are Buinovsky, Zholobov, and Kolodin. In the front row, at left, is Artyukhin, with Kugno and Demin at the far right. (Courtesy Eduard Buinovsky collection.)

Kapustin Yar missile launch ranges; an aircraft engineer (Kugno); and a navigator (Voronov), who, as a bomb crew-man, had taken part in H-bomb tests. All of the group were Academy graduates and were more senior than their 1960 colleagues, which created some tensions between the groups.

The Nelyubov incident

As the training of the new cosmonauts began, an incident involving three unflown members of the first selection resulted in a serious loss to the team. Following dinner in Moscow on 27 March 1963, Anikeyev, Filatyev and Nelyubov were returning to the cosmonaut training centre. They had apparently been drinking, and became involved in a heated argument with a military patrol at a railway station. The police commander wanted to place all three on report, but Nelyubov threatened to go over his head if he did. Not wishing to cause a scandal, the police (and officials at TsPK) agreed that the report would not be filed if all three apologised. Nelyubov categorically refused. One week after the report was filed, Kamanin dismissed all three from the team, with their dismissal notices signed on 17 April.

The departure of the three men was met with deep sadness in the team, as many felt that Filatyev and Anikeyev had taken the blame for Nelyubov's behaviour. Both men subsequently returned to Air Force duties. For Nelyubov, it was one setback too many. As one of the brightest and most qualified cosmonauts, his colleagues found his departure difficult to accept. There was talk of taking him back, based on his performance at a Far East air base, but this never materialised. He experienced a psychological crisis during the next three years, watching news of other members of his selection – some less able than himself – flying into space. He was also not believed when he told his fellow Air Force pilots that he had been a cosmonaut and was back-up for the famous Yuri Gagarin. Tragically, on 16 February 1966 he was killed by a passing train on a railway bridge at Ippolitovka station in the Far East.

He was drunk at the time. The official report hinted that his 'crisis of the soul' had led him to commit suicide.

Second group flight preparations

Throughout 1962, training of the women's group followed an intensive programme, which included theoretical and practical courses to qualify them for a Vostok mission. The major difference with this group was that the five were competing for one seat on one mission. There was no guarantee of other flights, and so, unlike the men's group, personal relationships focused more on competitiveness rather than on unity. The group completed centrifuge runs (enduring up to 10 g), as well as to up to eighty parachute jumps, parabolic flights in a Tu-104, and long sessions in the isolation chamber.

Only four took their examinations in November 1962. Kuznetsova was eliminated, as she had performed less than satisfactorily in the centrifuge and pressure chamber, and was absent with poor health. At the time top of the list was Ponomaryova, who displayed better skills in theory and practical tasks than had some of the male cosmonauts. However, Kamanin was concerned about her 'unsteady morals' of independence, self-assertiveness, and even over-confidence. Her main rival for the flight was Tereshkova, who was judged to be 'a model of good breeding'.

Solovyova was shown to be morally and physically sturdy, but had a personality that cut her off from the rest of the group, and she lacked visible 'activity in social tasks'. Yorkina was the 'weakest' of the four, but was shown to be improving and was expected to eventually be a good cosmonaut. In his journal, Kamanin recorded the strengths and weakness, on 29 November, adding: 'We must send Tereshkova

Tereshkova, studying for her historic mission, burns the midnight oil.

Tereshkova is monitored inside the altitude chamber during tests.

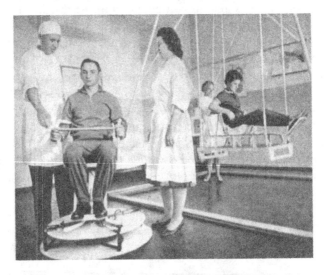

Bykovsky and Tereshkova undergo vertigo training before their dual mission.

into space first and her double will be Solovyova. Tereshkova, she is a Gagarin in a skirt.'

During their training, the flight profile for their mission was still being discussed. When they began training in March, their mission was scheduled for August 1962, but that month came and went with no decision about the date or content of the mission. By the time they qualified in November, the mission had slipped to the spring of 1963 at the earliest.

In January 1963 there were three options open to mission planners. There could be either a solo flight by a female cosmonaut lasting one to three days, or a female group flight with a one-day launch interval, but both landing on the same day (as had been the case with Vostok 3 and Vostok 4). The third option was a mixed flight of two Vostoks, with a man launching first for a 5–7-day mission, then the woman joining him in orbit for two days. Kamanin favoured the second option, as it was easier to prepare, and he had undergone a similar mission the year before.

Two training groups were formed for this mission. The four women (Ponomaryova, Solovyova, Tereshkova and Yorkina) would train for a female group flight, and three men (Bykovsky, Komarov and Volynov) would train for two or three long-duration solo flights of between 5–7 days later in 1963.

There was also the issue of hardware for the mission, as the design lifetime of spacecraft 3KA #7 (Vostok 5) and 3KA #8 (Vostok 6) would expire on 15 June 1963. In recognition of this, the Military Industrial Commission (VPK) issued a different plan. This would see the launch of 3KA #7 with one woman on board, and would keep the other spacecraft as a back-up if required. Should 'Vostok 5' prove successful, then capsule #8 would be relegated to the museum and so end the manned Vostok programme.

However, the 21 March meeting of the Central Committee opted for a mixed flight and recommended launch in August. Three weeks later, on 13 April, Kamanin and Korolyov agreed on a plan to launch the man on an eight-day mission onboard 'Vostok 5', and the woman for a 2–3-day flight on 'Vostok 6'. As the 'shelf life' of onboard equipment could not be extended into August, the flights were approved, at a meeting of the Central Committee on 29 April, for May–June 1963.

Also in April, the male training group for the solo missions was reassigned to train for a dedicated joint flight with the women; but in May, Komarov was medically disqualified and left the training group, to be replaced by Leonov and Khrunov.

Cosmonaut Irina Solovyova, back-up to Tereshkova, in a Vostok spacesuit.

The three women cosmonauts of the Vostok 6 training group at the pad: (*left to right*) Ponomareva, Solovyeva (back-up) and Tereshkova (prime).

Finally, on 10 May the crew assignments were announced. Bykovsky would fly Vostok 5, with Volynov as his back-up and Leonov in the support role. The installation of extra equipment onboard Vostok 5 required the extended flight to push the R-7's maximum payload capacity almost to the limit when carrying a Vostok spacecraft. As Bykovsky was 30 lbs lighter than Volynov, his selection may have been in part due to his lesser weight. Tereshkova would take the coveted Vostok 6 seat, with both Solovyova and Ponomaryova as her back-ups.

By this time Kamanin was becoming even more frustrated with the state of the planning of future missions. 'Because of the squabbling between various departments we make very poor use of our technical capabilities, hastily prepare flight programmes and a whole lot of other stupid things. Preparation should begin by giving the crew a flight programme, but we are doing the exact opposite. We first prepare the ships and their equipment, and then tailor the crew's programme to the configuration and equipment.'

The launch of Vostok 5
Bykovsky's launch was scheduled for 7 June, but due to a technical problem was delayed to 11 June before rollout. On 10 June the launch was remanifested to 14 June due to concerns over high solar activity. On 14 June Bykovsky was finally strapped into the spacecraft.

After a perceived problem with the ultrashortwave radio transmitters (which eventually proved groundless), the hatch was reopened and the launch delayed by thirty minutes due to a problem with the arming device on the cosmonaut's ejection seat. Then, close to the final minutes from launch, signals failed to indicate that the Block Ye upper stage was ready for launch. A hold was called for two to three hours while the problem was evaluated. The fault was found to be in the gyroscope device, and after a couple of hours the unit was replaced on the pad with a new one. A postponement would have meant emptying the rocket of fuel and sending it back to

Cosmonaut Valeri F. Bykovsky remains the holder of the solo spaceflight endurance record (119 hours) set on Vostok 5 in 1963.

the factory, delaying the mission until August – beyond the 'shelf life' of most of the hardware. A decision was reached that the Blok Ye was safe to fly.

Almost at the end of the day's launch window, and after the cosmonaut had lain in the capsule for six hours, Vostok 5 left the launch pad at 17.00 hrs local time. (15.00 hrs MT). At the last minute in the countdown, a cable connecting ground power sources failed to separate as planned. With just seconds to go, the decision to launch was given, and as the rocket left the pad the cable was simply torn from its socket.

Vostok 5's flight objectives were stated to include continued studies of the 'influence of different factors of spaceflight on the human organism'; to carry out extensive medico-biological research into conditions of a long flight (suggesting that it would probably surpass Nikolayev's record); and adjustments and improvements to the piloting system.

Vostok 5 was placed in an initial orbit of 112.4 × 146.0 miles, inclined at 64°.69, and with an orbital period of 88.4 minutes. Bykovsky's callsign was 'Yastreb' ('Hawk')

Bykovsky later commented on the experience of his first space launch, in enduring considerable acceleration forces. 'It was difficult, of course. I would not call it good or excellent. 'Satisfactory' is good enough. The space ship was placed into an orbit close to the one envisaged.'

The orbit was, in reality, slightly lower than that planned, due to a lower performance from the upper stage of the R-7. As a result, instead of a ten-day orbital decay ensuring an eight-day mission, the orbit would decay in just eight days, adjusting the planned mission to about five or six days.

Bykovsky realised he was becoming weightless as, 'the muscles of my arms felt unusually light. It was no effort at all to lift my arms, and if I picked up an object it felt as if it weighed nothing. One gets used to it of course, but not immediately.'

Five days of work

For the first two days, the cosmonaut submitted the now standard reports on the state of his spacecraft and his adaptation to weightlessness, and exchanged messages with Khruschev before beginning his work programme.

On all the previous Vostok missions there had not been time to include many observations or scientific investigations. For this flight, however, there were plans to include Earth observation and some simple experiments. M. Keldysh, the President of the Soviet Academy of Sciences, had proposed a formal plan on 17 May 1963. It included studies into the light intensity of the Earth's atmosphere at the horizon, the transparency of the atmosphere, using photography and subsequent photometry observations, and the structure of clouds. Some of these experiments were to be conducted by Bykovsky from Vostok 5, although very early in his flight he encountered difficulties in completing his tasks. When using the film camera to record the horizon data, he had one film cassette stuck in the camera and a second cassette without a roll of film inside it! When he tried to observe the Earth and the Sun's corona using special light filters, he was unable to see the corona, and in an experiment to record the change of his vision in orbit he found the equipment difficult to use.

One of the scientific experiments investigated the growth of peas outside the influence of Earth's gravity, while a second recorded the behaviour of liquids – the latter following on from experiments conducted by Popovich on Vostok 4. Accompanying the cosmonaut in orbit was a host of biological specimens, including cancer, fibroblast and amnion cells, frog ova and sperm, *drosophilia* insects, several plants, air-dried seeds, clorella algae and bacteria samples.

Added to Bykovsky's frustration with the experiments was the discovery of a pressure reduction – to just 10 atmospheres – in the nitrogen storage bottle for the gas manoeuvring system. To ensure the vehicle was capable of manual orientation for entry if required, a minimum of 5 atmospheres of pressure was a mission rule. To save what supplies he had left, Vostok 5 was placed in a slow roll, at a rate of one revolution every eight minutes.

'I am Seagull'

Only after Vostok 5 had entered orbit was the launch date for Vostok 6 chosen. The original plan had been to launch Vostok 6 five days after Vostok 5, and then to have both cosmonauts return on the same day.

The night before the launch, Tereshkova slept in the same bunk as had Gagarin had. Solovyova offered it to her, as she 'would also be a first'. On the morning of the launch the two women were woken at 7.00 am and completed thirty minutes of exercise before breakfast. Prior to suiting up, the doctors performed their final pre-flight medical check, which included administering a cleansing enema. After suiting up, and the short bus ride to the pad (accompanied by Gagarin, who would be the

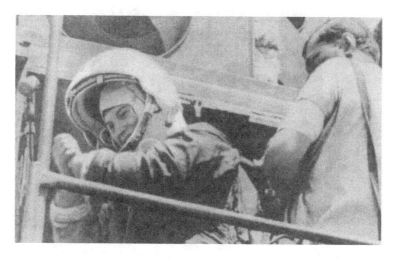

Tereshkova gives a confident clasped hand salute as she prepares to enter her spacecraft.

launch CapCom), they were greeted by Korolyov and other members of the State Commission. Korolyov expressed his disappointment that he could not be with Tereshkova in the spacecraft, but she told him that one day they might fly together to Mars.

'On 16 June 1963, at 12.30 p.m. Moscow Time, a spaceship, Vostok 6, was launched into orbit... piloted, for the first time in history, by a woman, citizen of the Soviet Union, Communist Comrade Valentina Vladimirovna Tereshkova,' the TASS statement announced. Tereshkova's callsign was 'Chaika' ('Seagull').

The new spacecraft was placed in an orbit of 113.7 × 144.7 miles, 64°.95, 88.4 minutes. The flight objectives were very similar to that of Vostok 5, apart from 'a comparative analysis of the influence of factors on the organism of a man and a woman'. Bykovsky had been in flying for about 45.5 hours at the time of the launch of Vostok 6.

'It is I, Seagull! Everything is fine. I see the horizon; it's a sky blue with a dark strip. How beautiful the Earth is!', Tereshkova called from orbit.

Valentina Tereshkova was the first woman in space, and her name was soon to become as famous as that of Yuri Gagarin. Unable to tell her family of her real role before the launch, she was anxious that her mother would not be worried during her 'special parachute training'. She wrote regular letters, but was unable to disclose the nature of her 'secret mission' even during visits home. The letters told how she was well and very busy, but some were delayed in the post. One arrived on the day of the flight, and when reports of her spaceflight appeared on television, her mother said that it could not be Valentina, as she had just sent her a letter explaining she was about to complete her parachute training and would be home soon. When Tereshkova's mother realised it was indeed her daughter in orbit she was very upset that she had been deceived, and apparently took a long time to forgive her.

Vostok 6's orbital plane differed to that of Vostok 5 by about 30°, which meant that the two spacecraft approached each other only for a few minutes, twice each

Valentina Tereshkova becomes the first woman in space, onboard Vostok 6, in June 1963.

orbit. The closest approach was only three miles, shortly after Vostok 6 entered orbit, and although both cosmonauts had seen the separated upper stage of their respective launch vehicles, neither reported a clear visual sighting of the other's spacecraft, although Tereshkova thought she might have glimpsed Vostok 5.

With Vostok 6 safely in orbit, the two cosmonauts established communications with each other and received further greetings from Khruschev. Moscow TV added live pictures of Tereshkova from space to those of Bykovsky that it had been showing for two days.

With a woman in space, the Soviet news machine took full advantage, proclaiming the event around the world as the latest development in the superior socialist system. Western coverage of the flight was divided between those who looked upon the flight as nothing more than a propaganda stunt, and those who marvelled at the scope and daring of the Soviet space programme. In any event, the news of a woman flying in space generated the international headlines and attention that two years earlier had been predicted by Kamanin.

Observations and experiments

Onboard Vostok 5, Bykovsky continued his experiments and observations while maintaining radio links with Vostok 6. During his observations of Earth he commented on the ease of distinguishing features on the surface, and how the different colours of the water surfaces were prominent. As he emerged from darkness during each orbit, he noted that, 'the curvature of the horizon exhibits a beautiful range of colours, with reddish hues predominating.'

Throughout the flight he continued to pay close attention to his physical conditioning, and supplemented the stretching exercises performed on earlier flights with a rubber strap for power exercises. He often demonstrated the effects of weightlessness to the television cameras, and described his eagerness to unstrap himself and float free in the cabin, which he did during his eighteenth, thirty-fourth, fiftieth and sixty-fourth orbits. He remained floating in the cabin for ninety minutes during one session – a complete orbit around the Earth 'I floated up to the portholes... and carried out observations of the ground. It is extremely peculiar. The slightest push sends you flying in the opposite direction, and with your eyes closed you cannot tell what your position actually is.'

Onboard Vostok 6, Tereshkova took film of the terrain and clouds on Earth, but found the camera film very difficult to remove. She was unable to reach the biological experiments, which prevented her from activating them, but by using light filters, was able to observe the horizon over the poles, and saw the Moon several times. She tried to identify the constellations, but had difficulty, and was unable to observe the solar corona.

From the photographs of the Earth's horizon obtained by the two Vostok vehicles, structures of the stratospheric layers were determined, including the recording of two aerosol layers at 7.1 and 12.11 miles (± 1 mile). Data on the structure of the atmosphere was compared with data recorded from the ground, from aircraft, and from unmanned balloons.

Space adaptation

New code-phrases had been adopted for this mission, so that there would be no misunderstanding about the condition of the cosmonauts when they were actually looking at thunderstorms as had happened on Vostok 4. A report of 'feeling excellent' signified there were no problems, allowing the mission to continue. 'Feeling well' really meant that the cosmonaut was having doubts about completing the assigned mission objectives. 'Feeling satisfactory' was a code for immediate termination of the mission. Reports of Tereshkova's 'satisfactory' condition throughout the flight gave rise to speculation and confusion about her actual condition.

Tereshkova was apparently not well during the first orbits as she adjusted to the new environment, but in the initial post-flight reports she stated: 'I stood up well to the state of weightlessness and quickly adjusted myself to it.' During later post-flight briefings, she reported vomiting only once – early on the third day – and this was caused by something she had eaten rather than the result of her adaptation to weightlessness.

Over the next few years, the uncertain state of her health during the flight generated much speculation, ranging from her suffering from a bad menstrual cycle in orbit, to being extremely sick after landing. During an interview more than thirty years later, Tereshkova revealed that, 'physicians checked on this aspect [menstruation]. I had been launched in the middle of my cycle, when menstrual bleeding would not start. After the landing, this aspect of a woman's physiology [also] came under medical observation.' Television pictures of her from 18 June 1963 had shown her tired in the face and looking a little weak.

The previous day, the decision was made to return Bykovsky on the fifth day instead of the sixth, because of his reduced orbit. Tereshkova would stay up for three days to return with Bykovsky, – barring any emergency. During a scheduled attempt at manual orientation of her spacecraft, Tereshkova attempted to use the gas-attitude system, but for reasons that are still unclear she was not able to do so. Ground controllers expressed disappointment that one of her major objectives had not been accomplished, and were concerned that she might not be able to manually orientate the Vostok for entry if the automatic systems failed.

After Gagarin, Titov and Nikolayev worked on instructions to help her overcome the problem, Tereshkova successfully orientated the spacecraft by manual control in a twenty-minute experiment on 19 June. This included holding the correct entry attitude for a full fifteen minutes. This reassured ground controllers about her performance.

At times, communications were difficult, either because something was wrong with her receiver or because she had selected the wrong channel. Western monitoring of the transmissions picked up calls from ground control to Vostok 5, trying to relay messages to Vostok 6: 'Yastreb, Yastreb – contact cosmonaut Tereshkova. We cannot contact her on the ground.' Bykovsky replied: 'I have tried myself to contact her without success, but I don't think there is reason to worry.' The problems seemed not to be too serious, as Kamanin recorded in his diary that Bykovsky later reported that communications with Vostok 6 were excellent, and that Tereshkova was singing songs for him! One explanation for the drop in communications was that she had fallen asleep!

Habitation

Bykovsky's flight onboard Vostok 5 presented the Soviets with the opportunity to evaluate the habitability of the spacecraft, and to obtain important data that could help in the design of succeeding spacecraft.

Bykovsky reported that instruments on the Vostok control panel were accessible, but were difficult to read. He also noted that when strapped into the seat it was difficult to reach the food rations and impossible to reach the medicine cabinet without unstrapping the harness.

Both cosmonauts reported that wearing the suit for such a long time was uncomfortable. Bykovsky reported that the suit helmet was weighing him down after four or five days, while Tereshkova found that by the second day she had a nagging

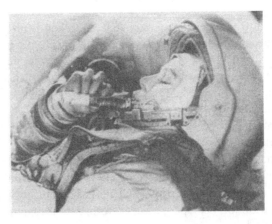

Tereshkova takes refreshment during training for her mission.

pain in her right knee, which became worse by the third day. Her helmet pressed against her shoulders and her left ear, and the biomedical sensors attached to the headband gave her itches and headaches, although the band itself was not a problem. In future, they suggested, it would be more satisfactory if the suits could be taken off – at least for part of the flight.

The cosmonauts used moist flannels to freshen their faces, and Tereshkova noted that the toilet wipes were too small and were not sufficiently moistened. They also requested that in future there should be something with which to clean their teeth.

Food supplies for the cosmonauts were similar to those on the first group flight, but more perishable food was also supplied, to be eaten for the first few days. For breakfast they had a caviar sandwich, coffee with milk (in a tube), fresh lemon sections, and a multi vitamin pill. Lunch was selected from roast beef chunks, fish, meatloaf, wheat bread, fish chunks, apple sections and fruit juice. Dinner included roast tongue chunks, bread, juice and a second multivitamin pill. Supper featured chicken fillets, filled pastry (Pirozhki) with rice and eggs or fruit fillings, and prune purée (in tubes).

Bykovsky's daily calorie content was 2,526 kcal, although the energy ration was changed so that on the first day he consumed 1,670 kcal, and on the final day he had 2,500 kcal. Tereshkova's daily rations amounted to 2,529 kcal.

Bykovsky's appetite remained good, but Tereshkova reported that she did not consume about 40% of her food. Unfortunately, it was not possible to confirm this post-flight, as she gave some away to those who greeted her after landing. This led doctors to suspect that she did not accurately report her condition in orbit. She certainly did not like the sweet dishes, which caused her to vomit on the third day, and added: 'The bread was dry, and so I didn't eat it. The juice and cutlet were pleasing. Towards the end of the flight I began to want some black bread, potatoes and onions.'

Bykovsky reported one amusing incident that happened on 18 June, when he reported a personal space achievement – the motion of his bowels in orbit! Due to radio interference, what should have been received as 'Stul' (motion of the bowels) was received as 'Stuk' (knocking). On the ground this caused 'lively discussion' concerning the cause of the knocking, where it had come from, whether anything struck the spacecraft, and whether it was a buzzing or a banging noise. Bykovsky explained that he was merely reporting his use of the sanitary appliance, and in reply he heard loud laughter from ground control.

The landing phase

On 19 June, both cosmonauts prepared to return to Earth. Vostok 6 re-entered first and caused some concern at ground control, as Tereshkova failed to indicate the correct working of the solar orientation system. She also remained silent throughout the re-entry phase, and did not report retro-fire or the separation of the capsule. She ejected at an altitude of four miles and, suspended from her parachute, and against regulations, she looked up, and was struck by a falling piece of metal, which cut her face, as she had opened her face-plate.

Bykovsky begins post-flight medical tests shortly after his five-day mission.

Bykovsky and Tereshkova being welcomed after their historic missions.

After landing she was offered fermented milk, cheese, flat cakes and bread, as locals celebrated the latest Soviet heroine. This was probably very welcome, but also ruined any post-flight analysis of her diet and food intake during the mission. Tereshkova had landed during her forty-ninth orbit, at 11.20 MT, 385 miles northeast of the city of Karaganda, in Kazakhstan. Her flight had lasted 2 days 22 hrs 50 min – just seven minutes less than Popovich in Vostok 4.

Bykovsky's entry was more eventful, as once again the sharik failed to separate cleanly from the instrument compartment – although it did not appear to trouble him. He landed without further incident at 14.06 pm on 19 June, 335 miles northwest of Karaganda, two orbits (three hours) after Tereshkova, and nearly 500 miles away. Both spacecraft landed on the same latitude of 51°. Bykovsky had flown for a record 4 days 23 hs 6 min, surpassing the flight of Nikolayev by almost a day, and that of Cooper (holder of the US endurance record) by more than three days. After flying 2,066,749.6 miles, Bykovsky set a new solo spaceflight endurance record that still stands thirty-eight years later, and seems unlikely to be broken.

Post-flight

The medical team, led by Yazdovsky, indicated that Tereshkova had performed poorly during the mission. Tereshkova countered this with claims of feeling well during the flight, although she had suffered from fatigue and a lack of sleep. Yazdovsky's report mentioned that she had felt ill on the thirty-second and forty-second orbits, had a weak appetite, had vomited, and had recorded 'weak cardiac activity'.

These reports reached Korolyov, who invited Tereshkova to a personal interview to discuss the details of her flight. The details of this meeting – which took place on 11 July – are not known, but her performance had apparently disappointed Korolyov. OKB-1's first deputy, Mishin, recalled that 'Tereshkova turned out to be at the edge of psychological stability.' Although neither Titov nor Tereshkova flew in space again, Tereshkova seemed to be singled out, possibly because she was a woman in a male-dominated system. Her performance was judged against the standards of the male cosmonauts, and although her flight was not the expected outstanding success, neither was it a complete failure. Tereshkova had demonstrated that a properly trained female could operate in space alongside the men, although it was to

Tereshkova receives the adulation of a flag-waving crowd, in a typical scene of cosmonaut post-flight celebrations in the 1960s.

Tereshkova and Nikolayev with their daughter Alyona.

be another nineteen years before a second woman would be presented with the chance to prove that it could be done again.

The media attention caused by Tereshkova's flight was extended by the high-profile 'celebrity' marriage of Tereshkova to Nikolayev on 3 November 1963. The event appeared to be almost an 'arranged' marriage. Nikolayev, 'the most eligible bachelor', had indeed been taken with the young trainee cosmonaut, and helped her during preparations for the flight, but the relationship did not appear to be leading to marriage. On 8 June 1964, Tereshkova gave birth to a healthy 6-lb 13-oz baby girl, Alyona. The medical team closely monitored this event, as Tereshkova was the first woman to give birth after a spaceflight.

After Vostok 5 and Vostok 6, there were no officially approved Vostok missions on the manifest, and Korolyov was more interested in developing the advanced spacecraft and boosters that he had been working on for some years. Although no more flights were to be made under the Vostok programme, there were several that had been planned to follow Tereshkova.

Cancelled Vostok missions

Korolyov was an obsessive worker, and was constantly putting in 18-hour days for several weeks without a day off. He had followed this regime for several years, and by late 1962 the effects were beginning to catch up with him. After the first Vostok group flight in 1962, he was hospitalised, with pain and internal bleeding, until 15 September. But during an enforced vacation at the holiday resort of Sochi he continued to work, spending hours on the telephone or with visitors, and making notes on further flights in the Vostok programme.

Initially, there were only eight Vostok—3KA (manned) vehicles approved, but during 1961 and 1962, both OKB-1 and the Air Force sought approval for a further ten spacecraft, and formulated plans for between ten and twelve manned Vostok missions. These plans received almost indifferent reception, and were once again opposed by Defence Minister Malinovskiy, who was not really interested in civilian space operations. Finally, in the spring of 1963, the Central Committee agreed to authorise four extra Vostok spacecraft to support Flight Operations in 1964.

An unofficial flight schedule was formed as late as 1963. It included a ten-day mission to a 'high altitude' of 373 miles, carrying one dog and a variety of other animals. There was also a solo manned flight of up to eight days, and a new, two-spacecraft group flight of up to ten days. The altitudes of the manned flights would be determined by the results of the canine flight, with both military and scientific experiments planned for the missions. These four missions were to be followed by the first flight of a new spacecraft (7-K) towards the end of 1964. On 17 September 1963 a new cosmonaut training group for these missions was formed.The group included Belyayev, Komarov, Shonin, Khrunov, Zaikin, Gorbatko, Volynov and Leonov, but no actual crews were formed for several months.

Other planned but unrealised Vostok missions have been identified:

Vostok 7 (April 1964) A high-altitude flight to the lower Van Allen radiation belts, for radiological and biological studies. Natural orbital decay after ten days would be used for recovery of the crew. An alternative Vostok 7 mission has been suggested to fill the gap between Vostok 6 in 1963 and the first Soyuz in 1966. This plan would have seen Boris Yegorov (a young medical doctor from Moscow) fly on a biomedical seven-day mission.

Vostok 8 (June 1964) The same mission profile as Vostok 7.

Vostok 9 (August 1964) and Vostok 10 (April 1965) Two high-altitude missions designed for extended scientific studies for up to ten days before the natural decay of the orbit. Proposed research areas included geophysical and astronomical research, photography of the solar corona, solar X-ray imaging, continued medical and biological research, studies of the effects of long-duration weightlessness on the human organism, dosimeters, and engineering tests of the ion-flow orientation sensors planned for the Soyuz spacecraft.

Vostok 11 (June 1965) and Vostok 12 (August 1965) EVA (spacewalking) tests. This would require the removal of the ejection seat and support equipment, attachment of an airlock, and installation of a braking rocket for soft landing, with a single cosmonaut onboard.

Vostok 13 (April 1966) A further high-altitude mission for extended scientific studies up to ten days, similar to the flights of Vostok 9 and Vostok 10.

It is probable that these missions existed only on paper, as no training or assignments of crew occurred.

Vostok design studies 5K, 7K and Sever

In the autumn of 1956 – a full twelve months before the launch of Sputnik – Korolyov proposed a long-range plan that envisaged the creation of a 'satellite space station', proximity operations, and the rendezvous and docking of two or more spacecraft.

Two years later, on 5 July 1958, Korolyov and Tikhonravov signed a plan for future space activity, which included heavier manned spacecraft that could not be launched by the R-7. These plans also included manned circumlunar missions and space stations. The launch vehicle required by these plans existed only on paper, but in the document Korolyov also proposed the techniques of Earth Orbital Rendezvous (EOR) first proposed in the studies of Tsiolkovsky earlier in the century. By using a fleet of R-7s, elements of these larger spacecraft could be joined together in Earth orbit, and could then be sent to the Moon or expanded into a large space station. In 1961–62, when deciding how to send Apollo to the Moon, the Americans also evaluated this technique, but opted instead for Lunar Orbital Rendezvous (LOR).

From 1959, Tikhonravov headed OKB-1 Project Department #9, which was tasked with solving the problems of space rendezvous. A special Space Assembly team (headed by K.S. Shustin) handled the problem of the joining of spacecraft (docking) in orbit. From 1960, the group began working on the problems in some detail, by designing and evaluating several types of hardware to achieve the rendezvous and docking goals.

During the early 1950s, Gleb Yu Maksimov had worked on satellite studies with Tikhonravov, and was now OKB-1's leading engineer working on an automated lunar and interplanetary spacecraft. From 1959 he conducted design studies for a 75-ton manned interplanetary spacecraft (launched by the proposed N1) that would be crewed by three cosmonauts launched in a modified Vostok capsule. This would lead to a manned fly-by of Mars, using the gravitational forces of the Red Planet and of Earth. This work also attracted the attention of several Vostok designers, including future cosmonaut Feoktistov, whose own plans pushed this proposal further to include a manned Mars landing. Designs of the Heavy Interplanetary Ship (TMK) continued into the mid-1960s, until Soviet attention was finally (and too late) turned towards reaching the Moon before the Americans could do so.

Work was also being conducted on a Heavy Orbital Station (TOS), which resembled the habitation elements of the Mars vehicle. The possibility of using the station to test elements of the Mars craft was not lost on Korolyov. 'Maybe the Heavy Interplanetary Station will be the Heavy Orbital Station during the first phase,' he wrote, 'contributing to the reliable debugging of all systems in the vicinity of Earth.'

From April 1959, Department 9 of OKB-1 continued to work on two types of mission, to follow Vostok, that would help develop techniques needed to construct larger space structures and reach more distant targets. *Sever (North)* was a multi-unit manned military space station, while the multi-stage Space Tug was capable of taking payloads from low Earth orbit (LEO) to high orbits or escape velocities. The spacecraft assigned to achieve this plan included:

Vostok-5K (Vostok 5) A two-man spacecraft weighing 5.6 tons, for delivering crews to the Sever military space station.

Vostok-5KA (Habitation) and *5KB (Instrument)* Modules for Sever, each weighing between 4.5 and 5.6 tons.

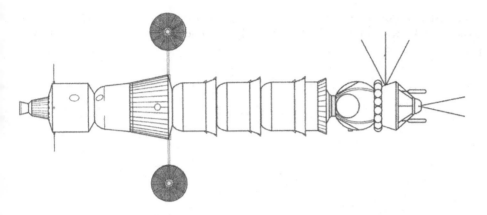

Vostok-7 (J)/Sever configuration, with the Vostok capsule on the right – the genesis of the design that evolved into Soyuz. (© 1998 Ralph Gibbons.)

Vostok-6K A second-generation Zenit military satellite.

Vostok-7K (Vostok 7, or Vostok J) A 5.8-ton manned spacecraft, with a crew consisting of an 'Assistant Spacecraft Assembler' and a 'Space Docker'. They would perform the final phases of rendezvous and docking of the Sever module and Space Tug units, as the evisaged system included only a half-automated method of rendezvous that needed a crew to complete the final docking.

Vostok-8K An upgraded second-generation Zenit spy-satellite.

Vostok 9K A 5.6-ton rocket unit for use in a multiple Space Tug. This tug could include two or more 9K units, depending upon its mission. It would be used to send the payload into geostationary Earth orbit (GEO) or on a lunar trajectory, following completion of the docking operations and separation of the 7K spacecraft.

Proposed payloads for 9K units included a 5.8-ton two-man circumlunar spacecraft (1L, or KL); a manned military spacecraft (5KM) weighing 5.6 ton, with orbital manoeuvring capability for inspector or interceptor missions, or a geosat communications satellite of 1.1–1.2 tons.

By 26 January 1962 these plans had evolved into the design of a four-module, 15–25-ton 'space train' for circumlunar spaceflight. This emerged as a completely new spacecraft that was summarised in a scientific/technical prospectus, signed by Korolyov on 10 March 1962, in wich the project was redesignated Soyuz (Union) for the first time. On 16 April 1962, Korolyov's plans for Soyuz were accepted by a Government Decree, which, as far as manned operations were concerned, also signalled the beginning of the end for the Vostok series.

There was, however, a considerable period between the end of the planned Vostok missions in 1963 and the beginning of the Soyuz programme in 1966, during which time the Americans would fly the versatile Gemini series. To help fill this gap, and to offer a chance to record further Soviet achievements over the

Americans, Korolyov assigned a team of fifty engineers the task of modifying the basic Vostok 3KA manned spacecraft, to fly 'extended' Vostok missions during the period 1964–1966.

First crew and first EVA

The year 1963 marked a watershed in several areas of space research. It saw the final flight of the American one-man Mercury spacecraft, and (although it was not known at the time) the end of the final single-seat Vostok missions. It was also the year that Korolyov appeared to realise that further flights of the original design, or even of the upgraded 'extended' version was delaying the far more advanced Soyuz programme. At the same time, Chelomei was making rapid advances with his alternative designs for military manned spacecraft.

During the flight of Vostok 5 and Vostok 6, Korolyov stated that he wished he could hand over the Vostok to the military, as this would allow him to concentrate on the development of Soyuz and his lunar exploration plans. There were plans for several more Vostok missions, but other events were beginning to supersede these ideas.

PROGRESS ON A LONG JOURNEY

On the political front, events were also changing. In October 1962, President Kennedy had called thousands of Reserve and National Guard troops to active duty in what became known as the Cuban Missile Crisis. His action was a response to Khruschev as deployment of intermediate-range missiles on Cuban soil a few hundred miles off the coast of Florida, a show of propaganda on the back of the success of the space programme, and a further demonstration of Soviet rocket power.

On 22 October, Kennedy addressed the American people about the crisis, and indicated evidence of the preparation of missile sites with no other purpose, 'than to provide a nuclear strike capability against the Western Hemisphere.' He demanded the withdrawal of Soviet forces from Cuba, and two days later – with the Bay of Pigs fiasco fresh in the mind – authorised the blocking of sea lanes around the Caribbean island. An uneasy stalemate was reached before Khruschev very publicly removed his missiles from Cuba and Kennedy quietly removed American missiles – which were aimed at Moscow – from Turkey.

Against this background there were arguments both for and against increased funding and further development of the American space programme, despite the technological advantages that space 'spin-offs' would bring every American citizen. On 17 May 1963, former President Eisenhower hailed the recent twenty-two-orbit flight of Gordon Cooper as 'A great step forward in the United States space programme.' That same day, an editorial in *Life* suggested: 'Now the United States can be foremost in another and greater adventure [than was Columbus], or abdicate its own greatness by not doing enough. The issue is much bigger than a Moon race. It makes sense for military, technology and prestige reasons.' The following day, during a visit to Alabama, President Kennedy addressed workers at the Redstone missile arsenal, explaining: 'I know there are lots of people now who say 'Why go any further in space?' I believe the United States of America is committed in this decade to be first in space, and the only way we are going to be first in space is to work hard.'

In the days following the flight of Tereshkova in June 1963, the Chairman of the US House Committee on Science and Astronautics, George P. Miller, stated: 'It does not surprise me that the Russians want to pull another spectacular by putting a woman in space. It shouldn't interfere with our programme. I don't want to downgrade their achievement, but it doesn't mean we have to follow suit.' Dr Eugene Konecci, NASA's Director of Biotechnology, pointed out that Tereshkova was not a trained pilot, and stressed the importance that the Soviets attached to the gathering of biomedical data about the effects of spaceflight: '[It] may be a preparation for flying scientists who are not trained pilots.'

Clare Boothe Luce, writing in an article in *Life*, continued to promote the magazine's image of the All-American-Boy character, proclaiming: 'The astronaut of today is the world's most prestigious idol. Once launched into space he holds in his hands something far more costly and precious than the millions of dollars-worth of equipment in his capsule. He holds the prestige and the honour of his country. He is the symbol of the way of life of his nation.' The same was true of the Soviet cosmonauts.

Partners in space?

By the summer of 1963, questions were being asked of the Kennedy administration about rumours of a collaboration between the US and USSR on manned lunar landing efforts. Recently returned from a tour of Soviet observatories, Sir Bernard Lovell, Director of Jodrell Bank radio observatory, said at a press conference on 16 July that he considered that 'The Americans are racing themselves (to the Moon).' The next day, in a Washington press conference, President Kennedy was asked about such a cooperation. 'We have said before we would be very interested in co-operation,' but he then added that the most that had been agreed, after weeks of discussion, was an exchange of weather data. The President recognised the hurdles against such a cooperative effort: '[It] would require a breaking down of a good many barriers of suspicion and distrust and hostility which exist between the Communist world and ourselves.'

On 20 September, Kennedy listed the results of US–USSR negotiations, in his address to the UN General Assembly, stating that 'there is room for cooperation.' Unfortunately, efforts to join the two programmes were not always well received by the American or Soviet press. An article in the Soviet magazine *Za Rubezhom* quoted Kennedy's offer as 'American propaganda'. In the first official response to the offer, on 3 October, the news agency Tass stated: 'Maybe it's too early to discuss now the question of what is better – to combine a Soviet rocket with an American spaceship, or to include an American in a Soviet space crew. But first steps in cooperation cannot but give satisfaction.'

On 19 October, Kennedy reflected on the 1962 Cuban crisis: 'A year ago it would have been easy to assume that all-out war was inevitable, that any agreement with the Soviets was impossible, and that an unlimited arms race was unavoidable. Today, it is equally easy for some to assume that the Cold War is over. The fact of the matter is, of course, [that] neither view is correct. We have made slight progress on a long journey.' But it took forty years of many more steps on that long journey before cooperation evolved into the reality of the International Space Station.

On 26 October 1963, progress came to an abrupt halt with comments by Premier Khruschev in an interview published in *Izvestia*: 'At the present time we do not plan flights of cosmonauts to the Moon. I have read a report that the Americans wish to land a man on the Moon by 1970. Well, let's wish them success.' This statement also led to a louder call for a reduction in the American space budget.

On 29 October, Vice President Johnson emphasised the US commitment to the peaceful exploration of space: 'Those who say that our purposes and [those] of communism in space exploration are the same, misread and misunderstand the history and meaning of our times. In 1957 [with the launching of Sputnik], the communist rulers of Russia refused to consider sharing the fruits of space research with other nations, refused to consider committing themselves to developing space for peaceful purposes. In the same year, the United States clearly stated our own national [peaceful] policy and purpose. We will not retreat from our national purpose.'

On 22 November 1963, President Kennedy visited in Dallas but was not to deliver his speech as intended. An assassin's bullet struck him down and within two hours Vice President Lyndon B. Johnson had become the 36th President of the United States. Kennedy's undelivered speech included the comments: 'The success of our leadership is dependent upon respect for our mission in the world, as well as our missiles. That is why we have regained the initiative in the exploration of outer space – making an annual effort greater than the combined total of all space activities undertaken during the 1950s, and making it clear to all that the United States of America has no intention of finishing second in space. This effort is expensive, but it pays its own way, for freedom and for America. For there is no longer any fear in the free world that a Communist lead in space will become a permanent assertion of supremacy and the basis of military superiority.'

Successor to Vostok

Just over a month after the flights of Vostok 5 and Vostok 6, during a large meeting of Soviet space leadership on 26 July 1963, discussions were held concerning what missions any future Vostok spacecraft might be able to accomplish. Four missions were identified that encompassed extended-duration missions, and these were designated Vostok 7 to Vostok 10.

At that time, Soyuz was expected to begin flying in early 1964, but in November 1963 Korolyov indicated that he had no funds to continue its development. Government support for the new programme finally came in December 1963, with the ultimate goal of piloted circumlunar spaceflights. Though the Decree indicated that the first flight would take place before the end of 1964, Korolyov knew (while publicly supporting this) that Soyuz would not be ready to fly until 1965 at the earliest.

News from the United States indicated that NASAs Gemini programme would begin two years of flight operations in 1964 (although it slipped to 1965). These would include ten two-man flights of up to fourteen days, rendezvous and docking, and activities outside the spacecraft in what NASA termed Extravehicular Activity – more commonly known as space-walking, or EVA.

Even with the four planned (but not yet authorised) Vostok 'extended missions', it was impossible for the programme to compete with the plans for Gemini. How the Soviets responded to this remains unclear, but in February, Korolyov received Party orders, stating that work on the single-seat Vostok would be suspended, and that he should prepare to convert one of them into a three-seat spacecraft to fly in 1964, before Gemini or Apollo could begin to overtake Soviet accomplishments.

Two theories of the origin of the authority for a three-person flight have been put forward. William P. Barry's 1996 thesis into the evolution of Soviet missile design bureaux and manned space policy of the 1960s indicated that the test flight of an early Saturn rocket in January 1964 presented the Americans with the capacity, for the first time, to orbit heavier payloads than those of the Soviets. This alarmed the Soviet leadership, who issued orders to do everything possible to retain their lead in space.

In March 1964, Khruschev told Korolyov to launch a three-man spacecraft before the end of the year. Korolyov – having already submitted plans to modify the capsule to carry two men back in 1961– was told by Khruschev that conversion to fly a third man must be made possible, and insisted that the new three-man mission must achieve a result in order to deflect attention from the two-man Gemini missions. Korolyov despaired. Having finally won the argument to build a spacecraft capable of taking cosmonauts around the Moon, and realising that Vostok was at the limit of its capabilities, he was now being asked to shelve both Vostok and Soyuz and squeeze in a new project ordered by the Kremlin. Furthermore, it was a project designed not to meet domestic goals, but as competition against an American mission.

However, the second theory – put forward by the Soviet space historian Asif Siddiqi – questioned whether Khruschev was personally involved in the direct running of the Soviet space programme, or that he was able to manipulate plans to achieve short-term goals. In his book *Challenge to Apollo* (2000), Siddiqi recalls that orders to change the design of Vostok to carry three cosmonauts came from the Kremlin, but did not specifically mention Khruschev. He suggests that Korolyov had thought of the three-man concept in February 1963, during evaluations to 'extend' the original Vostok design, and that when approached by the Kremlín to actually achieve it, he tried to argue that it would be difficult to convert the vehicle in such a short time.

One reason for this could have been that Korolyov was holding out for a much more important commitment from the Kremlin. Before he agreed to build the three-seat Vostok variant, he wanted a firmer commitment to the manned lunar programme. It is not known whether such a commitment was offered.

Siddiqi suggests that authority to press ahead with the design change came from someone in the Communist Party government – possibly Leonid I. Brezhnev, the Secretary of the Central Committee for Defence Industries and Space. The choice of Brezhnev was interesting, since he would take a more leading role in Soviet affairs in a few short months! According to Kamanin, Korolyov was not happy with the decision, but he also wanted to beat the Americans as often as possible.

Whatever the true origins of the conversion, Korolyov had less than six months to change a one-man capsule into a three-man spacecraft. On 13 March 1964 the official decision to adopt the proposal was discussed at a meeting of the Military Industrial Commission, which also called for a programme of four three-seat launches, with the first launch in August 1964.

It was also decided that the 'new' spacecraft would be given the name Voskhod (Sunrise), the details of which would remain secret, as had those of the Vostok craft. It was hoped that this would suggest to the West that the Soviets had a completely new and improved spacecraft that could compete with Gemini, and would disguise the fact that it was nothing more than an adapted Vostok.

Before he would be allowed to develop Soyuz, Korolyov was also assigned a secondary target for the Voskhod. A second mission would include an EVA by a cosmonaut, *before* the Americans completed their planned EVA in early 1965. To achieve this, Korolyov would have to adapt the system designed for the now cancelled Vostok extended missions. In this system, a pressure-suited canine test-subject was to have participated in an 'EVA' study in an extending 'compartment' from the side of the capsule. These original space-walking experiments were to be called Vykhod (Exit), and this name would be reactivated for the new plans.

The decree approving both the Voskhod and Vykhod missions was issued on 13 April 1964. Two variants of the spacecraft would be built, both adapted from the original 3KA spacecraft. For the Voskhod missions, three 3KV spacecraft would be developed and for the Vykhod, two 3KD spacecraft would be built. Each manned flight would be preceded by an unmanned flight carrying an animal payload, and the third 3KV vehicle would remain as a spare.

Vostok becomes Voskhod

Yevgeni Frolov, a leading designer in OKB-1, led the team assigned to convert the spacecraft, with the sole purpose of achieving the flight of three cosmonauts. The first consideration was the problem of fitting three cosmonauts inside the capsule,which had limited volume, even for just one space-suited cosmonaut and the ejector seat. Feoktistov – another leading designer – proposed that the only solution was to eliminate both the spacesuit and the ejection seat from the design. This would save considerable weight, but would also present a serious safety issue.

Arguments continued between the doctors (in favour of wearing suits) and the engineers (who pointed out the physical impossibility of fitting three men inside the sharik spherical descent capsule, unpressurised – let alone pressurised – suits). In the end it was reluctantly agreed that the only way would be to launch three men without pressure suits, and to rely on the structural integrity of the capsule to retain pressure and atmosphere. This had been shown to be reliable enough during the Vostok missions, but there was still a risk. This risk did not manifest itself in the forthcoming missions, but would haunt the Soviets during the Soyuz programme. In 1971, three cosmonauts were returning from Salyut 1 onboard Soyuz 11. A valve opened prematurely, and the cosmonauts died because they were not wearing pressure suits.

The omission of the ejector seat was a far more serious safety risk. OnVostok, the ejector seat enabled the cosmonaut to escape from the capsule in the event of a launch failure. During the craft's descent, the cosmonaut ejected from the capsule, to prevent injury from the hard and potentially dangerous impact with the ground. By removing this system there could be no escape during launch, and the crew would suffer an extremely hard landing at the end of the mission. This would necessitate an alternative method of protecting the health of the crew.

At the time of Voskhod's development, studies were available from the former 'extended Vostok' missions and from the forthcoming Soyuz programme, and some of these plans were re-examined to determine wether they could be incorporated into the Voskhod design. Korolyov first assigned KB-2 Plant 81 the task of modifying the launch escape tower (intended for Soyuz) for use on Voskhod.

The cancelled Vostok missions were due to have a parachute-reactive descent system, in which the parachute canopies would be supplemented by a solid propellant motor that would fire shortly before 'dust-down', to soften the landing impact. This system had already been successfully employed by the military for landing armoured vehicles and other logistics by parachute. To evaluate this method for a returning space capsule, Korolyov was forced to recycle Titov's old capsule from the OKB-1 museum and use it for drop tests, as there were no spare capsules. Every new sharik that came off the production lines was assigned to the Zenit reconnaissance programme. There had been ten previous tests of the Voskhod parachute system, but they probably all had a scaled-down descent apparatus. The test of the Titov descent module – widely thought to have taken place on 6 September in the Crimea – was the first full-sized demonstration, but it was not successful. The parachute hatch failed to open, and Titov's historic but thankfully unoccupied capsule smashed into the ground and was destroyed.

The two additional cosmonauts planned for Voskhod also presented a problem for the duration of the mission and the capabilities of the life support system. Vostok's capacity to support a cosmonaut for up to ten days was sufficient for natural orbital decay to ensure the safe recovery of the cosmonaut before failure of the life support system. On Voskhod, without a complete redesign, the life support system was incapable of sustaining three men for ten days. The most that could be hoped for was a short 24-hour flight, but if the braking engine failed then the crew

Voskhod in half of the launch shroud during launch processing at Tyuratam.

would perish before natural decay occurred. For Voskhod, a second, back-up engine would have to be installed in case the first engine failed.

These modifications had been in the minds of Vostok designers for some years and they merely adopted them for the new mission. For some years it was assumed that Voskhod was a hastily assembled project, whereas it had in fact benefited from the lessons learned on Vostok and some new concepts being developed for the next generation of manned spacecraft.

Essentially, the dimensions of Voskhod resembled those of the Vostok, at 16.4 feet long and 7.9 feet maximum diameter. The combined weight of the sharik, the instrument module and additional equipment was 11,730 lbs. Inside the capsule, the three Elbrus couches (incorporating shock absorbers to help cushion the landing) were arranged side by side, but because of the width restrictions the central couch had to be moved forward slightly, producing a triangular configuration. As a result, the control panel and couches were moved 90° from the position that they were in in Vostok, so that they were 'sideways' in the capsule.

Removing the ejector seat also removed the post-flight cosmonaut survival pack, which had to be replaced by three separate units fitted into the capsule. A new TV system, operating at ten frames per second, replaced the original 25 frames per second system to observe the crew. The movie camera was removed, but a second TV camera that was crew-controlled was placed outside the vehicle. Finally, new radio systems were installed, together with an upgraded landing beacon.

The Voskhod flight profile
Due to the additional launch weight, Voskhod required the improved 11A57 version of the R-7. This vehicle had an upper stage that employed the RD-0108 engine developed by OKB-154 for the R-9A missile, with a vacuum thrust of 30.4 tons and a 240-second burn time. The version of the R-7 used was originally developed for the unmanned Zenit reconnaissance class of spacecraft. The upper stage, at 29.5 feet, was longer than that used for Vostok, and had a dry mass of 3.35 tons and a propellant mass of 95.70 tons. The overall launch mass of the vehicle was 307.26 tons and it stood 147.3 feet tall.

The launch sequence was similar to that for Vostok, except for the change of escape method in the event of a launch failure; but it soon became evident that the escape tower would not be ready in time for Voskhod. The draft plan issued by OKB-1, dated August 1964, stated that in the first 45 seconds of the flight, rescue of the crew would be 'difficult', which in reality meant it would have been impossible. Following the 45-second point, recovery could be attempted following a nominal Vostok recovery profile. Up to 501 seconds into the mission, a landing would be within the territory of the Soviet Union, but afterwards, if orbital insertion was not achieved, the landing would be outside the USSR.

In orbit the crew would have very little to do, due to the short duration of the flight, the restrictions in volume inside the spacecraft, and launch weight limitations. The exterior TV camera could record the separation of the upper stage, and views of Earth and the Moon, but little else. The inside of the sharik was padded to help reduce noise levels inside the spacecraft during launch, and was cramped, so there would be no room for moving around and only limited space to operate equipment.

Initially, the crew would also have supplies sufficient for up to one, or maybe two, days. Re-entry would be achieved by using a new orientation system, based on ion sensors that were able to use the ionised layer of Earth's atmosphere to provide alignment for the spacecraft during the night-time pass of the orbit – a capability that was not available during the Vostok missions.

Using the TDU-1 braking engine for two seconds would be enough to de-orbit the sharik and, after entry, it would be recovered by a three-tier parachute system of an drogue parachute, a braking parachute, and a pair of primary canopies. To activate the solid propellant braking rocket in the parachute harness, a small probe extended 3.9 feet below the capsule. Upon striking the ground it initiated a single firing of the solid rocket motor that reduced the descent velocity from 26–32 ft/sec to a more comfortable 0.5 ft/sec The spring-loaded suspension couches would absorb the g-loads encountered at landing.

Korolyov's cosmonauts

The six Vostok missions had established that it was possible to fly into space, perform basic functions there, and return in a good state of health. The Air Force had fought against plans to select doctors or scientists in the early cosmonaut selections for a single-seat spacecraft that needed a pilot to 'fly' it. Voskhod – a highly automated spacecraft – hardly warranted two pilots, let alone three, and there were several organisations involved in the space programme that wished to assign some of their own employees to cosmonaut training. These included the Academy of Sciences, the Ministry of Health, and the Ministry of Aviation Production.

Korolyov began to suggest that the time was right to send space researchers, scientists, physicians, engineers, and even journalists into space. Initially, the medical services of the Soviet Ministry of Defence was opposed to the idea, but was forced to screen up to thirty 'passengers' and thirty physicians in April 1964. From these, six were to receive approval for cosmonaut training.

During April and May 1964, thirty-six candidates were selected from the Academy of Sciences and the Ministry of Health, with fourteen passing the medical board examination. On 28 May, ten candidates (eight from the Ministry of Health and two from the Academy of Sciences) were evaluated by the credential committee, which then nominated five for training:

Dr (Eng.) Georgi P. Katys (Academy of Sciences) *b.* 31 August 1926, age 37
AF Major Vasily G. Lazarev (NII-7) *b.* 23 February 1928, age 36
Dr. (MD) Boris I. Polyakov (IMBP) *b.* ?
AF Capt. (MD) Alexei V. Sorokin (TsPK) *b.* 30 March 1931, age 33
Dr (MD) Boris B. Yegorov (IMBP) *b.* 26 November 1937, age 27
Lazarev was also a doctor, as well as being a pilot.

Cosmonaut Alexei V. Sorokin, the back-up doctor for Voskhod 1, who was considered for the 1965 Air Force selection but was not selected.

From the fourteen OKB-1 candidates submitted on 17 May, eight engineers were medically qualified:

Konstantin P. Feoktistov, *b.* 7 February 1926, age 38
Georgi M. Grechko, *b.* 25 May 1931, age 32
Valeri N. Kubasov, *b.* 7 January 1935, age 29
Oleg G. Makarov, *b.* 6 January 1933, age 31
A.M. Sidorov, *b.* ?
Vladislav N. Volkov, *b.* 23 November 1935, age 28
Valeri A Yazdovsky, *b.* 8 August 1930, age 33
V.P. Zaitsev, *b.* ?

Only Feoktistov was nominated for spaceflight training, and only after a personal appeal from Korolyov. He was restricted to spaceflights of only one day due to an earlier spinal injury that prevented him doing any parachute training. In 1961, he had asked to be selected for cosmonaut training, but was rejected because of the military monopoly of pilot cosmonauts for single-seat spacecraft. When Voskhod was first proposed, Feoktistov was among those who argued that it would be unsafe. But Korolyov convinced him otherwise, by saying that if they could design the craft to fly three people, one of the seats would be offered to an engineer from OKB-1. That was too good an offer to refuse, and so Feoktistov was one of the team that started to work on the vehicle that could take him into space.

A training group was formed consisting of nine Air Force Pilots and Engineers (Bykovsky, Titov, Popovich, Komarov, Volynov, Leonov, Demin, Shonin and Belyayev); a very experienced test pilot from the Tupolev Bureau of the Ministry of Aviation Production (Vladimir N. Benderov *b.* 4 August 1924, age 39); four physicians (Lazarev, Yegorov, Polyakov and Sorokin), a scientific researcher (Katys); and Feoktistov.

Cosmonaut G. Katys onboard an aircraft during training for Voskhod 1.

Cosmonaut Boris Yegorov, the doctor on Voskhod 1, discusses his evaluation reports with Yuri Gagarin after a training session.

Both Benderov and Polyakov suffered bad reactions in the centrifuges and were rejected, and the Air Force group was reduced to two (the most experienced pair, with back-up experience from Vostok). Seven finalists (Komarov, Volynov, Katys, Feoktistov, Lazarev, Sorokin and Yegorov) remained to train for the first Voskhod.

Cosmos 47
Several technical problems hampered the preparations for the first launch of an unmanned Voskhod, planned for 5 September 1964. (The manned flight was expected by the middle of the same month.) Problems with the telemetry system, the fixing of the support couches, and qualification of the parachute descent system engines delayed the launches into October. A successful test of the recovery system was finally completed on 5 October, and on the following day the first 3KV spacecraft was launched at 10.12 a.m. MT. Once in orbit, the spacecraft was designated Cosmos 47, to disguise its real mission. The unmanned Voskhod entered a 108 × 237.9 miles, 64°.62, 90.1-minute orbit. During the 24-hour flight, ground controllers tested the onboard systems to verify them for the manned flight, planned to take place a few days later.

Originally there was to be dogs onboard but instead, mannequins occupied the seats on the mission. Why the dogs were not flown is unclear, although one possibility could have been the difficulty of adequately restraining three dogs in separate containers inside the capsule; and there would also have been additional payload weight. It is also unclear what benefit flying mannequins would have offered, as the landing loads would be marginally better than those on the Zenit photoreconnaissance satellite recoveries since Voskhod carried the additional landing rocket in the parachute system.

There were no reported anomalies during the mission, which was completed after sixteen orbits in 24 hrs and 18 min on 7 October. The landing, at 10.32 am MT, was notable for a strong wind dragging the capsule 500 feet from the touchdown point. This was not a concern for the manned flight, as the crew could jettison the canopy from within the vehicle.

A pilot, an engineer and a doctor

On 9 October, forty-eight hours after the recovery of Cosmos 47, the State Commission formally assigned Komarov, Feoktistov and Yegorov to the flight, with Volynov, Katys and Lazarev assigned as the back-up crew. With the exception of Volynov and Komarov, all the others would be stood down from their temporary cosmonaut status after the mission.

During the night of 11 October, the three cosmonauts occupied the cosmonaut cottages. Feoktistov took the bunk that Gagarin and Tereshkova had previously taken, while Komarov chose the bunk that he should have had as back-up to Popovich before he was replaced by Volynov, due to ill health. Yegorov slept in an adjoining room.

On the morning of 12 October, the three cosmonauts completed the now familiar launch-day ritual of exercise, showers, medical examinations and breakfast, before boarding the bus to the pad. Dressed this time, not in the orange space pressure suits and white helmets, but in lightweight grey woollen trousers and shirts with light blue

The Voskhod 1 prime crew at the pad at Tyuratam prior to the launch: (*left to right*) Komarov, Feoktistov and Yegorov.

jackets and communication caps, they looked as though they were about to board an aircraft rather than take a flight into space.

Members of the State Commission, including Korolyov and Gagarin, were at the pad, where formal presentations were made. The three cosmonauts then took the elevator to the spacecraft, where they removed their jackets and boots. Yegorov climbed into the Voskhod first, and took the innermost couch. He was followed by Feoktistov, who took the central couch, slightly forward; and finally Komarov, who occupied the outermost couch, nearest the hatch.

Launch occurred at 10.30.01 MT, and was nervously watched by the controllers through T + 523 seconds, at which point orbital velocity was reached without incident. Korolyov was reported to be shaking – not from the cold, but from worrying that the crew would be lost in a launch mishap. Voskhod entered an orbit of 110.2 × 253.5 miles, 64°.9, × 90.1 minutes, as planned, with the crew using the call-sign 'Rubin' ('Ruby'). Inside Voskhod, the temperature was reported as extremely comfortable; but in reality, during the first six orbits the temperature climbed from 15° C to 21° C, and humidity rose from 45% to 60%. The life support system was obviously having difficulty in maintaining a bearable environment with three men in close proximity, even in the first few hours.

The Tass announcement on the flight hinted at an impressive mission, with 'three men launched into space... by means of a powerful, new launch vehicle.' The stated aims of the flight were, 'to test the new multi-seat piloted spaceship; to check the capacity for work and interaction during spaceflight of a group of cosmonauts consisting of specialists in different fields of science and technology; to carry out scientific, physical and technical investigations on conditions of spaceflight; to continue the study of the effects of different factors of spaceflight on the human organism; and to carry out *extended* medical and biological research in *conditions of a long flight*; with the help of instruments on board the spacecraft and with the direct participation of a scientific worker, a cosmonaut and space doctor.'

In the West, it was not only the size of the crew that was unexpected, but also that a scientist and a doctor had been included. American astronaut Scott Carpenter stated that he would not have been surprised if the Soviets had sent up two cosmonauts, as that was expected with Gemini on the horizon – but not three. 'The Russians seem to do this, what you don't expect them to do,' he said. America was not planning three-man flights on Apollo until at least the end of 1966, but these would be pilots, not scientists, and all three would have full pressure suits for launch and entry, and not ride into space in training suits!

As the crew completed their first orbit, they reported that they were 'Feeling fine. The assignment will be carried out.' Each of the crew had specific tasks to complete. Komarov would be in charge of the control and operation of the spacecraft and report on its functioning to ground control, Feoktistov would handle a number of visual and photographic tasks, and Yegorov would conduct an impressive array of medical experiments on the flight. During the third and fourth orbit, Yegorov took blood pressure and respiration readings of the other cosmonauts.

The crew had also talked to Premier Khruschev during the third orbit, and took their first meal during the fourth orbit. Although Komarov was in command of the

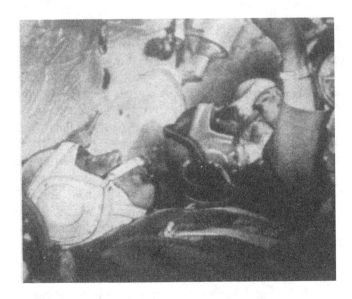

The Voskhod 1 crew undergoes simulator training. Komarov is in the foreground, Feoktistov is in the centre seat, and Yegorov is in the inner couch.

mission, it was Yegorov's task to schedule the cycle of work, rest and meals. Food supplies on the flight were stored in a large container, and the meals had a total energy value of about 3,600 kcal each for the whole flight. The selection of food was the same as carried on Vostoks 3–6, but was expanded a little. Appetites remained at satisfactory level, although neither Feoktistov nor Yegorov drank much fluid during the flight. (Upon landing, both felt immediately thirsty, and resorted to extracting liquid from the spacecraft water supplies.)

Each cosmonaut rested in turn, so that at least one was always alert to monitor spacecraft systems and communicate with the ground as they continued recording data. During the fifth and sixth orbits, Feoktistov observed the cloud cover and carried out photometric studies. Meanwhile, Yegorov conducted physiological studies into the state of the cardiovascular system and the functioning of the vestibular system, including the taking of pressure and blood samples from himself and Feoktistov. Although assigned to medical studies inside the crew compartment, Yegorov took the opportunity to look out of the porthole, and marvelled at the mountain ranges far below him. As a mountain climber, he was particularly looking forward to seeing Mount Kilimanjaro, in Africa.

During the sixth and seventh orbits, Komarov operated the ion velocity-vector adjusters ('postroityeli'), or electrostatic ion engines, which the Soviets reported was the first such test in orbital flight. This worked by creating an electrostatic field that accelerated alkali metal ions to produce a high-velocity ion flux that turned the spacecraft.

From the eighth and thirteenth orbit, the spacecraft was not over Soviet territory, but the crew continued their observations and experiments. During the flight they passed on their greetings to the Soviet team at the Olympic Games in Tokyo, and to the North Vietnamese Vietcong fighting the Americans in the growing war in south-east Asia.

Feoktistov continued his photographic observation of the Earth's surface, and the polar aurora over Australia, which, he recalled after the flight, was 'the most impressive phenomenon we saw'. He also noted luminescent particles that had been observed from Vostok and Mercury spacecraft, and observed the behaviour of liquids in weightless conditions, following on from experiments flown on Vostok.

Yegorov (the first medically trained person to fly in space) was the busiest of the three, with a range of medical experiments. Although these were minor in comparison to those conducted on the later flights, the fact that a doctor would conduct them in space added to the results and the publicity.

Yegorov's research included observation of the condition and behaviour of the crew, research into pain and tendon reflexes, and psycho-physiological tests that determined the quickness and accuracy of the crew in processing data from charts. He also measured arterial pressure, determined the thresholds of sensitivity to stimuli, and conducted observations of oral activity, as well as performing an experiment that required a rhythmic pressing of the hand on a constant force (which was recorded on a dynamograph) for sixty seconds.

The most extensive series of tests involved visual experiments, completing a predetermined sequence of eye movements before and after moving the head ten times, while recording electro-oculograms (which on later flights was found to induce space sickness in some cases), and periodically closing and opening of the eye while recording electroencephalograms. The crew also completed an experiment that investigated the coordination of head, eye, and hand movement while drawing a series of figures (four complex spirals, four figures of the number '6' and a signature), with their eyes both open and closed.

Biological specimens were also carried, and were similar to samples taken on Vostok missions: cells, plants, seeds, algae and bacteria.

Perhaps of most benefit to Yegorov and the doctors were the sensations that two of the crew felt shortly after entering orbit. Beginning on the second orbit, Yegorov began to feel warmth in his head as fluid shifted around the body, and explained that he felt as though he was bent over and facing downward. Feoktistov also felt he was upside down. Neither effect hindered their work capacity, but neither could they float out of the couch to find out, and both men experienced the sensations for most of the flight. This was apparently a result of their abbreviated training programme.

TV was broadcast from the crew compartment during the sixth and seventh orbits, and there was a small scare when telemetry indicated that Yegorov's pulse had fallen to only 46 beats/min when he slept – but after confirming that there was nothing wrong, it was found that his pulse was only 68 beats/min when he was awake!

A landing beyond belief

The flight was scheduled to last just over 24 hours, but when Korolyov was preparing to order the re-entry, Komarov put in a request to extend the flight a further 24 hours, having reported that they had seen, 'many interesting things.'

Korolyov's reply was a quote from Shakespeare's *Hamlet*: 'There are more things in Heaven and Earth, Horatio, than are dreamt of in your philosophy.' We shall go, nevertheless, by the programme.'

With thoughts of a troubled qualification programme (in which one spacecraft was smashed in the test of the descent system), all on the ground waited anxiously for news of a safe landing. The braking engine fired first time, and the descent into the atmosphere occurred without incident. It was an uneventful, if bumpy, descent, with slight vibrations felt as Voskhod dropped towards Earth. The crew heard the parachute container hatch open at 4.3 miles, the firing of the soft landing rockets (to slow the touchdown) was reportedly not felt by the crew, and after landing, Komarov jettisoned the parachutes to prevent the capsule being dragged across the field in which they landed. He opened the hatch, to be greeted by snow flurries and a cold wind.

When news reached the control room that the spacecraft had been seen on the ground with all three cosmonauts outside waving, huge applause erupted. Korolyov was greatly relieved: 'Is it really true that the crew has returned from space without a single scratch?' He could hardly believe that the converted Vostok had actually carried three men into space, as it had been asked to do.

The Voskhod touched down at 10.47 am MT, 193 miles north-east of the town of Kustany, after a flight of 24 hs 17 min, having completed sixteen orbits and 416,203 miles. Following the landing, the cosmonauts were scheduled to speak with Khruschev on the telephone, but they were informed that they were to return immediately to Tyuratam.

From the cosmodrome they spoke to Khruschev at his dacha on the Black Sea coast, where he was on holiday. The next day, following initial post-flight debriefing, news broke around the world of a change of leadership at the Kremlin. While Khruschev was away, a special meeting of the Central Committee of the Communist Party 'voted' him out of office, to be replaced by a trio of leaders. The three new men in power were Nikolai Podgorny (as President), Aleksei Kosygin (as Chairman of the Council of Ministers, and thought to be the instigator of the coup), and Leonid Brezhnev (as First secretary of the Central Committee). The once powerful but now deposed Khruschev was 'retired'. In fact, the telephone call to the crew was his last public act, and he later said that his colleagues were trying to take the telephone from him due to the excitement. Khruschev retired from public view, and died, aged 77, on 11 September 1971.

During the ninety-minute press conference held at Tyuratam on 15 October, Komarov indicated that touchdown was so soft that there was no indentation in the field, and that Voskhod was a much more comfortable spacecraft than was Vostok (disguising its true origins by stating that the new spacecraft was a 'small space station'). At the 21 October press conference held at Moscow University, Academician M. Keldysh countered reports of the flight being shortened by stating that the flight of Voskhod, 'was planned for 24 hours, but the general resources of the spaceship's apparatus were much greater.'

On 19 October the three cosmonauts returned to a heros' welcome in Moscow – Komarov in full military uniform, flanked by the two civilians in three-quarter coats

The Voskhod 1 crew takes 'the long walk' at Vnukovo airport, Moscow, on 23 October 1964, ten days after completing their mission: (*left to right*) Feoktistov, Komarov and Yegorov. Ceremonies were delayed due to the overthrow of Premier Khruschev.

and trilby hats, marching along a red carpet. They were met by the new Soviet leadership, who had kept them waiting for this celebration. If the crew had any personal comments on the political events that surrounded their mission, they concealed them as they were greeted by the new leadership. One interpretation of Korolyov's quote from *Hamlet* was to attribute it to the changing political leadership in Moscow.

Western reports of the landing expressed surprise, and suggested that something had gone wrong with the mission, especially as the last four Vostok missions had lasted several days. They were unaware that the temperature control system was experiencing difficulties. Others suggested that it was terminated early to allow the overthrow of Khruschev, and some blamed the apparent space sickness of either Feoktistov or Yegorov – especially since the doctor was supposed to be conducting research on a 'long flight'. At that time a flight of 24 hours was a long mission, and other reports countered that the first Gemini was planned to last only a few hours, that the week-long spaceflight had still to be achieved, and that this was the first flight of a new spacecraft. What was not known was that with a three-man crew the Voskhod could have flown for only 24 hours.

Twelve months after the loss of President Kennedy, the Soviet Union also had new leadership. The change would affect the Soviet space programme, and would have consequences equally as important as those resulting from the influence of the White House and Congress on the American space programme. The new era on the political front also signalled the end of the era of space spectaculars and flights of the manned Vostok capsules.

VOSKHOD 2

Early plans for EVA
Konstantin Tsiolkovsky was one of the first to suggest that suitably protected humans could exit a vehicle in the vacuum of space and perform a space-walk. Before and after World War II, significant progress was made in the development of pressure suits for high-altitude stratospheric flights, and these led to the concept of the space pressure garment for the Vostok series of missions.

When Korolyov proposed to follow the series of one-seat spacecraft with a programme of extended missions, one of the early ideas was to place an animal (possibly a dog) in a pressurised garment (a space suit) inside a container that could be depressurised. After depressurisation of the container, the animal would be exposed to the vacuum of space to demonstrate extra vehicular activity, (EVA). This would have occurred on a mission named Vykhod (Exit), but the plan was cancelled before the mission was authorised

Then, in April 1964, a new series of flights was authorised that would fill the gap between the end of the Vostok missions and the beginning of the Soyuz flights planned for 1966. The first objective had been to orbit three men in a 3KV capsule (Voskhod) in 1964, before the American two-man Gemini was launched early in 1965. The next objective was to attempt to perform an EVA by a cosmonaut (using a different version of the spacecraft, designated 3KD) before the American stand-up EVA that was also planned for 1965.

Voskhod EVA plans
For a while, the name Vykhod was reactivated for the EVA flight, and it was not until early 1965 that the name Voskhod 2 first appeared in Kamanin's diaries. In the government resolution of 13 April 1964, there were five capsules proposed for the EVA mission, but it is not known how many of these were to be manned.

By May/June 1964, there were plans to launch only two 3KD versions. One would be unmanned (Cosmos 57) as a precursor for the manned vehicle (Voskhod 2). The Soviets had known of plans for an American EVA since 1961, when Gemini was first proposed; but it was not until January 1964 that a stand-up EVA (during which the Gemini pilot would 'stand' on the seat with his head and shoulders out of the open hatch) was actually scheduled for an early Gemini mission. With the decision to attempt an EVA from Voskhod, the hardware had to be prepared and the cosmonauts trained for what would be the most challenging and demanding mission since Gagarin's launch.

Since the Voskhod was based on the Vostok design, there was no provision for a lone cosmonaut, wearing the Vostok pressure garment, to depressurise the cabin. Neither could they open the hatches, which were bolted to the structure of the sharik, and not hinged so that the crew could exit the vehicle directly from the crew compartment. It was also impossible to repressurise the spacecraft again, as there were insufficient supplies of oxygen and nitrogen on board. The hatch design did not allow for it, the ejector seat was in the way, and the suit was totally unsuitable for protection outside the vehicle. To attempt an EVA from a one-person spacecraft was also a very risky procedure. In addition, the electronics were designed to operate only in normal atmospheric temperatures and pressures, and not in the vacuum of space.

Soyuz – the spacecraft that would follow the Vostok series – would incorporate a separate compartment that could be sealed off from the crew descent module, which allowed it to be depressurised by using a side hatch for EVA. This hatch would be hinged so that it could be closed and resealed, allowing one or two cosmonauts to perform a space-walk while the rest of the crew remained in shirt-sleeves. However, it would be about two years before this spacecraft would be able to fly with a crew. What was needed was modification of the basic sharik design to achieve a short and relatively simple EVA, – essentially to prove that EVA was feasible, and to once more beat the Americans.

A method would need to be developed that would allow a cosmonaut (wearing a suitable protective garment) to transfer into a pressurised compartment that could be sealed off from the main vehicle and then depressurised, allowing exit through a hatch. After the EVA period, the cosmonaut would need to re-enter the compartment, be able to close and reseal the hatch, and repressurise before re-entering the vehicle to join the rest of the crew and continue the mission.

How to perform EVA from a Voskhod

To perform an EVA from Voskhod, a 'passageway' was devised. This was attached, folded, to the side of the spacecraft, and would be extended in orbit. The space-suited cosmonaut would transfer from the crew compartment through the side hatch, to enter the passageway or airlock. When the inner hatch was closed, the pressure could

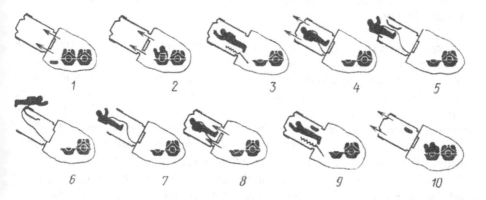

The Voskhod 2 EVA sequence. 1) Airlock extension; 2) Leonov prepares to enter airlock, and dons the back-pack while Belyayev checks the integrity of the airlock pressurisation; 3) Leonov opens the inner hatch and floats into the airlock; 4) the main hatch is closed, and airlock pressure is released; 5) Leonov opens the outer hatch, and floats outside to begin EVA; 6) Leonov during the EVA; 7) Leonov re-enters the airlock; 8) the outer hatch is closed, and the airlock repressurised; 9) the main hatch is opened, and Leonov returns to the cabin; 10) the inner hatch is closed, and the airlock is depressurised and released.

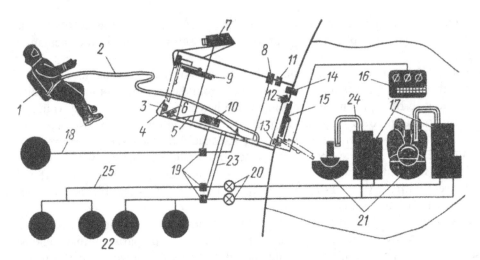

The Voskhod 2 EVA systems. 1) The cosmonaut's self-contained life support system (suit and back-pack); 2) the safety tether, with communications and telemetry wires; 3) lights; 4) airlock structure; 5) handrails; 6) cine-camera; 7) cine-camera; 8) airlock pressure relief valve; 9) airlock exit hatch; 10) duplicate exit system back-up control panel; 11) airlock pressure valve; 12) cine-camera; 13) lamp; 14) mechanism for equalising pressure in the airlock chamber and cabin; 15) interior airlock hatch (main spacecraft hatch); 16) exit system primary control panel; 17) spacecraft life support units for cosmonauts in spacesuits; 18) independent airlock pressurisation system; 19) electronically controlled valves; 20) gas pressure reducers; 21) cosmonaut couches; 22) spacesuit and cabin pressurisation system; 23) electrical cable connections; 24) umbilical connects between spacecraft life support system and cosmonaut pressure garment; 25) gas transfer line.

be lowered, and a second, outer hatch opened to the void of space. To return to the crew compartment, the cosmonaut would reverse the procedure.

The development and construction of the 'Volga' airlock took just nine months from mid-1964, and was the responsibility of Chief Engineer Boris Mikhailov, of the Engineering Plant 918 in Tomolino, Moscow. The airlock was a soft, inflatable structure, which was folded against the side of the sharik during launch (under a suitably modified launch shroud to take the protrusion), extended in orbit, and then jettisoned from the capsule after the end of the EVA.

The airlock was constructed of an upper rigid ring with an inward-opening outer hatch for the EVA. This was connected to the lower adapter ring that was bolted to the exterior of the Voskhod capsule, over the inward-opening crew entry/exit hatch. When extended, the cylindrical airlock was effectively a hermetically sealed double-walled rubber tube. Between the walls were forty inflatable sections (called 'aerobalky', or aerobooms) fabricated in three sections, any two of which would be sufficient to extend the airlock and maintain the internal pressurised volume against the external vacuum. However, if the outer hatch was opened, the structure could be weakened and could possibly have lost its rigidity, making it very difficult for the cosmonaut to re-enter the structure.

The air supply for inflation of the structure (which also served as a back-up air supply for the EVA cosmonaut) was located in four spherical air canisters mounted on the outside of the adapter ring at the base of the structure. The length of the collapsed structure was just 29.1 inches, while fully extended it measured 62.9 inches. When combined with the hatches, the structure measured 8.2 feet in length, with an internal diameter of 3.2 feet and an external diameter of 3.9 feet when fully deployed. It took seven minutes to inflate, and had an internal volume of 88.28 cubic feet.

Covering the airlock was a pressure-fabric envelope that also provided a thermal cover, and around the outer hatch was a handrail for the EVA cosmonaut. There were three 16-mm cameras (two inside and one outside) to record operations, with two lights provided to light the interior. The EVA crew-member operated the airlock from inside the crew module by means of an extra control panel and internal TV receiver added to the spacecraft controls. A duplicate set of controls was suspended from bungee cords inside the structure.

Five fully functional airlock units – as well as several mock-ups and training models – were constructed. The first two units were used for technical and functional testing, and both suffered undetermined damage. The third unit was for training the cosmonauts as a flight demonstration unit, while the fourth was the flight model attached to Voskhod 2. This unit was apparently dropped on 13 January 1965, but was quickly repaired and shipped to Baikonur for installation on the spacecraft. The fifth unit was the back-up, and after the flight it was moved to the Memorial Museum of Cosmonautics in Moscow. One of the other units is now at the Zvezda Museum.

'Exit' space suit

The development of the pressure garment for the EVA began in May or June 1964, and was also the responsibility of the Zvezda company. Called 'Berkut' ('Golden Eagle'), it was a multilayered construction that was a modified Sokol Vostok IVA suit with additional thermal protection layers. It featured the constant-wear undergarment, a liquid coolant layer, and an inner comfort layers, and on top of these was a full body undersuit of white nylon parachute cloth. This had aluminium umbilical interfaces that connected to the life support system and communications equipment. Over this, the cosmonaut wore the pressure layer with an additional white ribbed twill nylon oversuit (to reflect solar heat) that featured a dual front zip closer and detachable white cotton gloves with rubberised palms and finger pads. The boots were made of quilted white leather, and laced on top of these were a pair of overboots made of flexible waffle-textured insulating foil and white over-covering.

The helmet was white metal, featuring a hinged visor and hinged reflective sunshade. Sewn into the right sleeve was a pressure gauge display under a Velcro flap, while on the left forearm was a reflective mirror.

One of the problems of using an airlock system was that during depressurisation the suit would balloon to a rigid shape that would be impossible to bend. To alleviate this, a reduction in pressure of the suit would have to be incorporated into the design. In the early American spacecraft the astronauts breathed an oxygen-rich atmosphere, which was not dangerous if the pressure was reduced. In Vostok and

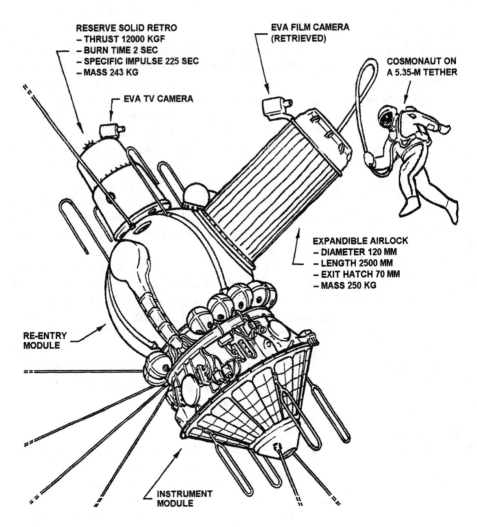

RESERVE SOLID RETRO
– THRUST 12000 KGF
– BURN TIME 2 SEC
– SPECIFIC IMPULSE 225 SEC
– MASS 243 KG

EVA FILM CAMERA
(RETRIEVED)

COSMONAUT ON
A 5.35-M TETHER

EVA TV CAMERA

EXPANDIBLE AIRLOCK
– DIAMETER 120 MM
– LENGTH 2500 MM
– EXIT HATCH 70 MM
– MASS 250 KG

RE-ENTRY
MODULE

INSTRUMENT
MODULE

An artist's impressions of the Voskhod 2 capsule and cosmonaut EVA. (© 1983 Dave Woods.)

Voskhod, however, an oxygen/nitrogen mix was used, and this added the potentially fatal problem of initiating the bends if the pressure dropped the prevention of which normally requires pre-breathing oxygen for a short period to cleanse the blood of the nitrogen. The Soviets obviated this problem by providing the Voskhod EVA suit with two operating pressures; – one at 0.4 atm (6 psia), and the second at 0.27 atm (4 psia). Using either of these levels for a short period of time was not normally thought to be dangerous, and provided additional mobility during the exit and entry of the airlock.

In choosing the airlock system, life support could not be supplied directly from the spacecraft to the cosmonaut via an umbilical. Instead, a back-pack was provided, with a 45-minute oxygen supply for both breathing and cooling, and with connections to the back-up supply in the adapter ring on the airlock. Oxygen was also vented into space by a relief valve that also carried excess heat and moisture, and exhaled carbon dioxide.

The added weight of the EVA suit, airlock and exit hatch meant that only two cosmonauts could fly an EVA Voskhod. The cramped cabin would also render it impossible for a three-person crew to move adequately inside the capsule allowing access to the airlock. It was stated that the second internal (intravehicular activity) cosmonaut was wearing a similar suit, and that a 'rescue' was possible, but it was highly unlikely that this could be effected with any degree of success. Therefore, the EVA cosmonaut was also provided with a 17.5-foot line that tethered him to the spacecraft at all times. This tether also contained the communications link through to the IVA crew-member, and thus to ground control.

Selecting a crew
The training group for the second manned Voskhod mission was formed at the same time as that of the first in April 1964, but they began Voskhod 2 training in early July. The training group consisted of six cosmonauts: Belyayev, Gorbatko, Khrunov, Leonov and Zaikin – all unflown cosmonauts from the 1960 AF selection – and Kolodin – from the 1963 AF selection. The training programme would give them about twelve months in which to perfect this exacting mission. The original assignment was three teams of two cosmonauts, in prime, back-up and support roles:

Command pilot: prime, Belyayev; back-up, Gorbatko; support, Zaikin.
Pilot (EVA): prime, Leonov; back-up, Khrunov; support, Kolodin.

In January 1965, concerns about the health of Gorbatko resulted in his removal from the training group, as he was suffering from tonsillitis. He was replaced by Zaikin as back-up to Belyayev. No replacement support Commander was assigned, but Khrunov also trained as an alternative to Belyayev. There were also concerns about the health of Belyayev, and Kamanin noted in his diary on 15 January 1965 that: 'the Khrunov–Leonov team is definitely stronger than the Belyayev–Leonov team.' However, since the EVA was the most demanding part of the mission, the decision had been made to train the strongest candidates – Leonov and Khrunov – for it. Once they had mastered this role, Khrunov undertook his command training, while Leonov gained further experience in his EVA tasks. During a 9 March meeting of the State Commission, further concern was raised over Belyayev, but since he and Leonov had trained as a crew for more than six months it was decided not to split them up so close to launch. Another decision made was that, despite Zaikin being the back-up Commander, only Khrunov would suit up, to replace either cosmonaut if required.

Training for EVA
Leonov recalled that he first saw the airlock chamber in the summer of 1964 at OKB-1, where it was attached to the side of a Vostok spacecraft. At the same visit to the design bureau, he was to participate in a strenuous two-hour simulation in an altitude chamber, while wearing the suit. The Soviets had adapted a flown Vostok

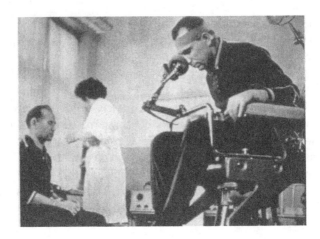

Cosmonauts Belyayev (right) and Leonov participate in medical tests as part of their Voskhod 2 training programme.

(Popovich's) to a Voskhod configuration for training. This capsule and airlock is now in the Zvezda Museum.

Training for the EVA included full-scale evaluation of the move from the spacecraft to the airlock, putting on the back-pack, closing the hatches, the exit and entry through the outer hatch, and the return to the capsule mock-up. These were all accomplished in 1 g environments before moving on to the parabolic simulations in the TU-104 aircraft using a spacecraft crew compartment and a wooden mock-up of the airlock. Leonov also completed altitude chamber work in the TBK-60 chamber. To condition him for the psychological 'shock' of exiting the confines of the Voskhod into the emptiness of space, he spent many days in the isolation chamber, and upon exit was taken immediately to an aircraft to complete a parachute jump from high altitude.

There was also a personal training programme to condition the cosmonauts who trained for the EVA. In preparing for the spaceflight and his EVA, Leonov recalled that in the year before his first spaceflight, he ran 311 miles, cycled 621 miles, skied 186 miles, and spent hundred of hours in the gymnasium.

Leonov's strength was exhibited during the parachute descent training, when his harness became caught on the seat structure. Using his bare hands, he bent the obstruction to release his harness, and landed safely. After the seat was recovered, the offending structure needed a hammer to straighten the metal back into shape!

Voskhod 3KV modifications

Although still based on the Vostok 3K design, the version used to support the EVA required a number of design modifications, in addition to the airlock, hatches, launch shroud protrusion and EVA support equipment previously described.

This included the installation of two Elbrus cosmonaut couches (again 90° to the entry/exit hatches), which could accommodate a spacesuited cosmonaut and allow for the inward opening of the hatch to the left. The Soviets also needed to supply two suits with air conditioning and life support inside the spacecraft, and there was also an emergency oxygen ventilation system for the suits in the event of the spacecraft life support system failing during recovery.

Launch preparations

During a visit to Tyuratam in September 1964, Khruschev observed Leonov exiting and entering the airlock on a training module of Voskhod 2 during a suited exercise. Korolyov had over-optimistically promised the leadership that the mission would be launched by November that year. By December 1964 the mission was scheduled for either March or April 1965 and, as with Voskhod 1, an unmanned precursor mission was to fly in January or February to test the airlock.

On 6 February the State Commission agreed to schedule the unmanned mission for between 14 and 16 February, followed by the manned mission between 25 and 27 February. The mission was to last just twenty-four hours, with the EVA occurring shortly after entering orbit. Should the EVA not take place, the plan allowed for the Voskhod to continue in orbital flight, on an alternative mission, for two to three days.

The commission confirmed the cosmonaut crews three days later, on 9 February. However, on 12 February, two days prior to the scheduled launch of the unmanned Cosmos 57, there was a delay of forty-eight hours in its launch. This pushed the manned launch into March, and since a Luna Moon-probe was scheduled to launch in the middle of that month, the required twenty days of launch preparations for the lunar mission saw the manned launch delayed again to the end of the month. On 19 February, the State Commission decided to go for a manned launch around 4/5 March, but this depended upon the results from Cosmos 57.

Cosmos 57

The unmanned version of Voskhod 2 was launched at 10.41 a.m. MT on 22 February 1965, initially without incident. The spacecraft separated and entered an orbit of 102.5 × 265.3 miles, 64°.74, 91.1 minutes. TV images revealed that the airlock was deployed, and the outer hatch opened and closed with the unit being repressurised without incident, providing confidence that the system would work on Voskhod 2. As Kamanin recorded: 'During the first orbit, the craft was observed via a special onboard television circuit in Simferopol and Moscow We gathered around the television set [at Tyuratam], though not really expecting to catch a glimpse of the craft since it was quite far from the proving ground. Quite unexpectedly, a distinct image of the front part of the airlock appeared on the screen, causing an outburst of joy among all present.' Kamanin then went off duty... and everything went terribly wrong.

As the unmanned spacecraft began its third orbit, well within range of three Soviet ground stations, all contact was suddenly lost. It slowly became clear that the automatic self-destruct system had triggered, destroying the spacecraft. It was tracked in 168 detectable pieces, which re-entered Earth's atmosphere between 31 March and 6 April 1965.

The cause of this accident was traced to duplicate signals being sent from two ground stations instead of one. The primary station (IP-7 at Yelisovo) transmitted command signals to the spacecraft, and the back-up station (IP-6 at Klyuchi) was to send signals only if commanded to do so by flight control. However, the Klyuchi station duplicated the first signal, which was read as a command to initiate the braking engine. The firing took place at the wrong moment and in the wrong direction, and the spacecraft therefore stayed in orbit, activating the automatic destruct system thirty minutes after the shut-down of the braking engine.

This would not have happened on a manned vehicle, as no self-destruct devices were carried. A year earlier, Kamanin had suggested that it would be better to launch the test flight with a single cosmonaut on board, as he could control the spacecraft; but he was overruled by Korolyov: 'He was troubled not by technical but by political considerations. Such a one-day test would not be another triumph in space.' The benefit of flying a crew was demonstrated by the loss of Cosmos 57, but according to Kamanin, in order to achieve this, 'you have to fly not for ballyhoo and nation-wide parades, but for the cause of spaceflight.'

As preparations continued for the manned launch between 15 – 20 March, the premature ending of the Cosmos 57 mission had left the subject of airlock ejection unanswered, although the system had been ground tested more than twenty times. The Soviets decided to continue ground testing up to the flight.

The second objective not achieved by Cosmos 57 was re-entry with the attachment ring on the side of the spacecraft, which changed the aerodynamic shape of the vehicle. The question of adequate parachute deployment also remained unanswered. In order to resolve these queries, a Zenit reconnaissance satellite of the Vostok design, which was undergoing launch preparation at the cosmodrome, was fitted with an attachment ring, and was launched as Cosmos 59 on 7 March.

The addition of the ring allowed for a test of the dynamics of the re-entry vehicle from orbit to landing, providing additional confirmation that the structure would not affect the aerodynamics of the descending sphere. Recovery had to wait one week, until its photography mission had been completed, but re-entry and landing indicated that there was no threat to a safe recovery of the crew, despite some loss of telemetric data.

Several days after the Cosmos 57 accident, a further drop test of a sharik ended in disaster when the parachute failed to open and the capsule smashed into the ground. The repeated failures caused some concern that bad workmanship, – or even sabotage – was to blame. As a result, the Committee on State Security (KGB) assigned a security guard to all preparations for Voskhod 2, which made everyone even more nervous.

Voskhod 2 is launched
Despite the recent setbacks, in their meeting on 16 March the State Commission set the launch of the manned flight to 18 March. The cosmonauts had flown to the cosmodrome on 9 March for final preparations for launch, where the crews were reaffirmed.

The night before launch, the crew stayed in the cosmonaut cottage and, following

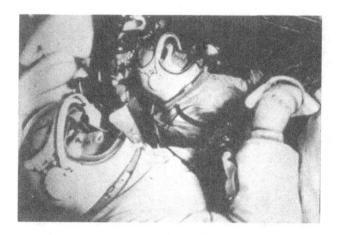

The Voskhod 2 crew complete a mission simulation before the flight. Belyayev is in the foreground, and Leonov is at the rear.

tradition, Leonov – who was expected to be the first to perform a space-walk – took the same bunk as had the first cosmonaut and the first woman in space.

On the morning of launch, the two spacesuited cosmonauts emerged from the bus to flurries of snow and a cold wind. At the top of the gantry they had to squeeze through the collapsed airlock and the hatch to gain entry to the capsule. The launch, at 10 a.m. MT on 18 March 1965, was highlighted by alarms in launch control, and was quite an ordeal for those on the ground, including Korolyov, who lit a cigarette – despite being a non-smoker! Voskhod 2 entered a 107.8 × 309.2 mile, 64°.79, 90.9 minutes orbit, at the same time setting a new altitude record for manned spaceflight. Gemini 3 – the first manned flight of the series, planned to perform the first manoeuvres by a crew – was set to launch in five days time; but Voskhod 2 was ready to steal some of the American thunder by setting its own new milestone.

As soon as orbit was confirmed, the crew prepared for the EVA, which could have taken place any time during the first six orbits while in range of tracking stations. The crew elected to attempt the EVA on the first orbit. Their call-sign was 'Almaz' ('Diamond').

Steps across the planet

The first Tass report of the mission gave the usual details about the launch, the crew, and the orbital data. Then, a short time later, the following statement was broadcast: 'At 11.30 a.m. Moscow time today, during the flight of the spaceship Voskhod 2, a man has left his spaceship in flight for the first time. Co-pilot Alexei Leonov, dressed in a special spacesuit with an autonomous life support system, stepped out into space carried out a series of planned experiments and observations, and returned safely to the ship.'

EVA preparations began as soon as the crew attained orbit, with the extension of the airlock from the side of Voskhod. Belyayev helped Leonov put on the life support system back-pack, and tried to pace the preparations as the eager Leonov

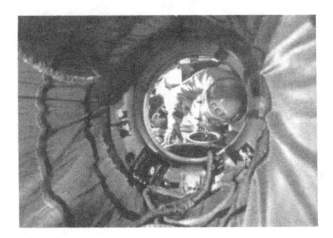

Leonov trains in the mock-up 'Volga' EVA airlock for his historic space-walk.

became impatient to step outside. Leonov then opened the inner hatch to his left, and with a brief 'good luck' from Belyayev, floated into the airlock and attached the safety tether to his suit. The hatch was closed, and the cosmonaut checked his suit for pressure integrity. Behind him, Belyayev, also wearing a full pressure suit, floated over to close the inner hatch and then returned to the controls to monitor his colleague's activities outside, by telemetry, radio, and TV.

At 11.34.51 hours MT, Leonov finally opened the hatch and prepared to exit. 'I gave a little push, and popped out of the hatch like a cork,' he explained later. It was the beginning of the second orbit as the spacecraft flew over the Soviet Union, as planned. Temporarily blinded by the intense sunlight, even through the protective visors, Leonov looked at the view below him. He could see across the Mediterranean, from the Straits of Gibraltar to the Caspian Sea. He pushed out to the full extent of his tether, – 17.5 feet – and found that it gave him tight control over his actions. The later American Gemini astronauts would not find matters so straightforward on their EVAs

Leonov's only tasks outside were to attach a cine-camera to the end of the airlock to record his actions, and to photograph Voskhod. He attached the camera on the hatch ring with no problem (he would recover it at the end of the EVA), and the exterior TV camera provided the world with ghostly images of the first man to walk in space. As he took off the lens cover, Leonov asked if he could send it into a new orbit, which he did by releasing it into space. However, when he tried to use the still camera on his chest, he found that the suit had ballooned and that he was unable to reach down to the shutter switch attached to the leg of the suit. He could not take the photographs of the Voskhod; but he was one of the foremost space artists, and he memorised the view before him in such detail that years later he was able to create his own images of his spacecraft in orbit and of his EVA.

Leonov later recalled his adventure, giving no indication of the difficulties he had encountered: 'As I pushed off, I felt as though the spaceship bounced off in the opposite direction. That's how it should have been, but the sensations were

The first EVA, by Alexei Leonov from Voskhod 2, 18 March 1965.

unfamiliar.' Swimming in space, he recalled, was not like swimming in water, when the liquid can be felt. 'In space you can float about as you like. I stretched out my arms and legs, and soared.' But, he added, 'It is true that my pressure suit resisted changes in the form of my body, so it required an effort to work.'

Then the artist in him emerged. 'I saw the Universe in all its grandeur. The view of the untwinkling stars on a dark velvet background changing to the velvet black of the abyssal sky was followed by views of the Earth. I saw floating beneath me great tracts of land. I recognised the Volga, [and] the mountain range of the Urals. Then I saw the Ob and Yenisei rivers as though I was swimming over a vast colourful map. There were clouds over the mountains, but over the coast there was wonderful sunny weather. Everything was clear for those who are familiar with brush and easel, it would be difficult to find a more majestic panorama than the one I beheld.'

Leonov discovered that by bending his arms or legs he reduced the internal volume of the suit, but the air pressure made any further bending increasingly difficult. When he tried to return the camera to the airlock, it stuck, and he had to exert considerable effort to push it down in front of him to ensure that the valuable film was retrieved.

Inside Voskhod, Belyayev held the spacecraft steady by using the attitude control system, and noted that he could feel the ship move as Leonov bounced against it, as his colleague had no means of controlling his actions. Leonov hit the spacecraft five times during his excursion, but encountered no problems. The Soviets stated that Belyayev could have exited the spacecraft to assist Leonov in an emergency, although it is questionable whether this would have been possible.

As Leonov twisted, soared and swam outside the spacecraft, the groundtrack covered 3,000 miles in ten minutes, until he was over the most northerly point of the orbit and about to move southwards across the Pacific Ocean. By then, Belyayev was ordering him back inside, to complete the excursion.

Initial reports indicated that Leonov could have stayed outside far longer, and that everything went as planned, with both cosmonauts feeling fine. It would be ten years before the truth emerged about the difficulties that he really encountered. The effort of trying to get the camera into the airlock made Leonov sweat profusely, and increased his body heat to more than the suit's system could accommodate. Despite his enormous strength, he was tiring quickly. He then found that his suit had ballooned so much that it would not fit into the airlock. He pushed harder and harder but to no avail, as his body temperature increased and sweat floated into his eyes and began to fog his helmet, which he was unable to clear. His attempts to solve the problem only made his heart beat faster, and he breathed harder, making matters worse by drawing oxygen from his own back-pack and the spacecraft's supply.

There was only one solution – and it was dangerous. He could reduce the pressure of the suit to allow him back into the airlock. Leonov lowered the pressure of the suit from 0.4 atmosphere to 0.3, but he still could not fit in the tight airlock hatch. He then lowered it to the safety limit of 0.27, but found that even this was not enough. Now the situation was becoming one of life or death at the moment of triumph. There could be no help. He was alone, and had to make a decision that could threaten his life, – to lower the pressure to 0.25 atmosphere. Then again, if he did not he would not be able to get back in, and would be dead anyway! With desperation combined with strength and determination, Leonov forced himself back into the airlock and closed the outer hatch.

More than thirty years later, Leonov revealed that he had had, in his helmet, a suicide pill which he could take if he had been stranded outside, leaving Belyayev to return without him! It also appears that he nearly suffered heat stroke, as his core body temperature climbed $3°.1$ (F) in twenty minutes. Leonov himself stated that he was 'up to his knees in sweat' – so much so that his suit sloshed around as he moved back into the cabin, to be greeted by a very happy and much relieved Belyayev.

The first EVA had lasted 23 min and 41 sec, with the cosmonaut outside for 12 min 9 sec. The world celebrated another Soviet space first, and Western newspapers once again highlighted that the US was 'playing catch-up'. The *New York Journal* headlined: 'A Walk In Space – And A Russian Takes It.' In London, the *Evening Standard* printed a cartoon of two astronauts leaping aboard a Gemini, with the caption 'Follow that cab!' Mstislav Keldysh, President of the Soviet Academy of Sciences, indicated that in future, cosmonauts would find EVA to be easy – although Gemini astronauts would soon discover that this was not quite true.

The recovery of Voskhod 2

Following the EVA, the airlock was jettisoned, and the crew settled down to completing the rest of their rather quiet flight programme. They reported tumbling by one revolution every 20 – 40 seconds, but were not bothered by it, and did not correct it until shortly before entry. More worrying was an over-saturation of oxygen in the atmosphere of about 45%, caused by the exit hatch not being completely sealed, and requiring the onboard systems to compensate. Consequently, the atmosphere was highly flammable, and both cosmonauts moved very carefully, as the slightest spark could have resulted in a flash fire and an explosion. Lowering the temperature inside reduced the humidity in the air and stabilised the atmosphere to safer levels.

After the EVA, Leonov's first duty was to write everything in the log-book, before both men continued taking still and motion photographs for the rest of the day. In a

Voskhod 2 lands in snow-covered forest many miles from the original landing area.

medical experiment they determined whether weightlessness had any effect on the colour perception of the human eye. Belyayev recorded a 26.1% reduction in his ability to view certain colours, while Leonov recorded a loss of 25%.

They then settled down to a well-earned meal (using the same type of supplies as Voskhod 1) and sleep period, before entry and landing planned for the following day, during orbit seventeen. In his diary entry, Kamanin noted that this was 'the most tense day in the history of our space programme to date.'

When the automatic guidance system failed to align the spacecraft correctly for entry, Voskhod 2 became stranded in orbit. Belyayev was ordered to initiate manual orientation for entry on either the eighteenth or twenty-second orbit, which meant that the landing area would then pass from beyond the normal flat steppes in Kazhakstan to an inhospitable area which was not usually considered for landings. However, as the cosmonauts tried to manually orientate the spacecraft, they found that, clad in their suits and strapped to the couches, they were unable to do so. To solve this (according to some accounts), Leonov floated below the couches and held on to a prone Belyayev 'lying' across the couches to reach the Vzor apparatus. It must have been very uncomfortable, and would have looked very strange.

Belyayev managed to successfully orientate the spacecraft, but in struggling to return to their seats it took an extra forty-six seconds to restore the centre of gravity before manually firing the braking engine. This extra forty-six seconds led to their overshooting the intended landing point by a further 1,242 miles, to land in the snow-covered forests 112 miles north-east of the town of Perm, in Siberia. Even though they were now coming home, their problems were not over, as once again the sharik failed to achieve a clean separation from the Instrument Module, which separated only during the heat of re-entry.

At 12.02 MT, Voskhod 2 crashed through the trees and landed, wedged between two fir trees, but safely in the deep snowdrifts of the forests of Siberia. Telegraph signals indicated that all was well, but it would take some time for the rescue teams to reach the two cosmonauts. Indeed, it was four hours before they had a visual sighting of the capsule and were able to report that both cosmonauts were apparently alive and well. One of the commanders of the search helicopters radioed that the landing site was 'on the forest road between the villages of Sorokovaya and Shchuchino, about nineteen miles south-west of the town of Berezniki. I see the red parachute and the two cosmonauts. There is deep snow all around.'

As there was no clearing nearby they were unable to land the search-and-rescue helicopters, although one of them landed three miles away, but could not then find the crew. Warm clothes and supplies were dropped, and the cosmonauts spent a cold and uncomfortable night in the capsule. On Radio Moscow, music was played for some time without announcing any news of the landing. At first the music was quite light, but gradually the mood became more sombre–which was usually the sign of bad news. Eventually, however, came the good news that the cosmonauts were down, and were safe.

Temperatures dropped to –5° in the area, and, according to several accounts, the cosmonauts had to retreat to the safety of the sharik as wolves surrounded the landing site. The next morning, search parties took four hours to ski through the dense forest to reach the cosmonauts. It was too dangerous to try to airlift them from the site by helicopter, so they had to spend a second night in the forest – this time with about twenty rescuers. The next morning they skied more than a mile to begin

Leonov – the first man to walk in space, on Voskhod 2 – returns to a more domestic assignment at his apartment at Star City in 1965.

the long journey back to Tyuratam, and eventually arrived forty-eight hours after landing.

To disguise the dilemma that they faced, the official explanation to the world's media was that the Voskhod had landed in the prescribed area, and that the two cosmonauts had chosen to rest for two days after their challenging mission. Full details of their experiences were not revealed until several days after the compulsory post-flight briefing.

At Tyuratam, Korolyov and the other Chief Designers greeted the cosmonauts as heroes. Korolyov raised a toast to future success: 'Before us the Moon. Let us all work together with the great goal of conquering the Moon.'

A follow-on mission planned as Gemini flies

The last cosmonauts to fly a Vostok-type spacecraft in orbit were back safe and sound – but only just – and while the Soviet Soyuz now suffered increasing delays, the American Gemini programme began to score success after success in 1965. Gemini 3 completed the first manoeuvres of a manned spacecraft in March, and the first American space-walk was performed from Gemini 4 in June. In August, Gemini 5 broke Bykovsky's endurance record of five days (and set a new record of eight days), only to be surpassed by the Gemini 7 crew, who flew fourteen days in December. Gemini 6 also joined them in space for a few hours and at a distance of a few feet, in the first true space rendezvous experiment. The Americans were not just catching up; with each new flight, they were overtaking the Soviets.

There had been several plans for follow-on missions after Voskhod 2. As early as 28 October 1964, Kamanin indicated that two 12–15-day flights by unsuited single cosmonauts could be flown, as well as two 'specialist scientific research' missions and a second EVA experiment. Meanwhile, Korolyov had devised his own plan in February 1965, assigning missions to the five remaining Voskhod-type spacecraft (production serial numbers 5–9). These were planned as a 15–30-day flight with animals, targeted for July–August 1965 (3KV #5), and a fifteen-day artificial gravity experiment with a pilot and a scientist for September–October 1965 (KV #6). Then there would be a 15–18-day biomedical flight with a pilot and a doctor (with up to four days in 'artificial gravity') planned for March–April 1966 (3KV #7). The programme would be completed with two 3–5-day EVA missions (with two pilots on each, and including two or three EVAs up to 300 feet from the capsule) some time in 1966.

These ideas changed several times, but essentially they formed the planning for what was flown as a Cosmos long-duration animal flight (Cosmos 110), and what should have been Voskhod 3 and Voskhod 4 long-duration flights and Voskhod 5 and Voskhod 6 EVA missions.

Delays and frustration

For several months after Voskhod 2, the schedules and objectives for other Voskhod missions remained undecided. This caused even more friction between Korolyov and the Air Force before they gradually agreed to deliver spacecraft 5 in October, spacecraft 6 in November, and spacecraft 7 in December 1965. The other two would follow in February (#8) and March (#9) 1966. Any experiments proposed for these vehicles were to be supplied six weeks prior to delivery.

On 30 August the Air Force finalised its own schedule for any future Voskhod missions. There would be a two-man 15-day flight of Voskhod 3 – with artificial gravity and military experiments – in November 1965, and a 20–25 day Voskhod 4 mission, with one cosmonaut performing military experiments. This would possibly be in the spring, but no launch date had been decided. The Voskhod 5 and Voskhod 6 missions in the summer of 1966 would include 15-day flights with EVA and some military experiments.

This was a radical change from the biomedical flights that Korolyov wanted, and when he learned of these plans three weeks later, he was outraged. Even when small changes were agreed by the Air Force, it took several weeks to have them authorised, and saw only an increase of the Voskhod 3 mission to twenty days. In October, OKB-1 turned down the Air Force proposal, stating that 'tens of vehicles' would be required to fulfil their objectives. Korolyov tried to include biomedical experiments on the proposed flights, and attempted to drop the artificial gravity experiments to allow the inclusion of the medical experiments. In November, he decided not to build spacecraft 8 and 9, and to instead transfer the proposed EVA experiments to Soyuz.

The delays to future Voskhod flights that could have filled the gap until Soyuz flew were often blamed on the disputes between Korolyov and the Air Force over a firm flight plan. One day after the historic meeting of Gemini 6 and Gemini 7 in space, the Military Industrial Commission failed to achieve any further planning for Voskhod missions, but did agree to fly two Voskhods before the next Party congress in March 1966.

After a six hour meeting at Star City on 4 January 1966, Korolyov could present no more information on the prospect of new Voskhod missions. Just ten days later, Korolyov died during surgery, and the fate of further missions passed to his successor, Vasily Mishin. For several months the planning discussions dragged on, without resolution. There is no official date given for the ending of Voskhod, but by May 1966 the Air Force seemed to have abandoned all realistic hopes of follow-on missions.

From the planning, the flights of Voskhod 3–6 would have included:

Voskhod 3 It appears that Boris Volynov and G. Katys were considered to be the prime crew for the 15-day Voskhod 3 mission, and trained for it for several months. Volynov had already served as back-up on four occasions, and Katys had been back-up to Feoktistov on the first Voskhod. However, there were plans to fly World War II veteran Beregovoi instead of Volynov. (Beregovoi – born on 25 April 1921 – had been added to the cosmonaut team on 25 January 1964. He was an experienced test pilot and was already a Hero of the Soviet Union, due to his exploits during the war.) An Air Force training group of five cosmonauts (Volynov, Beregovoi, Shatalov Demin and Artyukhin) was formed in March 1965, with the assignment of the Voskhod 3 crew agreed on 17 April: Volynov–Katys (prime), Beregovoi–Demin (back-up), and Shatalov–Artyukhin (second back-up).

Katys was assigned purely for the gravity experiment that was cancelled in

Winter training for the Voskhod 3 crew. Volynov is in the foreground, and Shonin is reaching into the capsule.

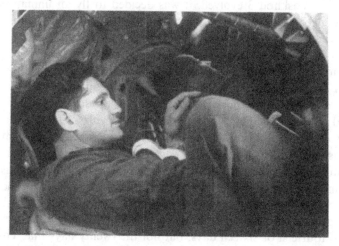

Cosmonaut Boris Volynov during Voskhod training. He was in training as Commander of Voskhod 3 before it was cancelled in 1966.

Cosmonaut Georgi T. Beregovoi, during a training session inside a Voskhod capsule.

October, when the mission was also extended to twenty days. It was decided to replace Katys with a stronger candidate, and when asked, Volynov chose Gorbatko, although the civilian scientist continued training for some time on the flight, with little hope of flying. By mid-January 1966, the crewing was changed again, to Volynov and Shonin (replacing Gorbatko), with Beregovoi and Shatalov as back-ups, and the second back-up crew of Gorbatko and Katys.

If it had flown, the mission would have attempted an elliptical orbit with high apogee of about 600 miles, to investigate the effects of the Van Allen radiation belt on the cosmonauts' health, and to set an altitude record for manned spaceflight – an objective of the cancelled Vostok follow-on flights. One military experiment would have entailed the visual observation of the launch of four missiles at different times and locations across the Soviet Union, and the tracking of their heat trails with the Svinets (Lead) instrument. The ill-fated Soyuz 11 crew finally performed this experiment onboard Salyut 1 in 1971.

As plans for Voskhod 3 evolved, it was decided to fly an unmanned precursor mission (using vehicle #5) with two dogs, on a thirty-day flight beginning in the first half of February. After the return of the dogs, Voskhod 3 would fly up to an eighteen-day flight, beginning between 10 and 20 March.

Several problems plagued the preparation of the missions, including the absence of a system for measuring water consumption. The only method available to the crew was to count the swallows of water that they drank – which was further complicated by the need to use the water for mixing with freeze-dried food. Another problem was the insulation of the inside of the capsule. A layer of foam rubber wall covering was used, which, although less flammable and more hygienic, released small fibres and dust particles when touched. These would irritate the mucous membranes of the cosmonauts, and the insulation was therefore removed. There was also continuing concern over the parachute recovery systems, which had shortly before failed four ground tests. Finally, the oxygen regeneration system was suitable for flights up to fourteen or sixteen days, but not the twenty days that Voskhod 3 was supposed to fly.

The prime and back-up crews passed their examinations on 28 February, and

Volynov (*right*) and Yegorov (*left*) undergo classroom studies as part of their Voskhod training programme.

As part of his Voskhod training, Shonin uses a camera on a high-altitude aircraft.

launch was planned for 20–22 April. Cosmos 110 flew quite successfully in February–March, and revealed that the spacecraft stood up well to the flight. However, on 27 March an 8K78 R-7 suffered a third-stage failure during an attempt to launch a Molniya comsat. This vehicle was almost identical to that used to launch

The Voskhod 3 crew (*left*) Shonin and (*right*) Volynov, in Red Square, Moscow.

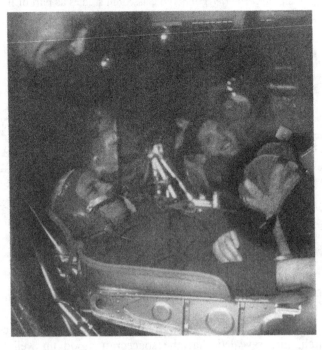

Inside a Voskhod crew compartment. In the centre couch is a smiling Boris Volynov, the back-up Commander of the first Voskhod crew.

Voskhod, apart from the addition of a fourth stage. As this new problem was investigated, the launch of Voskhod 3 was delayed, and the crews were sent on holiday. By May, the preliminary order was received to launch at the end of the month, but as Kamanin observed: 'We have already received so many orders like this that we don't believe this one will turn out to be final.' He also noted that the crews appeared to be tired during training, and although they tried to disguise it, the uncertainty must have been telling on them.

Following a meeting of the Military Industrial Commission on 10 May, it was reported that at last everything was ready to fly Voskhod 3 between 25 and 28 May. However, VPK Chairman Smirnov expressed the wish to cancel the flight on the grounds that it would offer nothing new, would also slow work on the already delayed Soyuz, and that 'a flight without orbital manoeuvres and docking will show the fact that we are lagging behind the United States, and will be seen by the public as proof of the Americans' superiority.'

There were arguments in favour of the mission on the grounds of setting a new endurance record, but there was no longer the unanimity between the members of the State Commission at their next meeting on 12 May. With the loss of Korolyov, there was no common will or sense of urgency to carry out the flight. Despite this, support for the mission was given for a launch between 22 and 28 May, providing the problem with the R-7 third stage could be resolved. By late May the problem remained, but Kamanin refused to believe that either the R-7 or the air regeneration system was the real cause of the delays. 'The main reason why the flight may not take place is that Smirnov and Pashkov have spoken out against it.'

On 16 May the flight was once again postponed (this time until mid-July), and the crews went on another holiday. By the time they returned, the focus had shifted to Soyuz. They resumed training through to November, when there was a final attempt to suggest that a flight with the first Soyuz could be attempted in order to once again upstage the Americans and their planned first Apollo flight for the first quarter of 1967. But this was interpreted more as wishful thinking than as real flight planning.

Voskhod 4 Apart from Voskhod 3, no other mission in the series came close to flying, and it is uncertain if any specific crew-training ever occurred. The Voskhod 4 mission was intended to be a fifteen-day biomedical flight, with a doctor in the crew, and it appears that several medically qualified candidates were suggested for the mission. These included Lazarev and Sorokin from the Air Force (and who were part of the Voskhod 1 training group), former candidates Yaroshenko, Ivanov and Voskresenskiy, and, from the Institute of Medical and Biological Problems (IMBP), Y. Ilyin, A. Kiselyov, Y. Senkevich and S. Nikolayev. Several of these underwent preliminary screening, as well as working on experiments for the proposed biomedical mission, but none seem to have been before the mandate commission to confirm them as 'real' cosmonauts.

It appears that when Korolyov confirmed the artificial gravity experiment in November 1965, he also supported the idea of including a doctor in the crew. But he also suggested that the gravity experiment and the doctor should fly on separate missions. Late in 1965, there was a suggestion to fly Voskhod 4 on a 20–25 day

Tereshkova in a Voskhod EVA suit. This could be a training photograph but is more probably a posed portrait.

military flight with one cosmonaut. Then, in March 1966, Beregovoi and Katys appeared to be assigned to the mission, but it is unlikely that any training was accomplished.

Voskhod 5 The idea of an EVA flight with a female cosmonaut appeared early in 1965, and received some crew assignments on 17 April, with Ponomaryova and Solovyova as prime crew, Zaikin and Khrunov as first back-up crew, and Shonin and Gorbatko as second back-up crew. It was also suggested that a manned manoeuvring unit should be used to increase the propaganda effect of such a mission.

There were arguments both for and against the idea of flying two women, with considerable opposition from the male cosmonauts – most notably, Gagarin. One argument was that Khrunov had already received far more EVA training as back-up to Leonov on Voskhod 2.

When the announcement was made, Kamanin indicated that the two unassigned women (Kuznetsova and Yorkina) would not be ready to fly for another five or seven years (1970–1972) – a decade after their selection for Vostok 6. The assignment of Ponomaryova as Commander brought objection from Tereshkova, Kuznetsova and Yorkina, who all thought she was unsuitable for the flight for the same reasons that she was not assigned to Vostok 6. Kamanin recognised this objection, but noted that she was the only suitable candidate for the role of Commander, and that Solovyova would be better trained for the EVA. He also warned the group that if the arguments and disagreements were not settled within four weeks, so that the pair could fly, the whole group would be disbanded.

Apparently the dispute was resolved – superficially, at least – but a second problem arose in that the team at Zvezda was categorically opposed to an all-female flight and refused to fabricate the EVA suits for them! During the final months of 1965, the idea of an EVA on the female flight received less and less support, in favour of a long-duration flight, or cancellation.

By early 1966 the State Commission had expressed its support behind the all-female flight, and ordered the resumption of training for a 15–20 day mission. This time, no male cosmonauts were assigned to the flight and, probably as part of the agreement to let the prime crew fly, Yorkina and Kuznetsova were assigned as the back-ups. It is unclear how long they trained before the mission was abandoned, as Ponomaryova recalled: 'Just as I imagined, our training ended very abruptly as they sent us off on holiday and everybody forgot all about it.' Ponomaryova and Solovyova however, underwent some training, and Ponomaryova's call sign was *'Silver Birch'*.

Voskhod 6 The original plan for this mission was to fly a two-week EVA mission crewed by members of the back-up crew for the original female Voskhod mission (Khrunov, Shonin, Gorbatko and Zaikin), who were still in training as a team in September 1965. The exact crewing is unclear, but by far the strongest candidate was Khrunov. Shonin, on the other hand, had registered unusual cardiac readings from centrifuge tests some years earlier, and also tended to perspire while performing even the lightest physical tasks. This was a major factor in that, for the proposed mission, the cosmonauts would spend the two weeks in suits. Some reports have indicated that the UPMK manoeuvring unit could have been assigned to this flight, but there is no evidence of training on such a unit. However, at least Khrunov and Gorbatko received EVA training during assignment to Voskhod 2.

Voskhod 7 This mission proposal evolved from discussions, in late 1964, between Korolyov and Raushenbakh on developing a method and mission profile to test an artificial gravity system in Earth orbit. The idea would have been to separate the system from the upper stage by about 15–30 feet and deploy a tether, where a solid fuel engine would fire (it is not clear from which location), and separate the two vehicles to the extent of the 3,000-foot tether. The two craft would then begin to

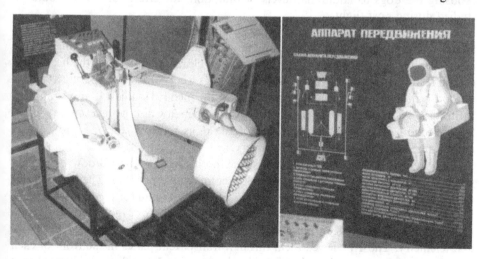

The proposed Voskhod astronaut manoeuvring unit at the Zvezda museum, with an accompanying explanatory diagram of how it worked and how it was worn.

rotate around a common axis, initially at 1°.5 per second to simulate 1/6 gravity loads – lunar gravity! It was planned to attempt to generate electricity from the interaction between the current-generating tether and the geomagnetic field of the Earth. A second demonstration of gravity would be at 1,000 feet and 12° per second. After two days tethered, the craft would separate, and then complete the rest of the 10–15 day mission.

Journalists fly Voskhod? Korolyov had expressed dissatisfaction at the level of post-flight reporting by the cosmonauts, who were pilots and engineers, not reporters. So he produced the idea of flying a journalist in space, – possibly onboard a Voskhod. Three journalists appear to have been identified in February 1965: Y. Golovanov, Yu Letunov and M. Rebrov. None of them were formally inducted into the cosmonaut team or even went through training. With the delays to the Voskhod series and the death of Korolyov, their mission was abandoned – assuming that it ever officially started.

Cosmos 110

Although there were no further manned flights of the Voskhod after Voskhod 2, one more man-related Voskhod mission – which was also the final Soviet orbital flight of canine passengers – was flown in February/March 1966. Designated Cosmos 110, the mission was led by a medical team that included former Voskhod 1 cosmonaut Dr Boris Yegorov, plus Vasily Parin and Vladimir Pravetsky.

Voskhod spacecraft #5 – carrying two dogs, Veterok (Breeze) and Ugolek (Blackie) – was launched on 22 February 1966, as a precursor for the manned Voskhod 3 mission. Initial orbital parameters were 108.7 × 318.1 miles, inclined at 51°.9 with a 95.3-minute orbital period. The spacecraft was then moved to a 118.0 × 548.0-mile orbit which ensured that it passed through the Van Allen radiation belts, exposing the dogs to additional levels of radiation, the effects of which would be studied during post-flight.

Each dog was located in a separate compartment inside the sharik, and was attached to a range of sensors recording biomedical data throughout the flight. This included pulse, respiration, arterial pressure and electrical activity of the heart, and certain responses of the nervous system. Both dogs were control-fed specific amounts of food via tubes into their stomach, as part of the medical studies. It also simplified the feeding system for two dogs over two weeks. Each dog also wore a sanitary 'coat' for collection of body waste, and one of the dogs received anti-radiation drugs, administered through a tube into an artery.

Data indicated that after a week in space, both dogs had settled down, and had stopped 'rocking their heads' in search of a comfortable position. Their heart rates settled to 70–120 beats/min for Veterok, and 60–90 beats/min for Ugolek, while their respiration per minute was 12–14 for Veterok and 18–21 for Ugolek.

As with the Vostok Korabl-Sputnik mission, both dogs were monitored via TV and telemetry. The TV viewed them through transparent plates in the front of each capsule. Ugolek was used as the control subject, while Veterok was subjected to a range of medicines.

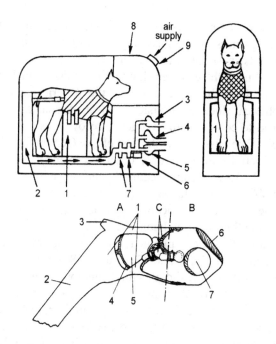

(*Top*) system for individual collection of solid and liquid wastes in dog cabin for 30-day biosatellite (Cosmos 110). 1) liquid waste container; 2) solid waste container; 3) main air fan; 4) back-up fan; 5) high-power fan for periodic cleaning; 6) device for automatic activation of fans; 7) filters; 8) transparent dome; 9) air sampler. (*Bottom*) sanitation device for dogs on short-duration spaceflight (Korabl-Sputniks). A) urine and faeces receiver; B) supporting 'coat'; C) harness strapping; 1) sheath; 2) tube; 3) opening for tail; 4) body/device seal; 5) opening for hind legs; 6) opening for legs; 7) opening for forelegs. (Courtesy NASA.)

The decision to land was taken on 15 March, about twenty days into the flight, as data showed that the support systems were in relatively good condition and the both dogs appeared to have satisfactorily withstood nearly three weeks in space. The landing, during the 330th orbit, took place at 17.15 hours MT, in slightly foggy conditions, 130 miles south-east of Saratov, more than 37 miles from the planned landing point. Within one hour, a team of seven parachutists had located the capsule, but it was not until the next morning that the full recovery team was able to land. Upon initial examination, both dogs had uncoordinated movements but seemed satisfactory, and appeared to have suffered no visible effects from flying through the radiation belts.

However, due to their inability to move for twenty-two days, the dogs' physical condition had worsened as the flight progressed. The recovery on 16 March (earlier than planned) was successful, but it was found that both dogs were suffering from extensive quantities of calcium in their blood and urine, and it was to be a further ten days before they were fully recovered. Two control dogs underwent the same test confinement, experiments and feeding regimes on Earth, to provide direct comparison with the flight pair.

The Soviets were surprised and alarmed at the poor condition of the dogs, as it indicated serious consequences for planned long-duration missions. It would be four years before cosmonauts in space could test protective measures developed as a result of this flight, – during the eighteen-day Soyuz 9 mission in June 1970.

The landing occurred on the same day that the American Gemini 8 mission achieved the first physical docking in space with an Agena target (although the astronauts were forced to complete an emergency recovery as they almost tumbled out of control in orbit). The mission was the first of five Gemini flights in 1966, when no Soviet cosmonaut left Earth. (The only other occasion that that has occurred since 1961 was in 1972.) Each Gemini featured rendezvous or docking and, from Gemini 9 (June), EVA activity, during which each of the four pilots reported problems with fatigue, maintaining location, and exhaustion, none of which had been made public by the Soviets after Leonov's EVA.

High-apogee missions that were planned for Voskhod were achieved by the Americans during Gemini 10 (July) and Gemini 11 (September). Gemini 11 also performed tethered operations, as did the final mission, Gemini 12 (November), which also completed a successful demonstration of EVA activities.

As America moved from the highly successful Gemini programme to the Apollo programme in 1967, so the Soviets abandoned the Voskhod series for cosmonauts and concentrated on Soyuz. Both programmes were striving for the Moon, and both would encounter tragedy and setback. But the Vostok capsules continued their missions in secret – an unmanned legacy for Korolyov, who was not there to see the move towards the Moon.

Loss of the Chief Designer

For several years, Korolyov's health had been deteriorating. He made several trips to the hospital, and needed to convalesce, but he refused to stop working or to slow down. Some of the frustrations and delays caused him to argue with other designers and members of his team, as he tried to forge ahead with Voskhod and plans for Soyuz that could take the first cosmonauts to the Moon – hopefully, ahead of the Americans.

In December 1965 he was diagnosed with a bleeding polyp in the straight intestine, and was prescribed surgery as soon as possible. On 4 January, – the day before he entered hospital, – he was still at his desk, working late into the night. 'I will die right here at this desk!' was his frequent comment when people suggested he should go home.

On 5 January 1966 he was again in hospital for what was expected to be routine surgery for the removal of a polyp from his rectum. On 11 January the Minister of Health, Academician Boris V. Petrovsky, performed a histological analysis by removing part of the polyp for examination, but the resulting bleeding was difficult to stem. Korolyov celebrated his 59th birthday on 12 January, and the operation was set for 14 January. Petrovsky thought that the operation would be simple and straightforward, and scheduled a separate operation on another patient immediately afterwards. But he was wrong. As the surgeon tried to remove the polyp, Korolyov haemorrhaged so severely that the bleeding could not be stopped. Petrovsky

After Korolyov's funeral on 17 January 1966, two guards stand beside his photograph and his awards beneath the nameplate on the Wall. (Courtesy Natalya Korolovya (his daughter), via Jim Harford.)

therefore cut into Korolyov's abdomen, which revealed a previously undiscovered cancerous tumour. The attempt to save him took more than eight hours, but it was to no avail, and the Chief Designer of the Soviet space programme died on the operating table. Had he lived, he would probably have done so for only a few months, as the tumour would have killed him.

The news of Korolyov's death spread immediately and Kamanin, upon hearing of the tragedy, recalled: 'Like an avalanche, this terrible misfortune came down upon us rapidly and unexpectedly. Korolyov was the main author and organiser of all our space successes. His personal contribution to cosmonautics is limitless. He could have done much more, but left us when his talent was in bloom.'

Only after his death did Korolyov receive the worldwide credit due to him, in his obituary in *Pravda* on 16 January 1966. His state funeral on 17 January, during which he was entombed in the Kremlin Wall, was a sombre affair. Mourners, from the highest political officer to the citizens of Moscow. all paid tribute to the great man of space exploration. Several of them spoke of his talent, ideals and burning desire, and the final speaker was Yuri Gagarin, who was little more than two years away from his own untimely death in an aircraft crash in March 1968. Gagarin – always the favourite of Korolyov – offered his view of the lost Chief Designer: 'The name of Sergei Pavolovich is linked with a whole epoch in the history of mankind – the *first* flights of the artificial satellite, the *first* flights to the Moon and to the planets, the *first* flights by human beings in space, and the *first* emergence of a human being into free space.'

The death of Korolyov marked the end of an era in the Soviet space programme – an era of outstanding success balanced by tremendous risk, inspired by one unique genius. It also signified a new direction which, after the loss of the Moon, matured into the space station programme of Salyut and Mir, and which now forms an integral part of the International Space Station programme four decades after Korolyov and Gagarin blazed a trail into space.

A fitting comment came from American astronaut Bill Shepard at the launch of the first resident crew from the Baikonur Cosmodrome (Tyuratam) to the ISS in November 2000. At lift-off, he echoed Gagarin's words, Korolyov's drive, and Tsiolkovsky's dreams, by beginning a new era of manned spaceflight with *'Poyekhali!'* – *'Off we go!'*

The legacy

A total of eight Object 3Ks (manned versions) carried cosmonauts under the Vostok and Voskhod designations, and there were a further ten precursor missions for man-related flights between 1960 and 1966. Therefore, eighteen 3K vehicles were assigned to the manned programme of the USSR from 1960 to 1966, with at least a further five missions planned but not flown.

UNMANNED PROGRAMMES USING THE VOSTOK BUS

Although the manned programme ended in 1966, unmanned military versions of the Vostok-type spacecraft had begun flying in 1961 under the code-name Cosmos. Between 11 December 1961 and 31 December 2000, more than 775 spacecraft of the 'Vostok' design were launched by variants of the R-7 rocket, under the Zenit, Resurs, Bion, Efir, Energiya or Foton designations. Of all these launches, there were only twenty-five known launch failures (3.2%). This long programme kept the 'Vostok' design flying missions into space for forty years after it was first designed. The following is a brief summary of the development and variants flown under these programmes.

The Zenit photoreconnaissance programme

History Work on the photoreconnaissance series of Soviet satellites has its origins in the Government decree of January 1956, which was followed by two years of design studies. The development was in parallel with those for the first artificial satellites selected for the IGY of 1957–1958 (as discussed in Chapter 2), which resulted in the launch of Sputnik on 4 October 1957. In addition to the scientific and propaganda benefits that satellites would generate, the Soviet government also realised that satellites could provide a valuable strategic platform for remote aerial reconnaissance of the United States and other countries which were not part of the Soviet bloc régime.

Such a military orientated project was already under development at Korolyov's OKB-1, and was called Object OD-1 (Orientirovanny D, or Orientated D). It was originally designed as an instrument module with a cone-shaped instrument section, to be launched on the 8A92 version of the R-7 ICBM. Korolyov planned to begin the orbital flights in 1958/59, and estimated the satellite weights to be not less than 3,300 lbs, including the final stage of the launch vehicle.

The early version of the satellite featured an instrument section that contained photographic and special radio equipment, as well as the support hardware to ensure independent operation in orbit. There would also be a separate recoverable capsule for the return of exposed film. The instrument section varied in shape from cylinders with a conical or spherical base, to a wide variety of complicated forms. The design of the instrument unit was dictated by the smaller size of the return capsule, which meant that all the elements not needed for entry and recovery were placed in a larger module. Incorporated into this design was provision for improvements, to extend orbital lifetime and to upgrade onboard equipment as new technology was developed.

The attached capsule was conical in shape, and contained the exposed film cassettes, with equipment necessary to ensure it functioned during re-entry and landing. A detachable engine – which served as a braking engine to initiate de-orbit after competing its mission – was installed on the capsule. During the design stages a variety of sizes were evaluated, ranging from 1.5 tons to 4.5 tons. Even at this early stage, the proposed photographic equipment included a camera with a focal length of about 3.2 feet.

By 1959, two versions of the reconnaissance satellite were under evaluation, both to be launched by the R-7. OD-1 had a mass of 1.5–2 tons, and was capable of being launched without upper stages, while OD-2 was a heavier version, at around 4.9 tons, which required the use of an upper stage. In early 1957 a two-stage version of the R-7 (8A92) was proposed, based on the original ICBM version (8K71). This was planned to be capable of lifting a 3,300–3,750-lb payload into orbit. Other improvements to upgrade the R-7 included increasing the specific impulse of the RD-107 and RD-108 engines, installing improved radio systems, and upgrading its control system.

The 8A92 was to be used to launch OD-1 satellites, while OD-2 would require an even larger version of the R-7. During 1957, OKB-1 Project Department 3 proposed a third stage for the R-7, designated Blok Ye, which would be used for lunar probes. This variant was also adapted to launch the heavier Object D-2. Two versions of this launcher (8A92 and 8A93) were developed, with different engines in the upper stage. The 8A92 version featured a 5-ton thrust LOX–kerosene engine, jointly developed by OKB-1 and Kosberg's OKB-154, with a payload capacity of 5.3 tons in low Earth orbit (LEO). The 8A93 version employed a 10-ton thrust LOX-UDMH engine (designated RD-109) developed by Glushko's OKB-456, with a payload capacity of 4.7 tons in LEO. The 8A93 was to be the primary launch vehicle, with the 8A92 as back-up.

During 1958, OKB-1 Project Department 9 was developing the concept of a manned spacecraft, also under the designation OD-2. In this design, a 7.5-foot spherical descent module would be attached to an unpressurised instrument module. The design could also be used as an unmanned version for systems testing and flights of biological payloads prior to manned flight, and for replacing the cosmonaut with photoreconnaissance equipment.

In November 1958 the Council of Chief Designers decided to adopt the Department 9 proposal in favour of the original OD-2 reconnaissance satellite design. During the same month, a decision was reached to abandon the OD-1 proposal and to absorb all design efforts for unmanned reconnaissance satellites into the common design for the manned spacecraft. The Government Decree to authorise the OKB-1 reconnaissance satellite programme came on 25 May 1959, under the generic named of 'Vostok' ('East'), with Vostok-2 (Object 2K) and Vostok-4 (Object 4K) assigned to the unmanned reconnaissance satellite programme.

Over the next twelve months, the name 'Zenit' ('Zenith') was adopted for the military satellite project, to distinguish it from the manned versions. Vostok-2 became Zenit 2, while Vostok-4 was renamed Zenit 4 and was planned to carry high-resolution photographic equipment.

Difficulties with developing the RD-109 engine for launch vehicle 8A93 led to the abandonment of the launcher project. Instead, it was initially decided to use the 8K72, developed for the Vostok manned launches. Then, once inertial control and telemetry sub-systems on the R-7 had been improved (to give a higher assurance in placing objects in orbit), the 8A92 would be used for Zenit 2 missions. The larger Zenit 4 would require the 11A57, which, in 1976, was developed for the lunar

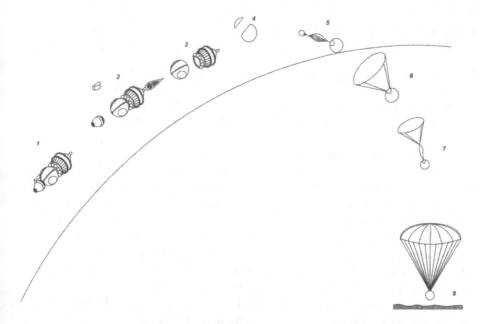

Recovery sequence of Zenit reconnaissance satellites (*left to right*) 1) the capsule in orbit, 2) retro-engine fires, and ejection of Nauka module; 3) the sharik separates for re-entry; 4) parachute cover ejected; 5) drogue parachute deployed; 6) main parachute deployment; 7) descent to ground; 8) landing in the Soviet Union. (© 1998 Ralph Gibbons.)

programme and had been used to launch Voskhod spacecraft. In 1967 the 8A92 was retired from the reconnaissance programme, and until 1972 all launches from either Tyuratam or Plesetsk were by 11A57, which, in 1976, was also retired. The 11A511M ('Soyuz-M') version of the R-7 was the Zenit launch vehicle for two years from 1972, until it was replaced by the 11A511U ('Soyuz U') variant, which had also been introduced in 1972. This continued to be the sole launch vehicle for Zenit-class satellites through 1999.

First generation Flights of the Vostok-derived spacecraft bus under the Zenit 2 and Zenit 4 series are usually termed 'first generation' in Western terminology, as these spacecraft did not include facilities for orbital manoeuvring, and usually had only an eight-day life-time. The non-manoeuvring Zenit 2M is also designated 'first generation', although it had an increased flight time of twelve days. On some of these missions, a 'Nauka' science package was attached, and was discarded in orbit prior to recovery.

Second generation Flights of Zenit 4M featured the ability to manoeuvre in orbit, and it flew for about thirteen days.

Third generation The Zenit 4M transmitted in morse code, while the Zenit 4MK series transmitted two-tone telemetry. It is therefore reasonable to assume that the 4MK represented a new generation of satellites that would later include Zenit 6 (flying in specific orbital regimes) and Zenit 8, on missions lasting up to three weeks.

Fourth generation These satellites were introduced on a new satellite bus, not related to the Vostok design. They were designated Yantar, and are not summarised here.

Scientific and remote sensing objectives In addition to flying the Zenit-class missions, Vostok-type buses were also used in a series of scientific missions that included remote sensing (Fram, Resurs), science (Energiya, Efir), biomedical research (Bion), materials processing (Foton), and commercial publicity (Resurs-500).

Photoreconnaissance – Zenit class
Zenit 2 The first series of reconnaissance satellites (programme code 11F61) began with a launch failure on 11 December 1961. The next launch placed the first Zenit in orbit, where it was given the cover name Cosmos 4. In this series there were eighty-one launches, seventy-four of which were successful. The final launch of the series was Cosmos 344, launched on 12 May 1970. There were also thirteen launches under the test development programme (up to and including Cosmos 20), of which three were launch failures. The spacecraft were launched from the Tyuratam and Plesetsk cosmodromes into orbital inclinations of between $51°.2$ and $81°.2$.

The major differences between the Zenit and the manned Vostok were the lack of the life support system, catapult seat system, manual controls, radio-telephone links and TV systems. Those systems that required very little change included the retro-rocket engine, orbital radio-control systems, radiotelemetry, and the landing system. Systems that required some adaptation from the original Vostok design were the thermoregulation system, the descent apparatus, the instrument section, and the

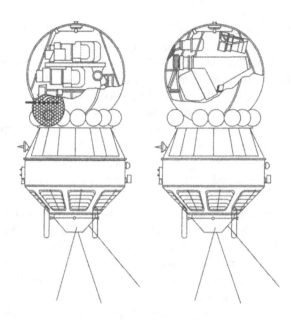

(*Left*) Zenit 2 and (*right*) Zenit 4 spy satellites. (© 1998 Ralph Gibbons.)

electrical supply system. The new systems introduced for Zenit-2 missions included the photoreconnaissance, photo-TV and special radio apparatus, command computer facilities, control systems for onboard complexes, and new radio lines. The instrument module on Zenit was slightly longer (8.8 feet) than on Vostok (7.3 feet), with a short cylindrical section added. The overall length of the Zenit was 15.7 feet, compared to Vostok's 14.4 feet. The diameter of the spherical descent module remained unchanged.

The photoreconnaissance equipment was housed behind one of the portholes in the sharik, and since the spacecraft orbited with its long axis parallel to its velocity vector, the result was an alignment at right angles to the vehicle's direction, so that it pointed 'downwards' towards Earth. The cameras carried on the original Zenit were one SA-20 with a focal length of 0.65 feet, the Fort-2R assembly consisting of three SA-20 and one SA-10 cameras (focal length 3 feet), and the Kust (Shrub) radio reconnaissance equipment. Some of the Zenit 2 spacecraft also carried experiments and test apparatus for scientific and meteorological research. In the event of landing outside the Soviet recovery zone, each spacecraft featured an automatic destruction system to prevent the hardware and data falling into foreign hands.

The photographic equipment remained inside the capsule during landing and, unlike the cosmonauts on Vostoks 1–6, did not eject. The Soviets stated that elements of the payload and the descent modules could be used on more than one launch, to reduce processing and launch costs.

Zenit 2M The upgrade to the basic Zenit-2 design received the military name Gektor (programme code 11F690), and covered eighty-six launches, of which there were eighty-one successes. The first launch took place on 21 March 1968 (Cosmos 208), and the final launch took place on 16 February 1979 (the latter being the last of the five failures). Orbital inclination of these satellites was between 51°.7 and 81°.3.

These launches initially used the 11A57 launch vehicle, but beginning with Cosmos 834 in June 1976, the launcher was changed to the 11A511U launcher, probably as a result of a further increase in payload mass over the original Zenit 2 design. The increased orbital lifetime to twelve days was probably due to increased camera resolution, or an increase in consumables and camera supplies carried on board. A few of the satellites were again used to test systems under development for later programmes. When carrying the Nauka science module, the satellite had the appearance of the publicised Foton spacecraft (as seen in the accompanying illustration).

The Nauka modules carried on Zenit 2 satellites were mainly flown on the Zenit 2M series, beginning with Cosmos 208, launched on 21 March 1968 and ending with Cosmos 973 on 27 December 1977. (One Nauka was flown on Cosmos 309 (Zenit 2) in 1969, and two were flown on Cosmos 1102 and Cosmos 1106 (Zenit 2M) in 1979.) These packages carried a range of instruments that could be mounted inside or outside the capsule, on a small sub-satellite, or on a deployable small scientific satellite. Fields of research usually focused on gamma flux, cosmic rays, solar radiation and microwaves.

Zenit 4 This was the final photoreconnaisance satellite that was developed at Korolyov's OKB-1. From 1964 the programme, including Zenit 2, was transferred to OKB-1 Branch 3 in Kuibyshev (headed by Kozlov), where all further development would be concentrated. The Zenit 4 programme (code 11F69) consisted of seventy-six launches, of which seventy-two were successful. The first launch took place on 16 November 1963 (Cosmos 22), and the programme ended on 7 August 1970 (Cosmos 355). All of them used the 11A57 launch vehicle. The programme was flown alongside Zenit 2, not as a replacement, as both versions flew until 1970. Zenit 2 could be classed as low-resolution or area survey missions, while the Zenit 4 series was of a 'closer look' nature, offering high resolution with improved optical devices. Orbital inclinations from Tyuratam varied between 51°.8 and 81°.4.

Zenit 4M These were the first manoeuvrable photoreconnaissance satellites flown by the USSR. (programme code 11F691; code name Rotor). There were fifty-nine launches, of which fifty-six were successful, between 31 October 1968 (Cosmos 251) and 25 July 1974 (Cosmos 667). Changes to this series included the addition of shielding for protection against molecular heating of the instrument unit, as shown on the representative diagram. The drawing also depicts the KDU orbital manoeuvring module on the descent capsule, and aerodynamic compensators. The orbital inclinations varied from 51°.6 to 81°.3.

Zenit 4MK – Zenit 4MKM These were improved versions of the Zenit 4M class, employing 'two-tone' transmissions. The 4MK (programme code 11F692; code name Germes) was replaced by the improved 4MKM (11F692M; code name Gerakl) in the late 1970s, but no details have been released to distinguish one series from the other. The first launch occurred on 23 December 1969 (Cosmos 317), with the final launch

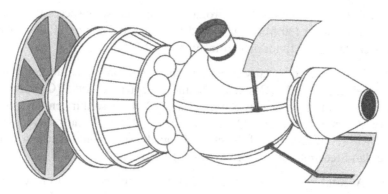

The Zenit 4M configuration. (© 1998 Ralph Gibbons.)

on 29 June 1984 (Cosmos 1580). There were 156 launches, and only three were failures. Orbital inclinations ranged from 62°.8 to 81°.3.

Zenit 6 This was a series of area survey satellites with high perigee. (217–223 miles) and appears to have been a replacement for earlier non-manoeuvring versions. The Zenit 6 series (11F645 – Argon) commenced with a launch on 23 November 1976 (Cosmos 867) and ended fifty-two launches later on 1 June 1984 (Cosmos 1568), with no launch failures. Orbital inclination ranged between 62°.8 and 72°.9.

Zenit 8 Western analyst Phil Clark has identified two versions of this series (17F116) of 'third generation' Zenit spacecraft, as follows:

Zenit 8A This group seems to have replaced the Zenit 6 series of area survey satellites, and consisted of thirty-eight launches between 11 June 1984 (Cosmos 1571) and 19 June 1990 (Cosmos 2083), with no failures. Orbital inclination for these satellites was between 62 .8 and 82°.6.

Zenit 8C This group appears to be a replacement for the Zenit 4MK series of 'close look' satellites. The series began with a launch on 4 September 1984 (Cosmos 1592) and ended with a launch on 7 June 1994 (Cosmos 2281). There were fifty-six launches, with no failures. Orbits were inclined at between 62°.8 and 82°.6.

Remote sensing – Zenit class

Disguised under the name 'Cosmos', there have been several series of remote sensing satellites, derived from the Zenit military satellite and Vostok-type spacecraft bus, and allowing recovery (though not always) of the payload. Objectives of these series have included topography and mapping, and the acquisition of photographic data. A sub-group of the military spacecraft has been identified (first launch, Cosmos 210, in April 1968, up to Cosmos 995 in March 1978) that provided photographic data on the break-up of arctic ice on the northern sea, supplementary to their military objectives.

Zenit-4MKM: Six launches, all successful. The first was on 10 October 1976 (Cosmos 859), and the last was on 5 July 1983 (Cosmos 1472). They were launched into orbits inclined between 65°.0 and 83°.3.

Zenit 2M/MKh (Code name Gektor-Priroda). Five launches, all successful. The first launch was Cosmos 741 (30 May 1975), and the last was Cosmos 1122 (17 August 1979). All were launched to an orbital inclination of 81°.3.

Zenit – 4MKT A group of twenty-seven launches (with only one failure) that flew between 25 September 1975 (Cosmos 771) and 6 September 1985 (Cosmos 1681). Launched into inclinations of between 81°.3 and 82°.4, these thirteen-day missions carried five still-image cameras. Designated Fram (11F635), they were used for photography in several regions of the electromagnetic spectrum, for use in geology, agriculture, land reclamation, oil prospecting, forestry, and fishing.

Zenit 4MT A total of thirty-three launches (with no failures) were completed between 27 December 1971 (Cosmos 470) and 3 August 1982 (Cosmos 1398). These satellites (11F629 – Orion) were thought to be dedicated to geodetic mapping, as first-generation cartographic satellites. This series was the first to use the new class Soyuz M launcher (11A511M). They were launched from Tyuratam and Plesetsk, into orbits with inclinations ranging from 62°.8 to 82°.3.

There were two further launches had remote sensing objectives announced but do not fit any previous named series:

Zenit 6 (11F645 – Argon) one launch, on 13 July 1983.

Zenit 8 (17F116 – Oblik- RS) one launch, on 27 July 1984 (Cosmos 1584).

Resurs-F1 This series featured an increase in orbital lifetime to twenty-five days. The Resurs (17F41, 14F40, and 14F43) appears to be a Zenit 4MK(M) photoreconnaissance satellite with a different set of cameras on board. It included the SA-33 sky camera, the SA-20M multispectral camera, and the SA-34 topographical camera. There were fifty-two launches between 5 September 1979 (Cosmos 1127) and 24 August 1993 (Resurs F 19), with only two failures. Orbital inclinations were between 81°.3 and 82°.6.

Resurs-F1M First launch on 18 November 1997. This satellite used a single SA-24 topographical camera and three SA-20M multispectral cameras, or an alternative cluster of three KFA-3000 cameras. It was launched into an orbit of 82°.3. A second launch occurred on 28 September 1999.

Resurs-F2 (17F42) This was a series of ten launches (with no failures) between 26 December 1987 (Cosmos 1906) and 26 September 1995 (Resurs F 20). It constituted a major upgrade of earlier Zenit-class satellites, with added solar panels for the generation of electrical power, as well as an increase of orbital lifetime to thirty days. The solar panels were discarded at retro-fire. Improved and upgraded onboard systems and new multi-spectral cameras were also part of the improvements. The satellites were launched into orbits inclined between 82°.1 and 82°.6. Further launches are planned, but are overdue.

Resurs-T (17F116 – Oblik/Resurs-T) A series of five launches which seem to have been upgrades of the Zenit 8 series. The first launch took place on 11 June 1986

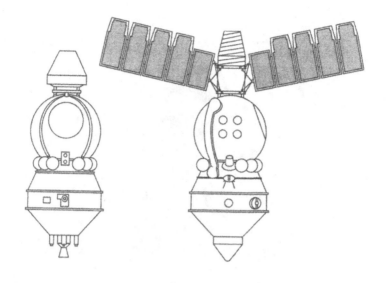

(*Left*) Resurs F-1, and (*right*) Resurs F-2. (© 1998 Ralph Gibbons.)

(Cosmos 1757), and the final launch took place on 22 July 1993 (Cosmos 2260). There were no failures, and they were launched into inclined orbits of 82°.3.

Scientific and application missions – Zenit class
A series of Zenit-derived spacecraft buses was used for a variety of scientific and application missions that fell outside the military reconnaissance/Earth resources series and the manned versions flown under Vostok and Voskhod. Using Vostok spacecraft technology, these vehicles have supplied additional information on life, materials, and space sciences, and with the Bion series applications for manned spaceflight.

Bion A series of twelve launches between 8 October 1970 (Cosmos 368) and 24 December 1996 (Bion 11), with no launch failures. The programme (12KS – Bion) had been headed by Dr Y. Illyin (former Voskhod candidate) at the IMBP. These were flown periodically to conduct experiments on a variety of research animals, to study the effects of space adaptation syndrome. They were launched into inclined orbits of between 62°.8 and 82°.4.

Efir The two launches of this series (36KS – Efir/SOKOL), both successful, took place on 10 March 1984 (Cosmos 1543) and 27 December 1985 (Cosmos 1713). Both were placed in into orbits of 62°.8. They were not intended for recovery, and therefore no heat protection was included. They studied high-energy cosmic radiation, and were a development of the Energiya series.

Energiya (13KS-Energiya) A pair of satellites launched on 7 April 1972 (Interkosmos 6) and 2 July 1978 (Cosmos 1026) into a 51°.7 inclined orbit. These spacecraft were used to study high-energy cosmic-ray particles. The Interkosmos 6 designation was the only Soviet bloc Interkosmos satellite to be recovered.

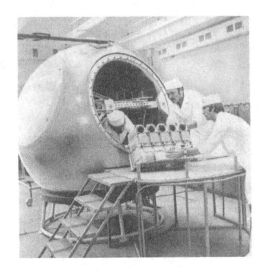

Biological experiments placed inside a Bion sharik based on the Zenit military reconnaissance spy satellite that was developed from the Vostok manned spacecraft.

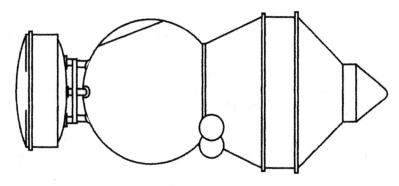

The Foton carrying the Nauka module – also representative of the Zenit 2M series and the Bion series. (© 1998 Ralph Gibbons.)

Foton (34KS – Foton) A continuing series of launches commencing on 16 April 1985 (Cosmos 1645), they were unmanned materials-processing research platforms. They were similar in configuration to the Bion series, and could transmit to Earth the results of the microgravity materials production completed in orbit. The Paton Institute developed a range of equipment flown on these spacecraft, including the Splav furnace, which was also used on Salyut manned space stations. All launches were placed into 62°.8 inclined orbits. The most recent launch took place on 9 September 1999 (Foton 12), and a Foton-M series is planned.

Resurs-500 This was a single launch, on November 15 1992, into an 82°.6 inclined orbit, as a one-off commercial flight, It carried space memorabilia in celebration of the 500th anniversary of the European rediscovery of North America, and as part of the 1992 International Space Year celebrations. The 'payload' was to be sold in

America after recovery. After almost a week in orbit, the descent module splashed down 200 miles south-west of Seattle, Washington. It missed the planned landing zone by sixteen miles, but the module was recovered successfully, and was donated to the Seattle Museum of Flight. This was the first Vostok-type sharik intentionally planned to achieve a splash-down rather than a traditional 'dust-down' on land.

SUMMARY OF RESEARCH UNDERTAKEN DURING MANNED MISSIONS, 1961–1965

The primary objectives of the man-related missions were: 1) to provide a launch system that was capable of delivering a human cargo to low Earth orbit; 2) to provided a suitable spacecraft to support manned flight at those altitudes for a duration of between one orbit and several days; and 3) to effect a survivable recovery of the spacecraft and human payload by a ground landing, ideally within Soviet territory.

In attaining these primary objectives, the Vostok-type flights can be considered a total success. Supplementary objectives were also achieved, in flying female and non-pilot crew-members, dual spacecraft missions, multi-person crews, and EVA. The technical achievement of proving that human spaceflight by either males or females, was not only possible but was also survivable, was a major advance in human history and exploration.

Due to the nature of these missions, it was not possible to carry out an extensive scientific research programme. This would have to wait for later programmes that would build upon the groundwork of the Vostok series in complexity, mission duration, crew composition, multi-spacecraft flight, and EVA.

Despite the restrictions that the missions imposed, several areas of investigation *were* addressed – most notably, space medicine. The one area that the Vostok could not investigate was orbital manoeuvring and physical docking with a second vehicle in orbit, although, the group flights provided experience in several support areas for this technique.

Spaceflight preparations Development of the launch vehicle, spacecraft, launch systems and flight support infrastructure for human spaceflight, as well as the selection and training of the cosmonauts, represented new fields of aeronautical technology that only the United States had the capability to match. The gains from the pioneering Vostok era continued beyond the fall of the Soviet Union. What has also survived over forty years is the system of training a small group of cosmonauts for a specific mission, and then selecting the flight crew from that 'group of immediate preparedness'.

Flight operations The development of the Tyuratam/R-7/Vostok system formed the basis of flight operations that have continued for more than forty years. The main legacy of that system is its longevity. The same pads are used by the latest versions of the same launch vehicle, which launches a spacecraft (Soyuz) that first flew in 1966. Soyuz itself is based on design studies that evolved from the original Vostok series. The early manned missions also presented ground controllers with the opportunity to establish a flight control system (mission control) and a network of tracking and control stations. They tested methods of controlling more than one spacecraft in orbit at the same time, and these methods were expanded upon with the advent of Soyuz. There was also an effective method of supporting crew return – and in two cases, a pair of spacecraft – by establishing an airborne and ground-based search-and-rescue network across the Soviet Union. Four decades later, it is still employed in the Soyuz missions.

Hardware development Although limited in capability, the series also provided early experience in the development of hardware for future spaceflights. Spacecraft systems, communications, navigation, habitability and equipment were basic, but with increasing flight experience were improved upon over the original concepts. This also had application in developing improved systems for future programmes, based on the results – both good and bad – from the Vostok and Voskhod series, and to some degree from the unmanned Zenit programme. There were also major hurdles to overcome with the reliability of spacecraft systems such as the sharik separation system, the temperature control system, and the braking rocket system. The simple experiments performed with liquid in containers looked fun when seen on television, but had a more serious application in designing future fuel-supply systems.

Radiation studies The Soviets were aware of the radiation levels in space, and the missions offered them a chance to investigate the effects on a wide range of biological subjects, including bacteria, seeds, plants, insects, dogs and, of course, humans. Instruments carried to record the levels of radiation included dosimeters, nuclear photoemission recorders, scintillator (ion flash) counters and crew thermolumines-cent glasses. In addition to the spacecraft shielding and the scheduling of flights for when solar activity was low, the cosmonauts had personal radiation protection, including supplies of anti-radiation medicine and a personal shield. Test samples carried included *E. coli* and lysogenic bacteria, and studies were conducted on the effects of radiation on the reproduction of *drosophila* flies. No manned flights traversed the Van Allen belts.

The dosage absorbed by each cosmonaut (reported by NASA from Soviet-supplied data in 1975) was found to be as follows:

Flight	Cosmonaut	Radiation
Vostok	Gagarin	2 m/rad /orbit
Vostok 2	Titov	11 m/rad /day
Vostok 3	Nikolayev	62 m/rad total.
Vostok 4	Popovich	46 m/rad total
Vostok 5	Bykovsky	80 m/rad total
Vostok 6	Tereshkova	44 m/rad total
Voskhod	Komarov	30 m/rads total
	Feoktistov	30 m/rad total
	Yegorov	30 m/rad total
Voskhod 2	Belyayev	60 m/rad total
	Leonov	60 m/rad total

Biological studies Every flight carried a small payload of biological samples such as cancer and amino cells, frog ova and sperm, drosophilia insects, air dried plant seeds, clorella algae, and bacteria samples.

Medical studies The most extensive experiment and research area featured biomedical investigations on each of the eleven flown cosmonauts. These included physiological methods of electrocardiographpy, pneuography, kinetocardiography, electro-oculography, electroencephalography, cutaneous–galvanic response and kinetocardiography. Additional post-flight investigations studied the effects of spaceflight on the reproductive systems of Nikolayev and Tereshkova, after they married and had a child. On Voskhod 1, Dr Yegorov performed additional research on both himself and his two colleagues, including measuring the arterial pressure, use of a hand-held spirometer, a dynamometer, and drawing capillary blood, by finger stick, which was stored on filter paper for post-flight analysis. On Voskhod 2, following the EVA, a hand-held spirometer was used to measure lung capacity.

Body parameters of the first six cosmonauts were recorded as:

Cosmonaut	Respiration	Pulse
Gagarin	16–26 cycles/min	Awake, 90–180 beats/min
Titov	4–28 cycles/min	Awake, 80–156; asleep, 53–67 beats/min
Nikolayev	10–18 cycles/min	Awake, 60–120; asleep, 60–65 beats/min
Popovich	10–20 cycles/min	Awake 60–130; asleep, 60 beats/min
Bykovsky	15–24 cycles/min	Awake 60–106; asleep, 45–56 beats/min
Tereshkova	16–22 cycles/min	Awake 64–82; asleep, 52–60 beats/min

For Voskhod, the maximum parameters were:

Komarov	17 cycles/min	Awake, 86 beats/min
Feoktistov	19 cycles/min	Awake, 86 beats/min
Yegorov	23 cycles/min	Awake, 90 beats/min
Voskhod 2		
Belyayev	n/a	Awake approx 90 beats/min
Leonov	n/a	Awake approx 90 beats/min

During the EVA, Leonov's pulse at one point varied between 100 and 143 beats/min, while Belyayev's was recorded at 100 beats/min:

Nutrition Despite the limitations of the flights, the food and nutrition provided for the cosmonauts provided baseline data on the balance of calorie intake, work output, and energy levels. The daily food rations (ounces), nutrient and calorie content for each cosmonaut were:

Spacecraft	Proteins	Fats	Carohydrates	Daily ration (kcal)
Vostok 1	3.51	4.16	10.86	2,772
Vostok 2	3.51	4.16	10.86	2,772
Vostok 3	4.20	2.98	10.77	2,529
Vostok4	4.20	2.98	10.77	2,529
Vostok 5	3.70	2.76	11.72	2,526
Vostok 6	4.35	2.99	10.75	2,529
Voskhod 1	5.29	4.58	15.16	3,600
Voskhod 2	5.29	4.58	15.16	3,600

Habitability In order to support the flight crew, the life support system had to maintain an environment to support human life both for the duration of the mission and in the event of delayed entry under emergency conditions. The cabin atmosphere parameters were recorded as:

Spacecraft	Pressure	Humidity	Temperature	% O_2 in air
Vostok	750–770 mm Hg	62–71%	19–22° C	n/a
Vostok 2	740–760 mm Hg	30–70%	10–25° C	n/a
Vostok 3	755–775 mm Hg	65–75%	13–26° C	n/a
Vostok 4	755–775 mm Hg	65–75%	12–26° C	21%
Vostok 5	775–780 mm Hg	40–65%	12–20° C	21%
Vostok 6	754–770 mm Hg	34%	18–23.6° C	20%
Voskhod	760–780 mm Hg	45–60%	15–21° C	n/a
Voskhod 2	760– ? mm Hg	35–40%	18° C	45% high

The Voskhod 2 figures are approximate. The high-level percentage of oxygen in the cabin air resulted from a drop in air pressure due to the hatch not being completely sealed after the EVA.

Visual observations Despite the briefness of the missions and the restricted view from the crew compartment, the cosmonauts performed the first visual observations from Soviet spacecraft. These included studies of the Sun, the Moon, the constellations and the Earth. Details of the atmospheric layers at sunrise and sunset were supplemented with observations of geological features, cloud cover, oceans, and weather phenomena.

The first eleven cosmonauts in space, 1961–1965

Cosmonaut	No.	Selected	Call-sign	H of the SU Order of Lenin	Pilot Cosmonaut of the USSR	Flight duration h:m
Gagarin	1	1960	Kedr	1961 Apr 14	1961 Apr 14	01:48
Titov	2	1960	Orel	1962 Aug 9	1962 Aug 9	25:18
Nikolayev	3	1960	Sokol	1962 Aug 18	1962 Aug 18	94:22
Popovich	4	1960	Berkut	1962 Aug 18	1962 Aug 18	70:57
Bykovsky	5	1960	Yastreb	1963 Jun 22	1963 Jun 22	119:06
Tereshkova	6	1962	Chaika	1963 Jun 22	1963 Jun 22	70:50
Komarov	7	1960	Rubin-1	1964 Oct 19	1964 Oct 19	24:17
Feoktistov	8	1964	Rubin-2	1964 Oct 19	1964 Oct 19	24:17
Yegorov	9	1964	Rubin-3	1964 Oct 19	1964 Oct 19	24:17
Belyayev	10	1960	Almaz-1	1965 Mar 23	1965 Mar 23	26:02
Leonov	11	1960	Almaz-2	1965 Mar 23	1965 Mar 23	26:02

LANDING SITE COORDINATES

Vostok	51° N	43° E	(approximately)
Vostok 2	51° N	47° E	(approximately)
Vostok 3	48° 02' N	75° 45' E	
Vostok 4	48° 10' N	71° 51' E	
Vostok 5	53° 23' 45" N	67° 36' 41" E	
Vostok 6	53° 16' 18" N	80° 27' 34" E	
Voskhod	54° 02' 00" N	68° 08' 00" E	
Voskhod 2	59° 34' 03" N	55° 28' 00" E	

LOCATIONS OF SHARIKS

The present (2000) locations of the recovered shariks are:

Vostok	RKK Energia Museum, Korolyov, Russia
Vostok 2	Destroyed during 1964 ground tests of Voskhod soft-landing system
Vostok 3	Location unknown
Vostok 4	NPO Zvezda Museum, Moscow (used in mock-ups of Voskhod 2)
Vostok 5	Tsiolkovsky Museum, Kaluga, Russia
Vostok 6	RKK Energia Museum, Korolyov, Russia
Voskhod	RKK Energia Museum, Korolyov, Russia
Voskhod 2	RKK Energia Museum, Korolyov, Russia
1KP	RKK Energia Museum, Korolyov, Russia
3K	Tsiolkovsky Museum, Kaluga, Russia
3K	(was located at) VDKH , Moscow, Russia

Voskhod ?   Black Sea Orlyonok Children's Centre (former Pioneer Camp)
Voskhod 3   RKK Energia Museum, Korolyov, Russia
Cosmos 110   RKK Energia Museum, Korolyov, Russia

In addition, two of the stratospheric balloon gondolas (the USSR and Volga) are located at the Air Force Museum, in Monino, near Moscow.

THE SHARIK LEGACY

The true legacy of the Vostok design has been in the sustained use of the sharik descent capsule in a variety of forms since it was first conceived almost forty-five years ago. The idea of spherical capsules for atmospheric exploration was first used in the 1930s during the stratospheric balloon ascents, and alongside the Vostok and Zenit series the technology of using spherical and displaced centre of gravity capsules was also applied to Soviet unmanned programmes to explore the Moon (as well as in the sample-return capsules), Mars and Venus for almost three decades.

Moreover the original designs of a follow-on vehicle to the single-seat Vostok series featured the sharik descent module that eventually evolved into the Soyuz design. The Soyuz itself has had a long and distinguished service history since the first test-flights during the mid-1960s to the present day, and now forms an integral part of the International Space Station programme.

The Soviets also employed the sharik design in the Yantar-1KFT (11F660) series of topographical and mapping missions, when costs escalated. The series began with

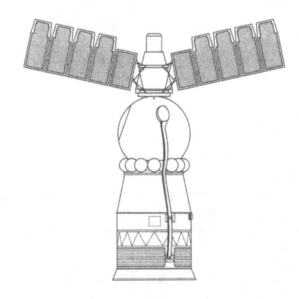

Kometa – one possible configuration of the Vostok/Soyuz hybrid. (© 1998 Ralph Gibbons.)

the launch of Cosmos 1246 on 18 February 1981, and were originally code-named Silvet (Cosmos 1246–1865), and then Kometa (beginning with Cosmos 1896) when the system became operational. The original design of a standard conical Yantar main descent vehicle was quickly exceeding the budget, and it was decided to adapt the Zenit-class sharik spherical descent module to the Yantar propulsion system (as shown in the accompanying illustration). This not only restrained the overall budget, but also reduced the overall mass of the vehicle.

The solar arrays were transferred from the Yantar service/instrument module to a conical attachment on top of the spherical descent craft, as was the case on the Zenit-class Resurs-F2 series. In the Resurs design the 'attachment' housed the in-orbit manoeuvring engine, while it is believed that on Yantar the propulsion systems were located in the service/instrument module. There have been nineteen launches in this series, with only one launch failure, and although they were not Vostok/Zenit-class vehicles, they contributed to an impressive record of using the spherical descent module.

Although Vostok is widely known as the vehicle that took the first men and first woman from the Soviet Union into space in a short, five-year period between 1961 and 1966, its legacy actually encompasses almost half a century of space exploration, and in the pages of history it will remain as the pioneering vehicle of human space exploration and discovery.

Conclusion

Forty years ago, Yuri Alekseyevich Gagarin was launched into space, and completed one orbit in just one hour and forty-eight minutes. Since that historic day, approximately 400 individuals from different countries have followed Gagarin's trail, on almost 250 missions. There have been more than 175 space-walks by 120 men and women. 24 American astronauts have completed nine flights to the Moon, and of them, twelve men have walked on its surface. More than fifteen Russians have each spent one year in space on long-duration missions on-board the Salyut and Mir space stations, and Dr Valeri I. Polyakov, on just one flight, logged 468 days in Earth orbit.

At the dawn of a new century, sixteen countries are working together in is creating the International Space Station (ISS). Every other month, a new human spaceflight leaves Earth to continue the building of the ISS towards a common goal, sharing the work, experience, cost, and long-term benefits.

Once the largest international construction project – the ISS – is completed after five years, it will have a mass of 470 tons, having been assembled from more than one hundred major parts launched on forty-five missions. There will have been more than seventy-five space-walks, totalling in excess of 1,000 hours outside the station, housing a crew of six or seven, rotating a permanent presence in orbit in 46,000 cubic feet of pressurised volume (about the same as two Boeing 747 jets) in six laboratories and four modules measuring 356 feet across and 290 feet long.

Four decades after Gagarin took that first trip beyond our atmosphere, Russians, Americans, Europeans and Asians are united in creating a human outpost in space. It will be the third brightest object in the night sky, and will remain in orbit for possibly twenty years.

All this seems to make Gagarin's mission a very very small step on the road, and in most records of human spaceflight experience, the flight of Gagarin and Vostok appears at the very bottom of the list of duration and experience. However, the flight of the first Vostok will be seen as one of the most defining moments of the evolution of the human race.

The dedication of the Soviet designers, led by Sergei Pavlovich Korolyov, was inspirational, imaginative, and very effective. They created the hardware, the

Cosmonauts lay flowers at the foot of Gagarin's memorial at Star City on Cosmonautics Day (12 April), which is celebrated every year.

facilities and an infrastructure in the greatest secrecy, with resources that were far fewer than those of the USA during that era. They conceptualised generations of spacecraft and launch vehicles, using a serialised production system. That system continues to be used to this day, in the manufacturing flow and for launches. The ability to launch from the Tyuratam (Baikonur) Cosmodrome, in all weathers, and at any time of the year in keeping with a previously planned schedule, is a lasting tribute to the sound engineering of the designers. To be still using the same facilities and systems almost fifty years later is a monument to the workmanship of the construction teams.

The space race was born at the height of the Cold War and, amazingly, survived its ending – a testament to the imagination of the human race. Since Gagarin, we have watched with amazement as mission after mission has been launched, each of them setting a new record, and all seemingly moving towards a defined common goal of permanent human habitation of the cosmos. We knew nothing of the in personal conflicts within the space hierarchy, the political interference, the industrial power-game, or the lack of clarity over what to do next. All we saw was one headline after another claiming a new space achievement.

Above all, Yuri Gagarin's feat will rank in history alongside the greatest voyages of exploration. Names like Captain Cook, Ferdinand Magellan, Vasco da Gama, Christopher Columbus, Wilbur and Orville Wright, Roald Amundson, Charles Lindberg and Edmund Hillary will always be linked to that of Gagarin. His name, and the date 12 April 1961, will forever be recorded in the history books as the time when the first human left Earth and began the exploration of a new ocean – outer space.

In the film of Arthur C. Clarke's novel *2010*, the name of the spacecraft is *Leonov* – after Alexei Leonov, the first human to walk in space. Perhaps in the near future, the international space community will agree that it would be fitting to name the International Space Station 'Gagarin', in tribute to the first to venture beyond our atmosphere. A further fitting tribute and legacy to those who have contributed to the exploration of space would be to name the major ISS components after fallen heroes of the conquest of the cosmos.

We trust that this book captures the sense of adventure and imagination that drove a small group of Russian explorers to achieve, against all odds, their dream of living and working in space. We also hope, above all, that the memory and achievement of one man on one short flight is appreciated and recognised as having changed human destiny and history. Preceding Gagarin was a period of human history that for thousands of years was tied to the Earth; but on 12 April 1961, human history truly began to encompass the stars.

Every single human who ventures into space will always follow the footsteps of Gagarin, as he was THE FIRST.

Vostok and Voskhod missions, 1961–1966

11F63 (Vostok 3) Production index for manned spacecraft

Name (Serial No.)	Type	Launch date	Landing date	Mass (lbs)	Period (min)	Inc. (deg)	Perigee (mi)	Apogee (mi)	Orbits	Duration (hh:mm)	Notes
Vostok	3KA (3)	1961 Apr 12	1961 Apr 12	10,416	89.1	60.7	108.7	187.6	1	01.48	1
Vostok 2	3KA (4)	1961 Aug 6	1961 Aug 7	10,430	88.6	64.9	110.6	159.6	17	25:18	2
Vostok 3	3KA (5)	1962 Aug 11	1962 Aug 15	10,410	88.5	65.0	113.7	155.9	64	94:22	3
Vostok 4	3KA (6)	1962 Aug 12	1962 Aug 15	10,423	88.5	65.0	111.8	157.8	48	70:57	4
Vostok 5	3KA (7)	1963 Jun 14	1963 Jun 19	10,405	88.2	64.9	108.7	137.9	81	119:06	5
Vostok 6	3KA (8)	1963 Jun 16	1963 Jun 19	10,390	88.3	65.0	113.7	144.7	48	70:50	6
Vostok (7)	3KA (modified) planned for April 1964	Mission to 375 miles by animals for 30 days cancelled in February 1964									
Vostok (8)	3KA (modified) planned for June 1964	Mission to a high apogee for 8 days cancelled in February 1964									
Vostok (9)	3KA (modified) planned for August 1964	Group flight to a high apogee for 10 days cancelled in February 1964									
Vostok (10)	3KA (modified) planned for April 1965	Group flight to a high apogee for 10 days cancelled in February 1964									
Vostok (11)	3KA (modified) planned for June 1965	Mission to include an EVA experiment by animals cancelled in February 1964									
Vostok (12)	3KA (modified) planned for August 1965	Mission to include an EVA cancelled in February 1964									
Vostok (13)?	3KA (modified) planned for April 1966	Mission to 745 miles (?) for 10 days cancelled in February 1964									
Cosmos 47	3KV (2)	06 Oct 1964	07 Oct 1964	11463	90.0	64.7	109.9	256.6	16	24:18	7
Voskhod	3KV (3)	12 Oct 1964	13 Oct 1964	11728	90.1	65.0	110.6	254.1	16	24:17	8
Cosmos 57	3KD (1)	22 Feb 1965	22 Feb 1965	12125	91.1	64.7	108.7	318.1	2 ?	03:02	9
Voskhod 2	3KD (4)	18 Mar 1965	19 Mar 1965	12526	90.9	65.0	107.5	307.5	18	26:02	10
Cosmos 110	3KV (5)	22 Feb 1966	16 Mar 1966	12566	95.3	51.9	116.2	561.7	330	522:00	11
Voskhod 3	3KV	Mission of 15 days with pilot and scientist to include artificial gravity experiment; cancelled November 1966(?)									
Voskhod 4	3KV	Mission of 15 days to include a doctor on a biological and artificial gravity mission; cancelled summer 1966									
Voskhod 5	3KD	Mission by female crew to include EVA; cancelled summer 1966									
Voskhod 6	3KD	Mission by male crew to include EVA possibly with UPMK manoeuvring unit; cancelled summer 1966									

Notes

1 Yuri Gagarin, *first* manned spaceflight

2 Gherman Titov *first* to spend 1 day in space
3 Andrian Nikolayev *first* to spent over 24 hours in space
4 Pavel Popovich performs *first* joint flight with Nikolayev in Vostok 3
5 Valeri Bykovsky set world endurance record for solo spaceflight that still stands 38 years later
6 Valentina Tereshkova *first* woman in space; performed joint flight with Bykovsky in Vostok 5.
7 Unmanned test of Voskhod spacecraft in orbit
8 Vladimir Komarov, Konstantin Feoktistov and Boris Yegorov become *first* multi-person crew and *first* three-person crew to fly in space; Feoktistov becomes *first* scientist/engineer to fly, and Yegorov *first* medical doctor
9 Unmanned test of Voskhod airlock, which was successful before vehicle exploded during second orbit
10 Pavel Belyayev and Alexei Leonov *first* two-person crew; Leonov *first* to perform EVA
11 Dogs Veterok and Ugolek on a long-duration biomedical mission as a precursor for a subsequently cancelled long-duration manned mission

Information released by Energia in 2000 revealed individual landing times of the Vostok cosmonauts and their shariks for Vostok 2 to Vostok 6. The data presented below recorded that the shariks landed after the cosmonauts, with an average difference of 10 minutes which, when applied to Gagarin's reported 108-minute flight, implies that he would have landed 98 minutes after launch. The five Voskhod cosmonauts landed inside their capsules.

Spacecraft	Cosmonaut	Pilot (hrs:min:sec)	Sharik	Difference
Vostok	Gagarin	01:38:00	01:48	Est. 10 min
Vostok 2	Titov	25:11:00	25:18	+ 06 min
Vostok 3	Nikolayev	94:09:59	94:22	+ 12 min
Vostok 4	Popovich	70:44:00	70:57	+ 13 min
Vostok 5	Bykovsky	118:56:41	119:06	+ 09 min
Vostok 6	Tereshkova	70:40:48	70:50	+ 09 min
Voskhod	Komarov	24:17:03	n/a	n/a
	Feoktistov			
	Yegorov			
Voskhod 2	Belyayev	26:02:17	n/a	n/a
	Leonov			

USSR/USA timeline for achievements in manned spaceflight

MR = Mercury Redstone; MA = Mercury Atlas; GT = Gemini Titan

First	USSR	USA	Time-span
Satellite in orbit	1957 Oct (Sputnik)	1958 Jan (Explorer 1)	+ 3 months
Animal sub-orbital	1950s ballistic altitude flights only	1961 Jan (MR 2 primate Ham)	n/a
Animal in orbit	1957 Nov (Sputnik 2 Dog Laika)	1961 Nov (MA 5 primate Enos)	+ 4 years
Unmanned spacecraft sub-orbital	None attempted	1960 Dec (MR 1A)	n/a
Unmanned spacecraft orbital	1960 May (Korabl-Sputnik 1)	1961 Sep (MA 4)	+ 1 year 4 months
Manned sub-orbital	None attempted	1961 May (MR 3 Shepard)	n/a
Manned orbital	1961 Apr (Vostok Gagarin)	1962 Feb (MA 6 Glenn)	+ 10 months
One day in orbit	1961 Aug (Vostok 2 Titov)	1963 May (MA 9 Cooper)	+ 1 year 9 months
Extended spaceflight	1962 Aug (Vostok 3 Nikolayev 3 days)	1965 Jun (GT 4 McDivitt/White 4 days)	+ 2 years 10 months
Long duration	1963 Jun (Vostok 5 Bykovsky 5 days)	1965 Aug (GT 5 Cooper/Conrad 8 days)	+ 2 years 2 months
Woman in orbit	1963 Jun (Vostok 6 Tereshkova 4 days)	1983 Jun (STS 7 Sally Ride 7 days)	+ 20 years
Multi-person spaceflight	1964 Oct (Voskhod three men)	1965 Mar (GT 3 two men)	+ 5 months
Solo endurance record	1963 Jun Vostok 5 (Bykovsky 119 hours)	1963 May Cooper (MA 9 34 hours)	Vostok 5 record for over 38 years
Extended duration	1970 Jun Soyuz 9 (18 days)	1965 Dec GT 7 (14 days)	GT 7 record held 4 years 6 months
Two-person crew	1965 Mar (Voskhod 2 Belyayev/Leonov)	1965 Mar (GT 3 Grissom/Young)	+ 1 week
Three-person crew	1964 Oct (Voskhod Komarov/Feoktistov/Yegorov)	1968 Oct (Apollo 7 Schirra/Eisele/Cunningham)	+ 4 years
EVA	1965 Mar (Voskhod 2 Leonov 20 min)	1965 Jun (GT 4 White 21 min)	+ 3 months
Civilian crew-member	1964 Oct (Voskhod Feoktistov /engineer)	1966 Mar (GT 8 Armstrong/ test pilot)	+ 1 year 5 months

Medical Doctor	1964 Oct (Voskhod Yegorov)	1973 May (Skylab 2 Kerwin)	+ 8 years 7 months
Duel spaceflight	1963 Aug (Vostok 4 and 5)	1965 Dec (GT 6 and 7)	+ 3 years 4 months
Rendezvous	1968 Oct Soyuz 3 and 2	1965 Dec (GT 6 and 7)	US record 2 years 10 months
Docking	1969 Jan Soyuz 4 and 5	1966 Mar (GT 8 and Agena 8)	US record 2 years 10 months
Gravity/tether experiment	None attempted	1966 Sep (GT 11 and Agena 11)	n/a
High manned apogee	1965 Mar (Voskhod 2 307.5 miles)	1966 Sep (GT 11 and Agena 11 643.1 miles)	1966 Feb (Cosmos 110 (dogs) 561.7 miles)

Appendix

THE COSMONAUTS

Ranks are those held at the time of appointment to the cosmonaut team.

Anikeyev, Ivan Nikolayevich (Senior-Lieutenant) was born on 12 February 1933 in Liski, Voronezh Raion. He attended the Stalin Naval Aviation School (graduating in 1956), before serving in the Air Fighter Division of the Northern Fleet. Anikeyev joined the cosmonaut team on 7 March 1960. He underwent some of the initial testing on the centrifuge, and was a capsule communicator on the Vostok 3 and Vostok 4 missions. In April 1963 he was dismissed from the team for becoming involved in a fight with the militia, after which he rejoined the Northern Fleet, in which he served as a fighter pilot and in several other positions, before retiring in 1975. He died of cancer on 8 August 1992, in Bezhetsk, Kalinin District.

Artyukhin, Yuri Petrovich (Engineer-Captain) was born on 22 June 1930 in the Moscow Raion. He initially tried to become a pilot, but was medically disqualified. He qualified as an Engineer in 1950. In 1952 he began a course at the Zhukovsky Air Force Engineering Academy, and graduated in 1958. He stayed on at the Academy as a senior Engineer until his selection as a cosmonaut on 10 January 1963. Artyukhin was a support cosmonaut on the Voskhod 3 mission, and went on to work on the Soyuz and then the lunar programme. He was paired with Voloshin on crew number 5, and served as a CapCom on the Soyuz missions during 1969 and 1970 and during the Salyut 1 mission. He then worked on the Almaz programme, flying with Popovich as his flight engineer on Soyuz 14/Salyut 3, in July 1974. Artyukhin became a Hero of the Soviet Union and Pilot Cosmonaut of the USSR. He then became head of Almaz training at the Cosmonaut training centre, and worked on four TKS missions between 1978 and 1981. He stood down from the team on January 1982, but continued working at the centre until he retired from the Air Force, as a Colonel of Engineers in 1987, after which he went to work at NPO Molniya as a staff member. Artyukhin died on 4 August 1998.

Belyayev, Pavel Ivanovich (Major) was born on 26 June 1925 in the Vologda Raion. In 1945 he graduated from the Stalin Naval Air School and went on to see action against the Japanese in the Far East. He continued to serve in various air units in the Pacific Fleet until 1956, when he entered the Red Banner Air Force Academy. After graduating in 1959, Belyayev became squadron leader of an air regiment in the Northern Fleet. He joined the cosmonaut team – as its eldest member – on 25 March 1960, and became its first commander. In August 1961, Belyayev broke his ankle, which put him out of action for more than a year. He supervised the selection of the 1963 group, and in April 1964 was selected to command the 1965 Voskhod 2 mission, after which he participated in various programmes, including Soyuz and Almaz. In 1967 he became Chief of Staff at the training centre, overseeing the selection of the 1967 group, and later went on to head a department within the centre. Belyayev died of peritonitis (following an operation for an ulcer) on 10 January 1970. He was a Hero of the Soviet Union and a Pilot Cosmonaut of the USSR.

Benderov, Vladimir Nikolayevich was born on 4 August 1924, and was a leading test pilot and Flight Engineer working for the Tupolev Design Bureau when he was considered as a potential candidate for Voskhod 1, planned for late 1964. The training group was formed on 24 May 1964, and Benderov was a member for the month of June, but then stood down. He went on to fly the MiG 21 test-bed craft. On 3 June 1973 he was killed when his Tu144 crashed at the Paris Air Show. He was the Flight Engineer on the exhibition flight. He is buried in Moscow.

Beregovoi, Georgy Timofeyevich (Colonel) was born on 15 April 1921 in the Poltava Raion of the Ukraine. He went to work in a steel plant before going to the Lugansk Air Force School, from which he graduated in 1941. He flew 185 combat missions against the Luftwaffe, and was shot down three times. In April 1944, Beregovoi became a Hero of the Soviet Union, and after the war he attended test-pilot school, where he flew sixty-three different types of aircraft. He has logged more than 2,500 flying hours. He also undertook correspondence course at the Red Banner Air Force Academy, graduating in 1956. In 1963 the heads of the manned programme decided that the team needed more experienced candidates, and Beregovoi joined the team on 17 January 1964. He underwent some initial training, and in April 1965 was selected as back-up commander of the Voskhod 3 mission with Demin. It was planned as a two-week mission, but the crew was changed on a number of occasions and the mission was eventually cancelled in 1966. Following this cancellation, Beregovoi was assigned to the Soyuz training group, and after the death of Komarov he was named as commander for the next test flight of Soyuz. This occurred in October 1968, when he flew Soyuz 3. Beregovoi retired from the team in April 1969, having been promoted to Major-General after his mission. In June 1972, he was appointed Director of the Cosmonaut Training Centre, where he remained in charge until January 1987. In 1977 he was promoted to the rank of Lieutenant General, and after his retirement he worked for the Soviet Academy of Sciences. Beregovoi died on 30 June 1995. He was a twice Hero of the Soviet Union and a Pilot Cosmonaut of the USSR.

Bondarenko, Valentin Vasilyevich (Senior Lieutenant) was born on 16 February 1937 in the city of Kharkov, now in the Ukraine. He attended the Armavir Higher Air Force Pilot (HAFP) School, graduated in 1957, and then became a fighter pilot. Bondarenko was selected to join the cosmonaut team on 25 March 1960, and was the youngest member of the group. He underwent basic cosmonaut training, including isolation tests. He was completing some medical work during one of these tests, only three weeks before Gagarin's flight, when a swab of cotton dipped in alcohol – which he was using to clean himself – caught fire. In the oxygen-rich atmosphere, Bondarenko suffered 90% burns. He was rushed to a hospital, but died of his injuries on 23 March 1961. He was the first selected cosmonaut to die.

Buinovsky, Eduard Ivanovich (Senior Engineer-Lieutenant) was born on 26 February 1936 in the Rostov Raion. He attended the Riga Higher Air Force Engineering School, graduating in 1958. He then joined the Strategic Rocket Forces, and was based at Baikonur, where he witnessed Gagarin's launch. Buinovsky joined the cosmonaut team on 10 January 1963, but during his initial training he failed a test on the centrifuge and was medically disqualified. He left the team on 11 December 1964, but continued to work on space control systems, eventually joining the Energia-Buran launch team. After the launch of Buran, he retired from the military with the rank of Colonel. He currently works for the Ministry of Foreign Affairs, in communications and control systems, and lives in Moscow.

Bykovsky, Valery Fyodorovich (Senior Lieutenant) was born on 2 August 1934 in Moscow. He attended the Kacha HAFP School, graduating in 1955, and then served in on air squadron as a pilot and instructor. Bykovsky joined the cosmonaut team on 9 March 1960, and was appointed to the top six group for accelerated training. He was awarded the Order of the Red Star. He was the back-up to Nikolayev on Vostok 3, and flew the Vostok 5 mission in 1963, breaking a number of records in the process. Bykovsky assisted on Voskhod 2, acting as training supervisor, and was then appointed commander of the Soyuz 2 mission, due to fly in 1967. He trained with Rukavishnikov for a Soyuz–Zond mission, and then undertook lunar training in the late 1960s. These missions were cancelled. He graduated from the Zhukovsky Air Force Academy in 1968, and then took up a number of training assignments, including ASTP training director. Bykovsky commanded the Soyuz 22 mission in 1976 and the DDR Intercosmos mission in 1978, and he also acted as the back-up commander for the Vietnamese Intercosmos flight in 1980. He left the team and stood down from the Air Force, with the rank of Colonel on 2 April 1988. Bykovsky worked in East Germany for some years before returning to Russia. After 38 years, he still holds the solo spaceflight endurance record. He is twice Hero of the Soviet Union and a Pilot Cosmonaut of the USSR.

Demin, Lev Stepanovich (Engineer Lieutenant-Colonel) was born on 11 January 1926, in Moscow. He worked in a defence plant in Siberia during the war. He also tried to become a pilot, but was rejected due to poor eyesight. He joined an Air Force Signals School and subsequently commanded a signal squadron. In 1949, Demin entered the Zhukovsky Air Force Engineering Academy, graduating in 1956. He stayed on at the

Academy, as well as working in a research institute that specialised in radar and communications. He joined the cosmonaut team on 10 January 1963, and was assigned to be the back-up pilot on Voskhod 3 before the mission was cancelled. He was then assigned to the Almaz programme, and trained on a number of crews, including as a crew-member for the 1972 military Salyut. In August 1974, Demin flew on the Soyuz 15 mission to Salyut 3 as the Flight Engineer, but the Soyuz failed to dock, and returned after two days. He continued to work at the cosmonaut training centre until he stood down from the team on 13 August 1983, with the rank of Colonel. He then went to work for a company which specialised in detecting resources in the sea. Demin retired in 1989, and died on 18 December 1998. He was a Hero of the Soviet Union and a Pilot Cosmonaut of the USSR.

Dobrovolsky, Georgy Timofeyevich (Major) was born on 1 June 1928 in Odessa, where he lived during the Nazi occupation. He graduated from the Chuguyev HAFP School in 1950, and then went on to serve in a number of air regiments in Germany and the Soviet Union. He rose to become a deputy squadron commander and a political officer. From 1957 to 1961, Dobrovolsky undertook a correspondence course at the Red Banner Air Force Academy. He joined the cosmonaut team on 10 January 1963, and worked on the Voskhod programme before joining the lunar training group, after which he worked on Almaz missions before joining the Salyut training group in January 1971. He had also served as a CapCom on various Soyuz missions from 1967 to 1970, and at the same time was the deputy commander of the team responsible for political affairs. Dobrovolsky served as third crew commander for Soyuz 10. He was also to be the back-up for Soyuz 11, but replaced Leonov only a few days before the mission. He spent twenty-three days onboard the station, setting a world duration record. On 30 June 1971, Dobrovolsky died during the decent of his Soyuz, when a valve opened on the craft and the air bled away. He was buried in the Kremlin Wall, and was posthumously honoured as Hero of the Soviet Union and Pilot Cosmonaut of the Soviet Union.

Feoktistov, Konstantin Petrovich was born on 7 February 1926 in the city of Voronezh. During the war he was a member of a partisan unit, and was captured. He was put in front of a firing squad, but was only wounded. In 1943 he entered the Bauman Higher Technical School, graduating in 1949, and then went to work in a weapons factory. In 1955, Feoktistov was working at the Korolyov design bureau, where he worked on the Sputnik programme and on other unmanned space vehicles. He was the leading engineer on the Vostok craft, and presented some of the lectures to the first cosmonaut team. He promoted the involvement of civilian engineers in the cosmonaut programme, which helped when candidates were being considered for the Voskhod 1 mission. Feoktistov was the candidate selected by OKB-1 as the Flight Engineer and was approved on 11 June 1964. He flew on Voskhod 1 that same year. On 23 May 1966 he was included in the first civilian cosmonaut group to be formed. He was involved in the design of Soyuz-T, Progress, and the civilian Salyut programme, and was the Flight Director for Salyut 4. His name was connected with a number of flights, but he was never called. In 1980 he was included as the Research Cosmonaut in a development Soyuz-T mission, but was excluded, having failed a medical. Feoktistov formally left the cosmonaut team on 28 October 1987, and continued to work at NPO Energiya; but he was very critical of the Buran programme. In 1990 he retired from the design bureau. He lives in Moscow, and is a Hero of the Soviet Union and a pilot Cosmonaut of the USSR.

Filatyev, Valentin Ignatyevich (Senior Lieutenant) was born on 21 January 1930 in the Tyumen Raion, Russia. He was originally a trained teacher, but joined the Air Force in 1951. He attended the Stalingrad HAF pilots School, graduating in 1955, and served as a fighter pilot in the Air Defence Force. Filatyev joined the cosmonaut team on 23 March 1960, and earned the nickname 'Gramps' because he looked older than the rest. In March 1963 he was dismissed from the team, having been involved in a fight with the militia. Filatyev returned to serve in the Air Defence Force until 1969, by which time he had been promoted to the rank of Major. He then worked in a state institute before returning to teaching. Filatyev died of cancer on 15 September 1990.

Filipchenko, Anatoly Vasilyevich (Major) was born on 26 February 1928 in the Voronezh Raion. He worked in a machine plant during the war, graduated from the Chuguyev HAFP School in 1950, and served initially as a fighter pilot in the Soviet Union and other Socialist bloc countries, rising to become an Air Force inspector. He also completed a correspondence course at the Red Banner Air Force Academy, graduating in 1961. Filipchenko joined the cosmonaut team on 9 January 1963. He was a CapCom for the Voskhod missions, and was initially assigned to the Spiral 'space plane' project under the command of Titov. He attended the Chkalov test-pilot school, and in 1967 graduated with the qualification of test pilot 3rd class. He joined the Soyuz group in 1968, was back-up commander for Soyuz 4 in 1969, and then commanded the Soyuz 7 mission in the same year. In 1970 he was back-up commander on the Soyuz 9 crew, and was also involved in the Kontact missions planned for 1970/71. He then worked in the training centre in a number of roles, including heading the Spiral group between 1970 and 1972, before working on the ASTP mission from 1972. He commanded the dress rehearsal mission of Soyuz 16 in 1974, and he then worked within the training centre, occupying a number of senior positions. He was promoted to the rank of Major-General in 1978, stood down from the team on 26 January 1982, and left the Air Force in 1988. Filipchenko went to work at the Pilyugin Research Institute from 1982 to 1993, before he retired. He is twice a Hero of the Soviet Union, and a Pilot Cosmonaut of the USSR.

Gagarin, Yuri Alexeyevich (Senior Lieutenant) was born on 9 March 1934 in the Smolensk Raion, west of Moscow. He initially went to a technical school, but learned to fly at a local club and was recommended to join the Air Force. He graduated from the Orenburg HAFP School in 1957, and went to serve with the northern fleet, flying MiGs. Gagarin joined the cosmonaut team on 9 March 1960, and was selected to join the top six accelerated training group in the summer of 1960. It became clear that he was the most likely candidate to be the first man in space. He was confirmed in that role only a few days before the flight, and on 12 April 1961 he became the first human to orbit the Earth. Following the mission, he was made commander of the cosmonaut team, but spent much of his time on public relations duties. He was a CapCom for Vostok 3, Vostok 4, Voskhod and Voskhod 2, and was made deputy Director of the cosmonaut training centre in December 1963. In June 1964 he was formally forbidden to fly, and after protesting for some years he managed to have the decision rescinded in 1966. He was then assigned as the back-up to Komarov on Soyuz 1, which flew in 1967. He also graduated from the Zhukovsky Air Force Engineering Academy in 1968, and was involved in the cosmonaut training for the lunar missions. Gagarin was killed in an air crash on 27 March 1968, while flying a MiG 15 with a very experienced pilot, Vladimir Seregin. The reason for the crash has been subject to a number of State Commissions, but it would seem that the aircraft was caught in a vortex from another aircraft, and went into a spin. Gagarin was a Hero of the Soviet Union, and was the first pilot Cosmonaut of the USSR. He was a Colonel in the Air Force.

Golovanov, Yaroslav Kirillovich was born on 2 June 1932, in Moscow. He graduated from the E.N. Bauman Moscow Higher Engineering School in 1956, and went to work on supersonic aircraft at a research institute. In 1958 he became a journalist for *Komsomolskaya Pravda*, working there until 1986, writing about the space programme. In 1965, Korolyov wanted to send a journalist into space, as he felt that the cosmonauts were unable to effectively communicate what they felt or saw. He selected two journalists, who underwent medical tests in July 1965. They never trained for a specific mission, and after Korolyov's death in early 1966, the plans for them to fly were dropped. Golovanov went on to write a number of key books about the space programme, including *Cosmonaut No 1*, in which the identities of the 1960 unflown cosmonauts were disclosed for the first time. He also wrote a key work on the life and career of Korolyov.

Gorbatko, Viktor Vasilyevich (Senior Lieutenant) was born on 3 December 1934, in the Krasnadar Territory of Russia. He went to the Bataisk HAFP School, graduating in 1956. He served with Khrunov in the same air regiment. Gorbatko joined the cosmonaut team on 9 March 1960. He completed his basic training, and was the original back-up commander for the Voskhod 2 mission, but was replaced by Zaikin when he became ill. He then went to work on the early Soyuz missions, backing up Khrunov on the Soyuz 2 mission in 1967, and then again on Soyuz 5 in 1969. He also graduated from the Zhukovsky Air Force Academy in 1968. Gorbatko made his first flight in 1969, as the research engineer on Soyuz 7. He was also a CapCom on Soyuz 9, and became Deputy Commander of the Almaz training group. He was appointed back-up commander for Soyuz 23 in 1976, and then flew as the commander of the Soyuz 24/Salyut 5 mission in early 1977. He then worked on the Interkosmos programme, initially backing up Bykovsky on Soyuz 31, and then commanded the Vietnamese Interkosmos mission in 1980. He went on to work in various directorates at the training centre, and left the cosmonaut team on 28 August 1982, with the rank of Major-General. He became a member of the Soviet Parliament, and worked at the Ministry of Defence and then at the Zhukovsky Air Force Engineering Academy. Gorbatko lives in Star City. He is twice Hero of the Soviet Union and a Pilot Cosmonaut of the USSR.

Gubarev, Alexei Alexandrovich (Major) was born on 29 March 1931, in the Kuybyshev Raion. He entered the military in 1950, and graduated from the Nikolayev Torpedo School in 1952. He was then assigned to the Pacific fleet, and flew a number of combat missions supporting Chinese and North Korean troops. In 1957 he went to the Red Banner Air Force Academy, graduating in 1961, and became a Commander of a mine torpedo regiment attached to the Black Sea Fleet. He joined the cosmonaut team on 10 January 1963, and was a CapCom for the Voskhod and early Soyuz missions. His first assignment was as a potential commander for a Military Soyuz 7K-VI mission, and he was then paired with Fartushny for an early Soyuz mission. He joined the Salyut programme in 1970, and was assigned as a support crew Commander to Salyut 1. In subsequent years he was a back-up crew commander for a number of civilian Salyuts and for Soyuz 12. Gubarev flew as Commander of the Soyuz 17/Salyut 4 mission in 1975, spending a month on board. His second mission took place in 1978, when he commanded the first Intercosmos mission, with the Czech cosmonaut Remek. He stood down from the team on 1 September 1981, after working at the training centre, and became Deputy Commander of the Chkalov Flight Research Institute. In 1985 he was promoted to Major-General, and he later left the Air Force to run a transport company. Gubarev still lives at Star City. He is twice Hero of the Soviet Union and a Pilot Cosmonaut of the USSR.

Gulyayev, Vladislav Ivanovich (Senior Engineer Lieutenant) was born on 31 May 1938, in Omsk. He attended the Kronstadt Higher Naval Engineering School, but on graduation was transferred to the Strategic Rocket Forces. He joined the cosmonaut team on 10 January 1963, and after completing his training he was assigned to the military Soyuz 7K-V1 programme as a Flight Engineer. He was also later on the Almaz programme. In the middle of 1967, he was involved in a car crash, which led to his retirement from the team on 6 March 1968. He continued to work at the training centre, rising to the rank of Colonel. Gulyayev died on 19 April 1990.

Ilyin, Yevgeni Aleksandrovich was born on 17 August 1937, in the city of Tula. He graduated from the S.M. Kirov Military Medical Academy in 1961, and was a physiologist as well as a doctor. He went to work at the Institute of Bio-Medical Problems, where much of the work in support of space medicine was being carried out. In 1965 the idea was put forward that one of the forthcoming Voskhod flights would explore biomedical issues, and so three doctors were selected to participate in the mission, which was planned to last two weeks. The trio undertook a five-day simulation in a Voskhod, but accomplished little else before the mission was cancelled in 1966. Ilyin considered himself a candidate cosmonaut. He continued to work at the Institute, went to Antarctica in the late 1960s, and became head of the biological Cosmos programme in the 1980s. He is now a Deputy Director at IMBP.

Kartashov. Anatoly Yakovelich (Captain) was born on 25 August 1932, in the Pervoy Sadovoye settlement, Sadovoye District, Voronez Raion. He attended the Chuguyev HAF Pilot School, and served as a fighter pilot in the Soviet Air Force between 1954 and 1960. He joined the cosmonaut team on 17 June 1960, underwent training, and earned a place in the initial six training group in July 1960. However, he failed a centifuge test when he developed haemorrhages along his spine, left the team in April 1962, and became a test pilot. Kartashov retired as a Colonel in 1985, and now lives in Kiev.

Katys, Georgy Petrovich was born on 31 August 1926, in Moscow. He graduated from the Moscow Auto-Mechanical Institute in 1949, and went on to do research work at the Bauman Higher Technical School. From 1953 he was involved in space research at the Academy of Sciences. He was referred for cosmonaut training in 1963, but was not considered. He joined the Voskhod 1 training group on 26 May 1964, and went on to serve as the back-up to Feoktistov. In 1965 he spent some months as the prime Flight Engineer for Voskhod 3, training with Volynov. He was replaced by an Air Force pilot as a result of the internal politics of the programme. Katys was named as a member of the Academy of Sciences group selected in 1967, but stood down from the team in 1970. He continued his work within the Institute of Automation of the Academy of Sciences and worked on the design of the Lunakhod series of space vehicles.

Khrunov, Yevgeny Vasilyevich (Senior Lieutenant) was born on 10 September 1933, in the Tula Raion. He graduated from the Bataisk HAF School in 1956, and served in an Air Guards regiment in Moldavia. He joined the Cosmonaut team on 9 March 1960, and was one of the first cosmonauts selected. He was a CapCom on Vostok 1, and was back-up to Leonov on Voskhod 2 in 1965. He went on to train for various Soyuz missions, and made his only spaceflight on Soyuz 5 in 1969, performing an EVA from one craft to another for the first time. Khrunov graduated from the Zhukovsky Higher Air Force Academy in 1968, and from the Lenin Military-Political Academy in 1974. He was involved in a number of other missions, including as original back-up Commander for Soyuz 7, and as CapCom and the training director of Salyut 3. In 1979 he was back-up commander to the Soviet–Cuban mission, and was a candidate to command the Romanian mission in 1980. He retired from the team on Christmas Day 1980, and went to work as a researcher in a military research institute. He then went on to work for the Committee for Foreign Economic Relations. Khrunov retired from the military in 1989, with the rank of Colonel. He died on 19 May 2000. He was a Hero of the Soviet Union and a pilot Cosmonaut of the USSR.

Kiselyov, Aleksandr Alekseyevich was born on 13 June 1935, in Moscow. He studied at the First Moscow Medical Institute, graduating in 1959. He then went to work at the Institute of Bio-Medical Problems in Moscow – the institute that was leading research into space medical issues. In 1965 there was a plan to fly a biomedical Voskhod flight for two weeks. Three doctors were identified as possible candidates, and although they conducted a five-day simulation flight in a Voskhod, they underwent minimum training before the mission was cancelled in 1966. Kiselyov returned to work at the Institute, and ultimately became the General Director of NPO Medinfo.

Kolodin, Pyotr Ivanovich (Senior Engineer-Lieutenant) was born on 23 September 1930, in the Zaporozhye Raion, Ukraine. He joined the military in 1946, attended a couple of artillery schools, and in 1952 entered the Govorov Military Engineering and Radio-technical Academy. After graduating in 1957, he joined the Strategic Rocket Forces, and served at Baikonur and Plesetsk. He was working at a rocket factory in Kharkov when he was informed of his selection to the cosmonaut team on 10 January 1963. Kolodin was assigned as an EVA crew-man for various early Voskhod and Soyuz missions, and in 1969 was assigned to the Salyut training group. He was back-up Research Engineer on Soyuz 7 in 1969 and Soyuz 10 in 1971, and was the Research Engineer on the original Soyuz 11 crew that was replaced only days before the mission in 1971. During the 1970s he was selected to serve on various other crews including the original Soyuz 26 crew, on which again he was replaced. He also served as a CapCom on a number of missions, including Salyut 6 and Salyut 7. In April 1983, Kolodin left the team and went to work as a CapCom at Mission Control. He retired with the rank of Colonel, and still lives at Star City.

Komarov, Vladimir Mikhailovich (Engineer-Captain) was born on 16 March 1927, in Moscow. He went to Bataisk HAF School, and graduated in 1949. He became a fighter pilot, based in the Grozny, before attending Zhukovsky Air Force Engineering Academy, from which he graduated in 1959. He was then assigned as a test engineer at the main Air Force Research Institute at Chkalovakaya, where he was when selected as a cosmonaut. He joined the team on 7 March 1960, but had some medical problems soon afterwards and had to fight to regain flight status. In 1961 he replaced Gagarin in the top six training group. He served as back-up to Popovich on Vostok 4, but was then diagnosed as suffering from an irregular heartbeat. In 1964 he was assigned to command the first multi-crewed flight, Voskhod. He was then made the cosmonaut member of the Soyuz design team, and in 1966 he was assigned to the first test flight of the craft. He launched on the Soyuz craft in April 1967. Immediately, the craft suffered major problems, and the mission was curtailed. On re-entry, the main parachute did not deploy, and Komarov was killed on 24 April 1967 when the Soyuz hit the ground. He was a Colonel in the Air Force, twice Hero of the Soviet Union, and a Pilot Cosmonaut of the USSR.

Kugno, Eduard Pavlovich (Senior Engineer-Lieutenant) was born on 27 June 1935, in Poltava, now part of the Ukraine. He attended the Kiev Higher Air Force Engineering School, from which he graduated in 1957. He was assigned to a squadron in the northern fleet, and was then reassigned to Leningrad, where he applied to join the cosmonaut team. He was selected on 11 January 1963, but was dismissed from the team on 17 June 1964. The official reason was his 'inability to tolerate weightlessness'. Kugno had said (at a meeting in early 1964) that a cult of personality was growing up around Khruschev. This caused real anger amongst the leadership of the cosmonaut team, who demanded his dismissal, together with another cosmonaut, Vorobyov. Kugno was not a Communist Party member at the time, and was duly dismissed. He continued to work in the Air Force in a number of engineering assignments, including a two-year posting to Algeria. He retired in 1990 as a Colonel. Kugno is a Master of Sport, with more than 800 parachute jumps to his name. He lives in Irkutsk.

Kuklin, Anatoly Petrovich (Major) was born on 3 January 1932, in the Chelyabinsk Raion, Russia. He entered the Volgograd HAF School, graduating in 1952. He served in various Air Guard regiments in the Soviet Union and in Germany, and attended the Red Banner Air Force Academy from 1957 to 1961. He joined the cosmonaut team on 11 January 1963, and in 1965 was assigned to the Spiral manned orbital spaceplane training group, with Gherman Titov and Filipchenko. He qualified as a test pilot First Class. In 1968 he was transferred to the lunar training group, and then went on to the Soyuz programme as a potential mission commander. He was the original back-up Commander on Soyuz 7. In 1970 he suffered a series of haemorrhages after a centrifuge test and was disqualified from mission training. Kuklin left the cosmonaut team as a Colonel in 1975, and served in the main staff directorate of the Air Force until he retired.

Kuznetsova (Pitskelauri), Tatiana Dmitryevna was born on 14 July 1941, in Gorky. She became a parachutist, setting a number of world records, and heard about the selection of a women's group in late 1961. She applied, and was selected on 28 February 1962. Kuznetsova was enrolled in the Air Force, initially as a private, but was then promoted to the rank of Junior Lieutenant. She was the youngest person (20 years of age) ever selected for space training. Kuznetsova was a support cosmonaut for the Tereshkova mission on Vostok 6 in 1963. She might have affected her own chances of a flight, as she married after having agreed to remain single until she had flown. She qualified as a cosmonaut in 1965, and in the same year was assigned as a back-up crew-member for the Voskhod 4 mission. This was to be a repeat of the Voskhod 2 mission, but with an all-female crew. The mission was cancelled in 1966. Kuznetsova graduated from the Zhukovsky Air Force Academy in 1969 – the same year that the women's group was disbanded. She continued to work at the cosmonaut training centre, helping cosmonauts with geophysical experiments. She is a Lieutenant-Colonel in the reserves, and still lives in Star City.

Lazarev, Vasily Grigoryevich (Major) was born on 23 February 1928, in the Altai Territory, Siberia. He attended a medical institute, with the intention of becoming a surgeon, but in 1952 transferred to study aviation medicine at the Saratov Medical Institute. After serving as a flight surgeon, he enrolled at the Chuguyev HAFP School, graduating as a pilot. In 1956 he was assigned to the flight test centre at Chkalov, and in 1959 he joined NII-7, the Institute of Aviation Medicine. He applied to join the cosmonaut team in both 1960 and 1963, but was unsuccessful. In 1962, he was assigned to the Volga high-altitude balloon project. This involved parachute jumps, in prototype spacesuits, from more than twenty miles high. This work came to the attention of Kamanin, commander of the manned space programme, who nominated him for inclusion in the first Voskhod crew. Lazarev served as back-up to Yegorov on this mission in 1964. He returned to work at the institute, but was considered for another Voskhod flight in 1965. He was selected to join the team on 17 January 1966, completed his training in 1967, and initially joined the Almaz training group before transferring to the Soyuz programme. He was assigned as back-up commander on Soyuz 9 in 1970, and was involved in the Kontact missions in 1970/71. He then joined the Salyut group, working as a prospective commander for civilian Salyut missions in 1972 and 73. These missions were cancelled. He flew as the Commander of the Soyuz 12 mission that tested the new Soyuz craft, and then acted as back-up to the Soyuz 17/Salyut 4 mission before being onboard the Soyuz 18 mission, which in 1975 experienced an explosion on the pad. The craft executed an 18-g re-entry. In 1976, Lazarev headed the Salyut training group within the cosmonaut team. He trained for a Soyuz-T mission in 1980/81, but failed his medical examination. He stood down from the team on 27 November 1985. He was a Colonel in the reserves. Lazarev died on 31 December 1990, from the effects of alcohol poisoning. He was a Hero of the Soviet Union and a Pilot Cosmonaut of the USSR.

Leonov, Alexei Arkhipovich (Lieutenant) was born on 30 May 1934, in Siberia. He graduated from the Chuguyev HAFP School in 1957, and served as a fighter pilot in various regiments in the Soviet Union and East Germany. He joined the cosmonaut team on 9 March 1960, and was the most junior officer in the selection. He was the assistant CapCom for Vostok 1, and was selected as the candidate to undertake the first EVA late in 1962. He trained for the mission, and was eventually paired with Belyayev in early 1964. On 18 March 1965 he was launched on Voskhod 2, as pilot. He spent ten minutes outside his spacecraft, experiencing some difficulty in returning to the craft at the end of the EVA. After the Voskhod 2 mission, he supervised EVA training for the early Soyuz crews, and in 1966 he was appointed commander of the lunar training group, and would have been the first Soviet cosmonaut on the Moon. Leonov trained with Makarov, but the whole programme was cancelled. He graduated from the Zhukovsky Air Force Engineering Academy in 1968, and was then named as Commander of the second crew to fly to the new Salyut station. A few days before the mission, a member of the crew was found to be ill, and the whole crew was replaced. Leonov would have commanded the ill-fated Soyuz 11 mission, having previously backed-up the Soyuz 10 mission. In 1972 and 1973 he was the prime Commander for the two Salyut stations that were subject to in-orbit failure. Also in 1973, Leonov was named to command the joint ASTP mission. He flew the Soyuz 19 mission in 1975, and was later promoted to the rank of Major-General. He later became Deputy Director at the cosmonaut training centre and the commander of the cosmonaut team. Leonov stood down from the cosmonaut team on 26 January 1982 and from then until 12 September 1991 he was the First Deputy Director at the training centre. He retired from the Air Force in 1992, and now administers a major Russian bank. He is twice Hero of the Soviet Union and a pilot cosmonaut of the USSR.

Letunov, Yuri Alexandrovich was born on 26 October 1926, in Tomsk. He graduated from the Moscow Juridical Institute in 1952, and joined Radio Moscow as a correspondent and then an editor. In 1965, Korolyov decided to select two journalists to train as cosmonauts, as he felt that current cosmonauts were unable to effectively describe their flights. Letunov underwent medical tests in July 1965, but after Korolyov's death in 1966 his interest in the mission waned. The journalists never trained for a specific mission. Letunov returned to work in radio and then in television, editing the major news programme 'Vremya', and later becoming a political observer. He died from a heart attack on 30 July 1984.

Matinchenko, Alexandr Nikolayevich (Engineer-Captain) was born on 4 September 1927, in the Voronezh Raion, Russia. He entered military service in 1943, and served as a radio operator in a bomber regiment. He entered the Balashov HAF School in 1948, and later served as a pilot in the Long Range Aviation Forces, flying a number of missions over Hungary in 1956. In 1957 he began a course at the Zhukovsky Air Force Engineering Academy, graduating in 1962. He was assigned, as a flight test engineer, to the military flight test centre at Chkalovskaya, where the cosmonauts were located while Star City was being built. He joined the cosmonaut team on 10 January 1963, completed his training in 1965, and worked on the Voskhod programme before training as a flight engineer on Soyuz and then working on the Almaz Military Space Station project. He stood down from the cosmonaut team on 19 February 1972, but continued to work in the Air Force, rising to the rank of Lieutenant-Colonel. He then worked as a crash investigator for the Ministry of Aviation Production. Matinchenko died on 18 June 1999.

Nelyubov, Grigori Grigoryevich (Senior-Lieutenant) was born on 9 April 1934, in Yevpatoriya, Crimea. He trained to fly at the Stalin Naval Air School, graduated in 1957, and then joined the Black Sea Fleet as a fighter pilot, flying MiG 19s. He joined the cosmonaut team on 7 March 1960, and undertook training for the Vostok programme, setting endurance records in the thermal chamber. He joined the first six group for accelerated training, and served as second back-up to Gagarin on Vostok 1. He was also assigned as a CapCom for the Vostok 3 and Vostok 4 missions. Nelyubov had what has been described as a 'strong personality'. In March 1963 he became involved in an incident with the militia and, together with two other cosmonaut colleagues, was dismissed from the team in May 1963. He was viewed as the ringleader of the trio, who apologised to the rest of the team. Nelyubov returned to the Air Force, and served in the Far East. He would tell people that he had been a cosmonaut, but he was not believed. On the 18 February 1966, while drunk, he was hit by a train and killed at Ippolitovka, Siberia.

Nikolayev, Andrian Grigoryevich (Senior Lieutenant) was born on 5 September 1929, in the village of Shorshely, Marinsky-Posad District, in the Chuvash Autonomous Republic of the former USSR. He attended the Marinsky-Posad Forestry Institute and became a lumberjack, after which he was drafted into the army in 1950. He began his military career training first as an aircraft gunner at the Kirovobad HAF school, and served on Tu-2 bomber crews for several months. He then trained as a pilot, graduating from the Frunze HAF School in December 1954, and for five years served as a MiG 15 pilot with the 401st Air Regiment, Air Defence Forces in the Moscow military district. He was selected with the first cosmonaut group in March 1960, having shown incredible resilience during the medical testing, and became Deputy Communist Party Secretary to the group. He was assigned to the Top Six training group, and served as back-up to Titov on Vostok 2 before flying on Vostok 3 in August 1962. Following his flight he assisted Tereshkova in her training for Vostok 6 before they married in November 1963. They divorced in 1982. Nikolayev was Commander of the first squad (civilian programme) of the cosmonaut team from 1966 to 1968, graduated from Zhukovsky AF Enginering Academy in 1968, and in the same year, after the death of Gagarin, he became the training deputy director of TsPK. He had been one of the first four cosmonauts assigned to Soyuz as early as 1965, and trained as back-up to the cancelled Soyuz 2 mission in 1967. A candidate for a planned circumlunar mission in 1968, he was instead assigned to command Soyuz 8, but failed his examinations and was reassigned as back-up to Shatalov. He flew a second mission as Commander of Soyuz 9 in 1970, during which he and Vitaly Sevastyanov set a new endurance record of 18 days, but returned in poor condition due in part to their lack of exercise. Retiring from active flight status in 1970, he was promoted to Major General and became the first Deputy Director of TsPK until retiring from military service in August 1992. He served as a deputy in the Great Soviet of the Russian Federation from 1991 to 1993 before becoming a deputy to Sevastyanov, who was then a member of the Russian State Duma. He is twice Hero of the Soviet Union and a Pilot Cosmonaut of the USSR.

Polyakov, Boris Ivanovich was one of a number of physicians identified as potential candidates for the 1964 Voskhod mission. He worked at the Institute for Bio-Medical Problems. The group was formed on 26 May 1964. Polyakov joined the team on 1 June 1964, and stayed for about a month. No other biographical details are known.

Ponomaryova, Valentina Leonidovna was born on 18 September 1933, in Moscow. She was interested in flying from a very early age, and enrolled at the Moscow Aviation Institute, graduating in 1959. While there, she learned to fly, and also became a parachutist. She married another student, Yuri Ponomarev, who later became a cosmonaut for NPO Energiya. In 1959 she went to work at the Institute of Applied Mathematics, and was encouraged to apply to the cosmonaut team by the Head of the Academy of Sciences. Ponomaryova joined the team in March 1962, and was the most experienced member of the group. The selection of the first female cosmonaut was in the hands of politicians, and Ponomaryova was appointed as second back-up to Tereshkova on the Vostok 6 mission in 1963. She was also enrolled in the Air Force. In 1965 she was assigned to command the Voskhod 4 mission that would repeat the Voskhod 2 flight, but with an all-female crew. The mission was cancelled in 1966, and the women's group was disbanded in 1969. Ponomaryova completed a post-graduate course at the Zhukovsky Air Force Academy. She continued to work at the training centre as an instructor until 1988. She is a Colonel in the reserves, and now works at an institute that is part of the Academy of Sciences.

Popovich, Pavel Romanovich (Captain) was born on 5 October 1930, in Uzin, near Kiev. He started work in Magnitogorsk as a building technician, and attended the Kacha HAFP School, graduating in 1954. He served as a pilot in an air regiment, flying MiG 19s. He was the first cosmonaut selected in March 1960, and was assigned to the top six training group. He also became the quartermaster of the group, and was famed for his singing. He was the CapCom for the Gagarin flight, based in Moscow, and flew on the Vostok 4 mission in 1962. Popovich became involved in a number of training positions, including the command of the cosmonaut team from 1964 to 1966, and was in charge of the early military Soyuz programmes. He graduated from the Zhukovsky Air Force Academy in 1968, and from 1966 to 1969 he trained as a commander for a lunar mission with Sevastyanov and then Grechko. After the cancellation of this programme he was placed in charge of the Almaz cosmonaut team, and was prime Commander for the 1972 Military Salyut, which broke up soon after launch. His second flight eventually came in July 1974, as Commander of the Soyuz 14 flight to Salyut 3. On his return he filled a number of training positions, including Deputy Director of the centre. In 1976 he was promoted to Major General. He left the Cosmonaut team on 26 January 1982 and the Air Force in 1989, and then went to work in a scientific and research institute. Popovich is twice Hero of the Soviet Union and a pilot Cosmonaut of the USSR, and now lives in Moscow.

Rafikov, Mars Zakirovich (Senior Lieutenant) was born on 30 September 1933, in Dzhalal-Abad, Kirghzia. He graduated from the Chuguyev HAF pilots School in 1954, and went on to serve in a fighter regiment, flying MiG 17s. After joining the cosmonaut team on 28 April 1960, he undertook cosmonaut training, but was dismissed from the team in March 1962 for being absent without leave and for being politically unreliable. He rejoined the Air Force and served in an air regiment, including a posting to Afghanistan. Rafikov left the Air Force in 1982, and went to live in Alma Ata (now Almaty). He died, after a heart attack, on 23 July 2000.

Senkevich, Yuri Aleksandrovich was born on 4 March 1937, in Mongolia. He attended the S.M. Kirov Military Medical Academy, graduating in 1960. He then joined the Institute of Bio-Medical Problems, which was the leading institute researching space medicine. In 1965 the idea was proposed that one of the forthcoming Voskhod flights would explore biomedical issues, and so three doctors were selected as potential candidates. They underwent a five-day simulation in a Voskhod craft, but carried out little other work, and the mission was cancelled in 1966. Senkevich returned to work at IMBP, and later participated as a crew-member on Thor Heyerdahl's Ra expedition, as well as presenting a travel programme on television. He became a department head at the Institute.

Shatalov, Vladimir Alexandrovich (Major) was born on 8 December 1927, in Petropavlovsk, Kazakhstan. He attended the Kacha HAFP School, graduating in 1949. He stayed at the school as an instructor, and then attended the Red Banner Air Force Academy, graduating in 1956. He had a number of assignments, rising to the post of senior inspector pilot. He joined the cosmonaut team on 10 January 1963, was a CapCom on both the Voskhod missions, and was assigned to the back-up crew for the Voskhod 3 mission. He trained with Beregovoi for more than a year, before the mission was cancelled. He then joined the Soyuz group, and acted as a CapCom on Soyuz 1. Shatalov was the cosmonaut who talked to Komarov during his re-entry. He was then back-up to Beregovoi on Soyuz 3 in 1968, and in 1969 he commanded the Soyuz 4 mission. He would then have been the back-up commander for Soyuz 8, also in 1969, but he replaced the Commander shortly before the mission. He was then assigned to command the first flight to the new Salyut station, reflecting his experience and seniority. His third flight, Soyuz 10, occurred in 1971, and he stood down from the team shortly afterwards, on 26 June 1971. Shatalov was promoted to the rank of Major-General, and replaced Colonel-General Kamanin as the Director of Cosmonaut Training for the High Command of the Soviet Air Force, – a post he held until January 1987. Shatalov became a Lieutenant-General in 1977, and was a member of the Supreme Soviet. In 1987 he became Director of the cosmonaut training centre, but was removed from his post in September 1991. He is now retired, and lives in Star City. He is twice Hero of the Soviet Union and a Pilot cosmonaut of the USSR.

Shonin, Georgy Stepanovich (Senior Lieutenant) was born on 3 August 1935, in the Voroshilovgrad Raion, Ukraine. He attended a number of pilot schools before graduating from the Stalin Naval Air School in 1957. He then joined an air regiment in the Northern Fleet. (Gagarin was in the same regiment.) Shonin joined the cosmonaut team on 7 March 1960. He was originally to be the back-up to Popovich in 1962, but suffered problems after a centrifuge test. In 1964 he started training for a Voskhod mission, being selected as the prime pilot for the Voskhod 3 mission that was subsequently cancelled. He then moved to work on the Soyuz programme, serving as back-up Commander on Soyuz 5. He also graduated from the Zhukovsky Air Force Academy in 1968, and in 1969 flew as the Commander of Soyuz 6, a mission that involved a welding experiment. In 1970 he was named as first Commander of the new Salyut station, but was replaced for disciplinary reasons. He then took up a position in the training centre, working on Almaz, ASTP, and the newly created Buran programme. He was promoted to the rank of Major-General in 1977, and left the cosmonaut team on 28 September 1979. He took up a number of senior positions within the Air Force, and then as a director of a scientific research institute, and retired with the rank of Lieutenant-General, in 1990. Shonin also wrote a book about the first selection in 1977. He died on 6 April 1997. He was a Hero of the Soviet Union and a Pilot Cosmonaut of the USSR.

Solovyova, Irina Bayanovna was born on 6 September 1937, in the Tula district. She attended the Ural Polytechnic Institute, graduating in 1957. While at the Institute, she joined her local air club and became a parachutist. She later became a member of the Soviet national team, setting many world records. She also became a Master of Sport, and has completed more than 2,200 jumps. She applied to become a cosmonaut, and was selected in March 1962. After completing her basic training in October 1962, she was enrolled into the Air Force. She was assigned as the first back-up to Tereshkova for the Vostok 6 mission of 1963, and in 1965 was selected as the pilot for the Voskhod 4 mission that was to repeat the Voskhod 2 flight, but with an all-female crew. The mission would have lasted ten days, and Solovyova would have walked in space, but it was cancelled in 1966. The women's group was disbanded in October 1969 and Solovyova graduated from the Zhukovsky Air Force Academy during the same year. She continued to work at Star City as a psychologist and as a scientist and also went on an expedition to Antarctica. She continued to parachute, setting more records and testing new equipment. Solovyova is a Colonel in the reserves.

Sorokin, Alexei Vasilyevich (Captain) was born on 30 March 1931, in the Kursk Raion. He attended a medical Institute from 1951 to 1955, and then entered a school for flight surgeons, graduating in 1957. He served as a hospital director in an air regiment, before joining the medical staff at the cosmonaut training centre in 1963, where he worked on the remaining Vostok flights. In 1964 he was nominated as a potential candidate for the first Voskhod flight, and served as the back-up to Yegorov. He was also considered for inclusion in the 1965 group, but ultimately was never enrolled in the team. During all these assignments, Sorokin continued to work at the training centre. This was where he served until his death on 23 January 1976.

Tereshkova, Valentina Vladimirovna was born on 6 March 1937, in the Yaroslavl Raion. On leaving school she went to work in a textile mill. She also joined a parachute club, and made more than a hundred jumps. She applied to join the cosmonaut team in late 1961, and unaware, at the time, that females were being recruited. Tereshkova was called to Moscow for an interview and was selected to join the group in March 1962. She completed her basic training by October 1962, and was selected to make the first flight by a woman. She was not necessarily the most experienced candidate, but the decision on who flew was political. Tereshkova came from a worker family, and this was a decisive factor. She flew on Vostok 6 in June 1963, and later that year she married Cosmonaut Nikolayev. They had one daughter, but they divorced in 1982. Tereshkova participated in training for Voskhod and early Soyuz flights, but there is no evidence that she trained for another mission. She remained within the cosmonaut team until March 1997, and reached the rank of Major-General in the Air Force. She graduated from the Zhukovsky Air Force Engineering Academy in 1969. Over the years, Tereshkova held a number of very responsible jobs. She was a member of the Supreme Soviet, President of the Soviets Women's Committee, President of the International Union of Soviet Friendship Societies, and, most recently, the Director of the Russian Centre for International Scientific and Cultural Cooperation. She lives in Star City, and is a Hero of the Soviet Union and a Pilot Cosmonaut of the USSR.

Titov, Gherman Stepanovich (Senior Lieutenant) was born on 11 September 1935, in the Altai Territory. He was always interested in becoming a pilot, and he therefore enrolled in the Stalingrad HAFP School, graduating in 1957. He was assigned to an Air Guards regiment in the Leningrad area. He joined the cosmonaut team on 9 March 1960, and was assigned to the top six group for accelerated training. In 1961 he was the back-up to Gagarin on Vostok, and later that year he flew on Vostok 2. He was the first person to suffer from space sickness, but was – and remains – the youngest person to fly in space. Titov undertook a number of public relations duties after the mission. In 1965, he was assigned to command the Spiral training group, working on a reusable space-plane. He attended the Chkalov test pilot school in 1967, and was awarded the title Test Pilot 3rd class. After the death of Gagarin in 1968 it became clear that the authorities were not about to allow Titov to fly again. He graduated from the Zhukovsky Air Force Academy in 1968, and retired from the team in July 1970. He graduated from the K.E. Voroshilov Military Staff Academy in 1972, and in the succeeding years, he was assigned to a number of increasingly senior posts within the Ministry of Defence. He rose to the rank of Colonel-General, and retired from the armed forces in 1991. He served two terms as a member of the Russian Duma (parliament). Titov was a Hero of the Soviet Union and a Pilot Cosmonaut of the USSR. He died on 20 March 2000, after suffering a heart attack in his sauna.

Varlamov, Valentin Stepanovich (Senior Lieutenant) was born on 15 August 1934, in the Penza Raion. He joined the Air Force, attending the Kacha HAF Pilots School and graduating in 1954. He then served as a fighter pilot. Varlamov joined the cosmonaut team on 28 April 1960 and was selected to join the initial six-person group for accelerated training. On 24 July 1960 he damaged the vertebrae in his neck while diving in a lake. He was medically disqualified, and left the team in April 1961. Varlamov continued working at the cosmonaut training centre as an instructor, and was a close friend to many of the cosmonauts. He died of a cerebral haemorrhage on 2 October 1980.

Volynov, Boris Valentinovich (Senior Lieutenant) was born on 18 December 1934, in Irkutsk. He attended the Stalingrad HAFP School, graduating in 1955, and served as a fighter pilot in the Air Defence Forces. He joined the cosmonaut team on 9 March 1960, and joined the Vostok training group in late 1961, replacing Gagarin and Titov, who were involved in post-mission work. He was back-up on Vostok 4 and Vostok 5, and also backed-up Komarov on Voskhod. He was then named as commander of the fourteen-day Voskhod 3 mission, but following one year's training with a number of different crewmen, the mission was cancelled in 1966. Volynov graduated from the Zhukovsky Air Force Academy in 1968. He was next assigned to the early Soyuz programme, acting as second back-up to Beregovoi on Soyuz 3, before commanding Soyuz 5 in 1969. He was then assigned to the Almaz military station, and was back-up Commander for Salyut 2 and Salyut 3 in 1973/4. Volynov commanded his second flight in 1976, on the Soyuz 21 mission to Salyut 5. He continued to work within the cosmonaut team, commanding the Air Force group from 1982 to 1990. He retired on 17 March 1990, as a Colonel in the reserves. He is twice Hero of the Soviet Union and a Pilot Cosmonaut of the USSR.

Vorobyov, Lev Vasilyevich (Major) was born on 24 February 1931, in the town of Borovichi, between Moscow and St Petersburg. He graduated from the Serov HAF School at Bataisk in 1952, and became an Air Force pilot based in the Urals, flying MiG 15s. He enrolled at the Red Banner Air Force Academy in 1957, graduating four years later, and was then assigned as a pilot-navigator on Sakhalin Island. He joined the cosmonaut team on 10 January 1963, and when he had completed his basic training he was assigned to the Almaz training programme. In March 1964, Vorobyov and another cosmonaut were critical of premier Khruschev during a meeting. Kugno, the second cosmonaut, was expelled from the team, but Vorobyov survived because he was a long-standing member of the Communist Party. In 1969 Vorobyov joined the Soyuz training group, and was teamed with civilian engineer Yazdovsky to fly the Soyuz 13 mission in late 1973. A couple of days before the flight, the crew was replaced, and Vorobyov was never given another opportunity to serve on a crew. He is described as 'a very principled man'. He retired from the team on 11 April 1974, but continued to work at the cosmonaut training centre, rising to the rank of Colonel. He is now retired.

Voronov, Anatoly Feodorovich (Captain) was born on 11 June 1930, in the Orenburg Raion. He originally trained as a teacher before going to Navigators' School, graduating in 1953. Voronov initially served in Long Range Aviation before transferring to an Air Guards Regiment based at the nuclear test range at Semipalatinsk. He took part in a number of nuclear tests. In 1957, he attended the Red Banner Air Force Academy, graduating in 1961. He was then attached to the flight research centre at Chkalovskaya as a Flight Test Navigator, and had logged more than 6,000 hours of flight time. He was selected to join the cosmonaut team on 10 January 1963, and after completing his initial training he was assigned to the early Soyuz programme before joining the lunar group. When these missions were cancelled, he transferred to Salyut 1 and joined a support crew to the Soyuz 11 mission of Gubarev and Sevastyanov. Voronov went on to work on Soyuz-T and the Salyut 6 station. He retired from the team on 25 May 1979, when he was a Lieutenant-Colonel in the reserves. He worked in the Priroda State Centre for the next fourteen years, and died, after a long illness, on 31 October 1993.

Yegorov, Boris Borisovich was born on 26 November 1937, in Moscow. He attended the Sechenov First Moscow Medical School, graduating in 1955. He wanted to specialise in eerospace medicine, and after graduation he worked in an institute that was conducting much of the medical testing of cosmonauts, including tests in the soundproof chamber. He was also part of a medical emergency team that would be parachuted to rescue a stranded cosmonaut. In 1964, he was named as the onboard physician for the Voskhod mission. In October 1964 he flew as the first medical doctor to work in space. Yegorov remained involved in space medicine, and helped supervise the Cosmos 110 mission in 1966. He became a professor and a director of a medical research institute under the Ministry of Health. He died on 19 September 1994. He was a Hero of the Soviet Union and a Pilot Cosmonaut of the USSR.

Yorkina (Sergeichik), Zhanna Dmitryevna was born on 6 May 1939, in Ryazan. She attended a pedagogical institute, graduating with a degree in English. In her spare time she was a parachutist, and she submitted an application to join the cosmonaut team via her local club. She joined the team on 28 February 1962, and was enrolled in the Air Force. She completed her basic training in late 1962 and was assigned as a support cosmonaut on the Vostok 6 flight of Tereshkova in 1963. In April 1965, after assisting with the Voskhod 2 mission, she was assigned as the back-up Commander of the planned Voskhod 4 mission that was to repeat the Voskhod 2 flight with an all-female crew. The mission was cancelled in 1966. Yorkina trained on Soyuz for a short time, but the women's group was disbanded in October 1969 – the same year in which she graduated from the Zhukovsky Air Force Academy. She continued to work at the training centre, and is now a Lieutenant-Colonel in the reserves.

Zaikin, Dmitri Alexeyevich (Senior Lieutenant) was born on 29 April 1932, in the Yekaterinovka part of the Rostov Raion. He attended various Air Force Schools before graduating from the Frunze HAF pilot school (with Nikolayev) in 1954. He then served as a fighter pilot. He joined the cosmonaut team on 25 March 1960, and was the back-up Commander to Belyayev on Voskhod 2. He also attended and graduated from the Zhukovsky Air Force Engineering Academy in 1968. Zaikin was working on the Soyuz programme when he was medically disqualified due to an ulcer, and he left the team in October 1969. He worked at the cosmonaut training centre until 1987, when he retired. He holds the rank of Colonel, and still lives in Star City.

Zholobov, Vitaly Mikhailovich (Senior Engineer-Lieutenant) was born on 18 June 1937, in the Ukraine. He attended the Azerbaijan Petrochemical Institute, graduating in 1959, and then joined the Strategic Rocket Forces, serving as a flight-test Engineer at Kasputin Yar. On 10 January 1963 he joined the cosmonaut team and, after completing his basic training in 1965, was initially assigned to the Zond programme. In 1968 he joined the Almaz group, and went on to be the back-up Flight Engineer to Artyukhin and then Demin on Salyut 3. Zholobov was the Flight Engineer on the Soyuz 21 mission that visited Salyut 5 in 1976 and was also CapCom on other military Salyut missions. He received a degree from the Lenin Military-Political Academy in 1974, and worked at Star City until he retired, as a Colonel, in 1981, after which he entered the reserves. Zholobov initially went to work in the Tyumen Raion, before going back to the Ukraine as a chief administrator in his home region. He is a Hero of the Soviet Union and a Pilot Cosmonaut of the USSR. He currently lives in Kiev.

COMMANDERS OF COSMONAUTS

Ranks are those held at the time of retirement.

Kamanin, Nikolai Petrovich (Colonel-General) was born on 18 October 1908. He joined the Air Force, training at the Borisoglebsh Higher Air Force Pilot School, and graduating in 1929. He also became the first Hero of the Soviet Union – for saving the crew and passengers from the ice-breaker Chukyushkin. He then entered the Zhukovsky Air Force Engineering Academy, leaving in 1938, and then commanded an air brigade in the Ukraine. From 1942 he commanded an air attack division, and after the war he became deputy chief of Aeroflot. In 1947 he headed the DOSAV, working towards the revival of aeroclubs and parachute schools. Many cosmonauts came via this route to join the Air Force or engineering schools. In 1956 he graduated from the General Staff Academy, and in 1958 he was selected to be a Deputy Chief of the Soviet Air Force, in charge of manned spaceflight. Kamanin had responsibility for all aspects of the manned programme, including the selection of the cosmonaut team. He worked closely with Korolyov, and was a major figure in the development of manned flight. At some point the post was renamed Director of Cosmonaut Training, and Kamanin remained in this post until June 1971, when

Shatalov replaced him. He retained his other responsibilities within the Air Force until his death in 1982.

Karpov, Yevgeniy Anatolyevich (Major-General) was born in 1921, in Kiev. He attended the Kirov Military Medical Academy, graduating in 1942, and was later appointed to an air division which was flying bombers on long-range missions to Germany. After the war he was appointed to the Institute of Aviation Medicine, where he developed a range of tests and knowledge which would be used to screen and test the first group of cosmonauts. Karpov was appointed the first Commander of the cosmonaut training centre on 24 February 1960, and in the press he was given the title 'Chief Doctor' (being a Colonel in the Medical Service). He served as Director for three years, and was replaced because the centre had acquired a special significance, and the higher leadership felt that it should be headed by an Air Force General with a reputation and fame comparable to that of the cosmonauts. He returned to the Institute and was promoted to Major-General in 1966. In 1973 he was appointed as a supervisor in the Ministry of Civil Aviation State Scientific Research Institute. He retired in 1978, and joined the Federation of Cosmonautics as one of its permanent staff. He went to live in Kiev, where he died.

Kuznetsov, Nikolai Fedorovich (Major-General) was born on 26 December 1916, in Petrograd (Leningrad). He trained at an Air Force school named after Kalinin, and later served in the Soviet Air Force. He served as a fighter pilot during World War II, and was honoured as a Hero of the Soviet Union on 1 May 1943, having shot down thirty-seven Fascist aircraft during the conflict. He then held a number of command positions, including a period in Korea during the early 1950s. He attended the General Staff Academy in 1956, and went on to command a pilot school. On 2 November 1963 he was appointed commander of the cosmonaut training centre. His deputy at the time was Yuri Gagarin. He remained in charge until May 1972, and after leaving that post he went to work at NPO Energiya until 1987, when he retired. Kuznetsov continued to live at Star City until his death on 5 March 2000.

Odintsov, Mikhail Petrovich (Lieutenant-General) was born in 1921, in the Perm Raion. He became one of the most outstanding fighter pilots of the great Patriotic War, and was twice awarded the title Hero of the Soviet Union – on 4 February 1944 and 27 July 1945. After the war he attended the Lenin Military-Political Academy and the General Staff Academy, and also held a number of senior positions in the Air Force. Odintsov was appointed commander of the cosmonaut training centre for only a few months in 1963. The reason for his departure is not known. He continued to serve in the Air Force until his retirement.

DESIGNERS

Barmin, Vladimir Pavlovich (Academician) was born on 17 March 1909, in Moscow. He graduated from the Baumann Higher Technical School in 1930, and became a specialist in refrigerated machines, including those used on Lenin's tomb. During World War II he was asked to work on the design of mobile rocket launch systems, including the Katusha rocket, and in 1945 he was sent to Germany to study the launch sites used by the Germans. Barmin was put in charge of designing and building the launch complexes for missiles, including mobile systems and space rockets, and all the assembly, integration, command and control buildings at Baikonour, Kasputin Yar and Plesetsk. He headed the General Machine Building Design Bureau (KBOM), and was a member of the State Commission that oversaw all aspects of the space programme. He was also a member of the Academy of Sciences, and honoured as a Hero of Socialist Labour, as well as being awarded the Order of Lenin six times. Barmin died on 17 July 1993.

Bushuyev, Konstantin Davidovich was born on 23 May 1914, in the Kaluga Raion. He initially worked in a smelting plant before attending the Moscow Aviation Institute, graduating in 1941. He went to work on the BI-1 jet fighter before becoming involved in rocket construction. Bushuyev became involved in the development of multi-stage liquid-fuel led rockets and ballistic missiles, and in 1954 he became deputy head of construction in OKB-1. He was also a corresponding member of the Academy of Sciences, and was honoured as a Hero of Socialist Labour. He died on 26 October 1978.

Chelomei, Vladimir Nikolayevich (Academician) was born on 30 June 1914. He studied at the Kiev Aviation Institute, graduating in 1939, and during the war he worked at the Central Aviation Motor Construction Institute, developing a pulse-jet engine that was tested on rocket planes. In 1944 he was Chief Designer at a factory that concentrated on the development and manufacture of winged rockets and missiles. In 1955 he specialised in winged space planes, and undertook early design work on the Spiral project. In 1959 he was appointed designer-general of the OKB-52. The title reflected the military work undertaken by the bureau, which was involved with the Proton booster and the Almaz space station (military Salyuts 2, 3 and 5). It also developed a manned vehicle called the TKS, and the related heavy module used so successfully on Salyut and Mir. Another project was the development of the SS-9 missile. The work of this design bureau is still viewed as secret. Chelomei was a member of the Academy of Sciences, and was twice honoured as a Hero of Socialist Labour twice. He was also a deputy to the Supreme Soviet. He died on 8 April 1984.

Chertok, Boris Yevseyevich was born on 1 March 1912, in Lodz (now in Poland). He graduated from the Moscow Power Engineering Institute in 1940, and specialised in aircraft control systems. In 1944 he moved to a rocket research institute, and went to Germany to study German rocket technology. He then went to work with Korolyov at OKB-1, where he held a number of development and research posts, involving control systems. Chertok was involved in the development of the Vostok launcher, and headed the team that worked on Vostok, Voskhod, Molniya communication satellites, Salyut, Mir and Energiya. He was a deputy to Mishin (Korolyov's successor), and still works at NPO Energiya. Chertok is a corresponding member of the Academy of Sciences, and was honoured as a Hero of Socialist Labour, as well as being twice awarded the Order of Lenin.

Glushko, Valentin Petrovich (Academician) was born on 2 August 1908, in Odessa. He studied physics and mathematics at Leningrad University, and after graduating, entered the field of rocket propulsion. In 1929 he became head of the Gas Dynamics Laboratory (GDL), based in Leningrad. Glushko developed the world's first electrical-rocket engine and the USSR's first liquid-propellant rocket engine, and worked on solid-rocket projectiles and rocket engines. In 1941 the GDL became a full Experimental Design Bureau. In 1946 Glushko attended a British test-firing of a V2, and during the next few years he worked with Korolyov (whom he had first met in 1932), putting his engines in Korolyov's rockets. He went on to provide the main-stage engines for all the Soviet Union's space launch vehicles, and was described in the press and other publications as the 'Chief Designer of Rocket Engines'. Glushko was a member of the state commission, and in 1974, – as part of a major reorganisation of design bureaux, – he was made the head of the NPO Energiya, – incorporating Korolyov's old design bureau – and succeeded Mishin as head. One of his first tasks was to cancel the manned lunar programme and the N1 launch vehicle. His title was General Designer of Manned Spaceships and Space Stations, and he oversaw the development of Salyut, Mir, Buran and Energiya. Glushko was a member of the Academy of Sciences, and twice Hero of Socialist Labour. He was also a member of the Central Committee of the Communist Party. He died on 10 January 1989.

Isayev, Alexei Mikhailovich was born on 24 October 1908. He began work on the development of a long-range bomber and then the BI-1 jet fighter, and in 1944 he became head of an experimental design bureau developing liquid-propellant rocket engines. He designed the engines of the Vostok, Voskhod and Soyuz manned craft and interplanetary probes. Isayev was a corresponding member of the Academy of Sciences, and was honoured as a Hero of Socialist Labour, and was three times awarded the Order of Lenin. He died on 18 June 1971.

Keldysh, Mstislav Vsevolodovich (Academician) was born in 1911, in Riga. He graduated from Moscow University in 1931, and joined the Central Aero-Hydrodynamics Institute, where he worked for many years and won many prizes. He became a member of the Academy of Sciences, and was the famous 'Chief Theoretician of Cosmonautics'. Keldysh carried out pioneering research into rocket dynamics and the mechanics of spaceflight, and was also involved in the development of the Soviet atomic bomb. He was a member of the State Commission, and became President of the Academy of Sciences in 1961. He was also honoured as Hero of Socialist Labour, three times. He died on 24 June 1978.

Kerimov, Kerim Aliyevich (Lieutenant-General) was born on 14 November 1917, in Baku. He graduated from the Dzerzhinsky Artillery Academy in 1944, and went to work at a rocket production plant. In 1946 he joined the Military of Defence Main Artillery Directorate, and worked with Korolyov on a number of projects involving military rockets. In 1965 he was appointed Chairman of the State Commission on

Flight Testing of Manned Space Complexes (which approves all manned flights, crews and planetary missions), to test the Soyuz craft. He remained the Chair of this commission until 1989.

Korolyov, Sergei Pavlovich (Academician) was born on 30 December 1917, in Zhitomir, Ukraine. He graduated from the Baumann Higher Technical School and went into the aircraft industry, becoming a designer in the GIRD group. He moved into the area of rocket research and development, but in 1938 was arrested and jailed. He worked in the mines in Siberia, and this affected his health for the rest of his life. In 1942, he was released, and went to work with Glushko, whom he had first met in 1932. He designed liquid-fuel rocket accelerators for aircraft, and in 1945 he was one of the leading designers sent to Germany to look at German rocket technology. For the next few years he worked as the Chief Designer of OKB-1, working on the Soviets Union's first ballistic missile. Korolyov's dream, however, was space travel, and he adapted the missile to be the launch vehicle for a series of space vehicles built by his design bureau. He led the design teams that built and launched Sputniks, Vostok, Voskhod, Molniya satellites, and a series of interplanetary probes. The Soviet people never knew Korolyov's name, and was always described as the 'Chief Designer of Carrier Rockets and Spacecraft'. Many of his colleagues went on to develop and design spacecraft for the next thirty years, several went on to lead their own design bureaux and achieve high honours, and some of the younger engineers became cosmonauts. Korolyov was a member of the Academy of Sciences and was twice honoured as a Hero of Socialist Labour, as well as being three times awarded the Order of Lenin. Korolyov was the inspiration behind the success of the Soviet space effort in the 1950s and 1960s. He died on 14 January 1966.

Kosberg, Semyon Arievich was born on 1 March 1903. He graduated from the Moscow Aviation Institute, and went to work at the GDL-OKB, where he became Chief Designer, working on liquid rocket engines. He designed a number of engines which were used on the upper stages of launch vehicles such as Vostok and Voskhod. Kosberg was honoured as a Hero of Socialist Labour. He died on 3 January 1965.

Kozlov, Dmitri Ivanovich was born on 1 October 1919, in Krasnodarskii Krai. He attended the Leningrad Military Mechanical Institute, but joined the Leningrad militia before completing his studies. He was wounded three times before being demobilised, after which he returned to the Institute to complete his studies in 1945. From 1946 he worked with Korolyov, as a design engineer. He was the leading designer of the R-7 launch vehicle, and in 1958 he headed the serial production of the rocket in Samara. He was a deputy Chief Designer in OKB-1, reporting to Korolyov until the bureau was reorganised in 1967, and he initially became First Deputy and then Chief Designer of the Central Specialised Design Bureau. This bureau is responsible for the manufacture of all Vostok, Molniya and Soyuz boosters, and also builds the Resurs, Foton, Bion and Fram satellites for both civilian and military use. Kozlov is a corresponding member of the Academy of Sciences, and has been honoured as a Hero of Socialist Labour, as well as four times being awarded the Order of Lenin.

Kuznetsov, Viktor Ivanovich (Academician) was born on 27 April 1913, in Moscow. He graduated from a polytechnic institute with an interest in engineering design, and in 1938 he began to work on gyroscopes. During World War II, the results of his work were used in ships, aircraft and tanks. In 1946 he became a Chief Designer, and was one of the group that visited Germany to examine German rocketry. He became one of the council of six designers, and specialised in control and command devices for missiles and spacecraft. His devices were used on Sputnik, Vostok, Voskhod, Soyuz, Buran, Salyut and Mir. He was head of NII-944. Kuznetsov was a member of the State Commission that oversaw all early space missions, and also headed the Commission that investigated the missile explosion at Baikonour in 1960. He was awarded many honours, including two Hero of Socialist Labour Gold Stars. He was also a member of the Academy of Sciences. Kuznetsov died on 22 March 1991.

Mishin, Vasili Pavlovich (Academician) was born on 5 January 1917. From 1935 he studied at a technical institute attached to the Central Aero-Hydrodynamics Institute. He also worked on the BI-1 jet aircraft, and then attended the Moscow Aviation Institute. Mishin then became involved in rocket research, and was one of the team that was sent to study German technology. In 1946 he went to work with Korolyov in OKB-1 – and worked with him for thirty years, eventually becoming his first deputy. Mishin was a specialist in control processes, and calculated the trajectory and ballistics for all early launches. When Korolyov died in 1966, Mishin succeeded him, and was the Chief Designer until 1974, when he was dismissed. He oversaw the development of the Soyuz, Lunar, Salyut and the N1 programmes. Mishin is a member of the Academy of Sciences, has been honoured as a Hero of Socialist Labour, and has twice been awarded the Order of Lenin.

Pilyugin, Nikolai Alekseyevich (Academician) was born on 18 May 1908, near St Petersburg. He graduated from the Baumann Higher Technical College in 1935, and began to specialise in aircraft instruments. During World War II he worked on autopilots, and in 1944 he began to work on rockets. He was one of the six designers who went to Germany to study rocketry, and he began to work with Korolyov on missile and rocket design. He also became the head of NII-885, in charge of automatic control systems on missiles and space equipment. Pilyugin's work was used on all manned craft and rockets, including Soyuz and Proton. He founded the Institute for Mechanics, and was also a member of the State Commission that oversaw all aspects of the space programme. He was a member of the Academy of Sciences and was twice honoured as a Hero of Socialist Labour. He died on 2 August 1982.

Rudnev, Konstantin Nikolayevich was born in June 1911, in Tula. He attended a mechanical institute, and then went to work in the defence industry. In 1952 he became the Deputy Minister for the Defence Industry, and in 1958 he was appointed Chairman of the Council of Ministers State Committee for Defence and Technology, – a post which he held until 1965. This is the famous State Commission, and he was in charge when all the key decisions concerning manned flight were made. In 1965 he became Minister for Instruments, Automation Equipment and Control Systems.

Rudnev was honoured as a Hero of Socialist Labour, and was six times awarded the Order of Lenin. He died on 13 August 1980.

Ryazanskiy, Mikhail Sergeyevich was born on 5 April 1909, in Petersburg. In 1935 he graduated from the Moscow Power Engineering Institute and became an electrical engineer. During World War II, he worked on long-range communications and on systems for the radio detection of aerial targets, and was one of the six Chief Designers who in 1946 were sent to examine German rocketry. Ryazanskiy was the head of NII-845, which specialised in the radio control of missiles and space technology. His systems were used onboard Vostok, Voskhod, Soyuz, interplanetary probes and all launch vehicles. He was a member of the State Commission that oversaw all aspects of the space programme. He was honoured as a Hero of Socialist Labour, and was five times awarded the Order of Lenin. He was also a corresponding member of the Academy of Sciences. Ryazanskiy died on 5 August 1987.

Serebin, Ivan Dmitriyevich was born on 25 February 1910, in Krasnador. He attended the Moscow State University, graduating as an engineer and worked in a factory, as well as becoming an organiser in the Communist Party. In 1942 he was appointed to a post in the Party's Central Committee, and he rose through the ranks. From 1958, as Head of the Defence Industries department, Serebin made a major contribution to solving problems connected with the creation of space technology and its application. He reported to Marshall Dmitri Ustinov. Serebin was a member of the Supreme Soviet, and was five times awarded the Order of Lenin. He died on 15 February 1981.

Severin, Guy I. was born in 1926, near Leningrad. During the war he served on one of the key fronts. Later, he attended the Moscow Aviation Institute, graduating in 1947. He then went to work at the LII.A research Institute, testing prototype aircraft. He became involved in designing ejector seats and pressure suits, and in 1964 the company Zvezda was formed, with Severin as its first General Designer. Zvezda has designed all Soviet space suits for use on Vostok, Voskhod and Soyuz craft, and on EVA operations. It also developed the Soviet MMU (including an early version, to be used on a Voskhod mission) and the airlock used on Voskhod 2. Zvezda also developed the seat system for all spacecraft and Buran. The design bureau also makes ejector seats and pressure suits for high-performance aircraft. Severin is a corresponding member of the Academy of Sciences, and a Hero of Socialist Labour.

Tikhonravov, Mikhail Klavdiyevich (Academician) was born in 1900, in Vladimir. In 1920 he attended the Zhukovsky Air Force Engineering Academy, and in 1932 he started work on engines at GIRD. In 1945 he was sent to study German rocketry, and later became a deputy to Korolyov in OKB-1, where he led a number of projects, including Sputnik, Vostok, Voskhod and Soyuz. Tikhonravov was a member of the Academy of Sciences, and was honoured as a Hero of Socialist Labour. He died on 4 March 1974.

Voskresensky, Leonid Aleksandrovich was born in 1913, and graduated from the Moscow Aviation Institute. He was a deputy Chief Designer to Korolyov in OKB-1, and was in charge of the flight-test department, including all the Vostok test-flights. Voskresensky was honoured as a Hero of Socialist Labour, and was twice awarded the Order of Lenin. He died in 1965.

Yangel, Mikhail Kuzmich (Academician) was born on 25 October 1911, in the Irkutsk Raion, into a family of German extraction. He graduated from the Moscow Aviation Institute in 1937, but his war record is not known. In 1946 he was a member of the team that was sent to study German rocketry. Yangel worked in Korolyov's OKB-1, and became one of his deputies. In 1954 he headed his own bureau, based in Dnepropetrovsk (OKB-586). (Brezhnev – a native of this place – represented the town in the Supreme Soviet.) This bureau was almost entirely committed to the design and construction of missiles, and among the missiles developed were the R-12 and the R-14. Yangel was present when an R-14 exploded on the pad in October 1960, killing a large number of technicians and engineers, including Marshall Nedelin. Nedelin was the commander of the rocket forces and a member of the State Commission. Yangel also designed the SS-7, RS-16 and RS-20 missiles, two of which were converted into space rockets – the SL-8 and the SL-11. He also headed design teams that developed the retro-rockets for Vostok and Luna craft and the early Cosmos series of satellites, and developed the Zenit space rocket which was used both independently and as a stage on the Energiya booster. Yangel was a member of the Academy of Sciences, and was twice honoured as a Hero of Socialist Labour, and four times awarded the Order of Lenin. He replaced Korolyov on the State Commission, becoming its Deputy Head, and became the dominant figure of the Soviet military and civilian rocket programmes. He died on 25 October 1971.

DESIGN BUREAUX ASSOCIATED WITH MANNED SPACEFLIGHT THROUGH 1966

Chief Designers' names are included in brackets

OKB-1 (Korolyov): *Piloted spacecraft; piloted lifting bodies; hypersonic spacecraft and space launch vehicles.*
This had originally been established as Department 3 of the NII-88 in Podlipki (later Kalinograd; now Korolyov). On 24 April 1950 the department was upgraded to Experimental Design Bureau No. 1 (still under the control of NII-88) but by 14 August 1956 the department had become independent of NII-88. In 1966, after the death of Korolyov, the OKB was headed by Mishin. It is now called Energiya.

OKB-52 (Chelomei) *Piloted lifting bodies and hypersonic spacecraft; piloted spacecraft; space launch vehicles and military space systems*
This group originated at Plant 51 in Tushino on 17 September 1944, and was headed by N.N. Polikarpov. During the next nine years the group worked on the

development of winged cruise missiles, before being dissolved in February 1953, when it became a division of OKB 155. A Specialist Design Group was re-established at Plant No. 500 in Tushino, and became OKB-52 on 26 August 1955, later moving to Reutov to work on naval cruise missiles. The bureau began working in the space and missile fields in 1959.

TsSKB (Kozlov) *Military photoreconnaissance satellites; military piloted spacecraft; space launch vehicle manufacturing and development*
This bureau was established in Kuybyshev on 23 July 1959, as a Specialist Design Department to supervise the manufacture of the R-7 and its derivative launch vehicles. In 1960 it became OKB-1 Branch 3, and inherited work on the Vostok-derived reconnaissance satellite programme in 1964. The bureau became independent on 30 July 1974.

OKB-155 (Mikoyan) *Piloted winged spacecraft; hypersonic spacecraft*
One of the earliest design bureaux, established in Moscow on 8 December 1939, as the Experimental Design Section (OKO). It became OKB-155 on 16 March 1942.

OKB-23 (Myasischev) *Piloted lifting bodies; space launch vehicles and piloted spacecraft*
This bureau, established in Fili on 24 March 1951, originally focused on the development of long-range bombers and cruise missiles before becoming part of OKB-52 on 3 October 1960. It remained in that role until it became a branch of Energiya on 30 June 1988.

OKB-10 (Reshetnev) *Piloted spacecraft complex elements; space launch vehicles*
Established on 4 June 1959 at Plant No 1001 in Krasnoyarsk 26, to supervise production of ICBMs for OKB-1. In December 1961 it separated from OKB-1 and became an independent bureau working on several satellite projects between 1962 and 1966.

OKB-256 (Tsybin) *Piloted winged and hypersonic spacecraft*
This OKB lasted from its creation in May 1955 until October 1959, when it was dissolved and absorbed into OKB-23. Its database on space-plane research transferred to OKB-155. Chief Designer Tsybin transferred to Korolyov's OKB-1 in 1960.

OKB-586 (Yangel) *Space launch vehicles; liquid and solid propellant engines*
On 9 May 1951 this bureau was formed from some of the transferred personnel of NII-088 OKB-1, who formed a Serial Design Bureau, supervising OKB-1 missiles located at Dnepropetrovsk Machine Building Plant No.586. On 10 April 1954 they were joined by a second contingent of engineers from NII-88, whereupon the plant received the designation OKB-586. It is now called Yuzhnoye.

OKB-456 (Glushko) *ICBM space launch vehicles; liquid-propellant rocket engines and nuclear rocket engines*
This bureau was established in Khimki in 1946 and became OKB Energo Mash in March 1965. Originally known as the Gas Dynamics Lab, it was incorporated into Energiya, and then became independent.

OKB-2 (Isayev) *Liquid-propellant rocket engines for long-range ballistic missiles; spacecraft; space launch vehicles; nuclear rocket engines (provided solid rocket engines for Soyuz escape rockets)*
On 4 February 1943, Isayev became Technical Department Leader of a larger department within the KB-D Plant No.293. The bureau was established from this department on 25 May 1943, and on 21 June 1943 Isayev was named Chief of Department Engineering at Plant No. 293. The following February this was merged with NII-3 to form the new NII-1, with Isayev being named Chief of Department at NII-1. The department was later transferred to NII-88, as the new SKB Departmet No.9. In 1952 this became OKB-2 of the NII-88, and finally received independent status as OKB-2 on 16 January 1959. It is now called Khimmash.

OKB-154 (Kosberg) *Liquid rocket engines for space launch vehicles*
This bureau was established in Moscow on 17 October 1941, before moving to Bedsk, and then, in April 1946, to Voronezh. It is now known as Khimavtomatiki

OKB-276 (Kuznetsov) *Liquid-propellant rocket engines for long-range ballistic missiles (LRBM) and space launch vehicles*
Established on 17 April 1946, first in Kuybyshev, then in Samara, where it became the Trud bureau.

OKB-165 (Lyulka) *Liquid propellant engines for space launch vehicles*
This bureau was established in Moscow, originally as a department in the NII-1, in March 1946.

OKB-? (Fakel) *Spacecraft thrusters*
This bureau was established in 1955, and became involved in the space industry from 1959, located in Kaliningrad and Königsberg.

OKB-300 (Turmanskiy) *Liquid propulsion rocket engines for spacecraft*
This bureau was established in 1943.

NII Mashinostroyeniya *Attitude control thrusters for spacecraft*
This bureau was established as a branch of NII-1 in 1958, and became independent in 1981.

GSKB SeptsMach (Barmin) *Design and development of space launch complexes*
This bureau was involved in the design of launch complexes during the early years, having been established on 30 June 1941.

GSKB (Petrov) *Design and development of launch and space complexes*
From August 1948 the primary focus of this bureau was the creation of launch complexes and, from 1963, space launch complexes.

Plant 918 (Zvezda) (Alexeyev/Severin) *High-altitude pressure suits; pilot rescue devices (ejector seats); space and EVA suits; manned manoeuvring units.*
This bureau was established in 1952, from designers and scientists of the Flight Research Institute (FRI), the Central Aero-Hydrodynamics Institute (TsAGI), and several leading aerospace design officers, including NI Umansky, I.L Makarov, A.I.

Boiko, A.S. Povitsky and V.G. Galperin. The bureau was headed by the Chief of the FRI Design and Production complex, S.M. Alexeyev. During the early 1950s the main focus was to supply life support systems for animals on high-altitude rocket flights, Sputnik 2, and then Vostok and Voskhod, including the EVA suit.

MAJOR CONTRACTORS OF VOSTOK MANNED SPACECRAFT

OKB-1 (Korolyov) served as the 'prime contractor' and integrator of the spacecraft and its components. The bureau also took responsibility for the orientation system, the guidance system of the braking engine, the thermoregulation system, the emergency system, and the supply of ground support and test facilities.

OKB-2 (Isayev) supplied the TDU retro-fire rocket system to bring the spacecraft out of orbit.

NII-88 (G.A. Tyulin) provided the Mir 2 automated system.

TsKB-598 (N.A. Vinogradov) was the sub-contractor for the Vzor optical orientation system and the Grif solar orientation system photoelectric sensors.

Plant 918 (S.M. Alekseyev) supplied the cosmonauts' space pressure suit, its associated air circulator and oxygen supply, the helmet, emergency provisions, the ejection seat system, and the mannequin and assemblies for all unmanned test flights.

LII- (N.S. Stroev) supplied the guidance unit.

OKB-124 (G.I. Voronin) supplied the oxygen regeneration system.

NII-137 (V. A. Kostrov) supplied the emergency destruction system, which was used only on the unmanned Korabl-Sputnik missions.

NII-695 (A.I. Gusev) supplied the Zarya radiotelemetry system.

NII-668 (A.S. Mnatsekanya) supplied the command radio system.

VNIIT (N.S. Lidovenko) supplied the electrical storage batteries.

OKB MEI (A.F. Bogomolov) supplied the radiotelemetry system Tra1-P1.

NII-380 (I.A. Rosselevich) supplied the Rubin (Ruby) radio control system and the Topaz television system.

GNIIA and SKTB Biofizpribor (A.V. Pokrovksiy) contracted to develop and supply the systems for life-signs monitoring and medical dosimeters.

NIEI PDS (F.D. Tkachev) supplied the parachute system for the re-entry capsule.

KGB (K.V. Bulyakov) provided security and propaganda for the radio broadcasts to be transmitted.

Red Mechanical Device Factory (N.M. Yegorov) supplied the Zritel cine-camera.

Index